Germ Cells

A subject collection from *Cold Spring Harbor Perspectives in Biology*

OTHER SUBJECT COLLECTIONS FROM COLD SPRING HARBOR PERSPECTIVES IN BIOLOGY

The Golgi

The Mammary Gland as an Experimental Model

The Biology of Lipids: Trafficking, Regulation, and Function

Auxin Signaling: From Synthesis to Systems Biology

The Nucleus

Neuronal Guidance: The Biology of Brain Wiring

Cell Biology of Bacteria

Cell–Cell Junctions

Generation and Interpretation of Morphogen Gradients

Immunoreceptor Signaling

NF-κB: A Network Hub Controlling Immunity, Inflammation, and Cancer

Symmetry Breaking in Biology

The Origins of Life

The p53 Family

Germ Cells

A subject collection from *Cold Spring Harbor Perspectives in Biology*

EDITED BY

Paolo Sassone-Corsi
University of California, Irvine

Margaret T. Fuller
Stanford University

Robert Braun
Jackson Laboratory

COLD SPRING HARBOR LABORATORY PRESS
Cold Spring Harbor, New York • www.cshlpress.com

Germ Cells

A Subject Collection from *Cold Spring Harbor Perspectives in Biology*
Articles online at www.cshperspectives.org

Printed in the United States of America

Executive Editor	Alex Gann
Managing Editor	Maria Smit
Project Manager	Barbara Acosta
Permissions Administrator	Carol Brown
Cover Designer	Denise Weiss
Publisher	John Inglis

Front cover artwork: Whole *Drosophila* ovary triple stained with DAPI (blue) to reveal all nuclei, anti-Vasa (pink) antibodies to reveal early germ cells, and phalloïdin to reveal actin filaments (green). Image courtesy of Florence Besse, Institute of Developmental Biology and Cancer, Nice, France.

Library of Congress Cataloging-in-Publication Data

Germ cells / edited by Paolo Sassone-Corsi, Margaret T. Fuller,
Robert Braun.
p. cm.
Includes bibliographical references and index.
ISBN 978-1-936113-51-4 (hardcover : alk. paper)
1. Germ cells. I. Sassone-Corsi, Paolo. II. Fuller, Margaret T.,
1951- III. Braun, Robert, 1955-
QL964.G43 2011
571.8′45--dc23

2011023813

10 9 8 7 6 5 4 3 2 1

All World Wide Web addresses are accurate to the best of our knowledge at the time of printing.

All Cold Spring Harbor Laboratory Press publications may be ordered directly from Cold Spring Harbor Laboratory Press, 500 Sunnyside Blvd., Woodbury, New York 11797-2924. Phone: 1-800-843-4388 in Continental U.S. and Canada. All other locations: (516) 422-4100. FAX: (516) 422-4097. E-mail: cshpress@cshl.edu. For a complete catalog of all Cold Spring Harbor Laboratory Press publications, visit our website at http://www.cshlpress.com/.

Contents

Preface

THE FAMOUS QUOTATION BY CHARLES DARWIN, "It is not the strongest of the species that survives, nor the most intelligent, but the one most responsive to change," beautifully illustrates the unique role of germ cells in the maintenance and evolution of the species. To be inherited, the changes to which Darwin refers need to occur in germ cells. For this, and many other reasons nicely developed in this book, germ cells have long fascinated a large number of researchers over many decades. Restricted for many years within the boundaries of descriptive science, the field of germ cell research has dramatically benefited from the development of molecular biology techniques, of genome-wide array approaches, and of cell- and tissue-specific gene ablation by homologous recombination, as well as from the increasing number of species in which experiments can be done. The past two decades have witnessed the accumulation of a spectacular amount of knowledge, including several discoveries that have fundamentally altered the manner in which we look at how germ cells develop and function.

The idea for this book originated from a desire to incorporate in a single volume not only the remarkable advances that have been made in the field but also the emerging concepts and hypotheses that currently drive thinking and experimentation in this field. In most cases, the authors have aimed to describe both facts and concepts arising from studies across multiple species. This approach reveals common themes, as well as intriguing examples of diversity. We, the editors, have worked with the authors to ensure that overarching ideas are linked among the chapters.

Germ cell development is unique in the way it generates the haploid cells responsible for perpetuating the species. The resulting germ cells are virtually totipotent; they will generate all possible cell types and tissues after fertilization. How is such a complex and elaborate differentiation program achieved? In what way is it unique with respect to the numerous programs of somatic cell differentiation? These questions have recently acquired even broader interest because of the excitement generated by stem cell biology and the therapies it promises. The intrinsic potential of germ cells to generate a whole new organism relies on very specific and fascinating molecular and cellular pathways operating during their development. Of course, the process of meiosis is one outstanding feature of germ cells that involves a series of highly specialized biochemical and metabolic processes. In addition, germ cell development requires choices between alternative pathways of gene expression, each generating a drastically different outcome. As an example, during mammalian spermatogenesis, spermatogonial stem cells face the choice of self-renewal or differentiation. Whichever pathway they choose, the task is formidable. On one hand, self-renewal is one of the most impressive examples of highly controlled and efficient cell proliferation, one that continues until old age in human males. On the other hand, cells unique in their structure are generated through the differentiation pathway requiring drastic biochemical reorganization and meiotic division. When things go wrong, the outcome can be dramatic, possibly leading to infertility on one hand or cancer on the other.

Two fundamental characteristics underlie the fascination of current germ cell research. First, germ cells must carry out all the basic functions of cells, ranging from transcription to division, but under the control of the germline developmental program, they do so in unusual and exceptionally interesting ways. Second, although the appearance of the exquisitely specialized cells that are gametes differs profoundly among species, in part to allow exploitation of different niches, strikingly unifying themes arise from the comparison of germ cell development across species. We strove to highlight both of these fundamental features in the organization of this book.

We hope this book will not only serve the aficionados in the field but also be accessible to students and researchers from other fields who wish to discover the fascinating world of germ cells. Indeed, part of our ambition is to motivate new generations of investigators to enter the field. The authors have contributed magnificently toward these goals, and we thank them for their participation.

Finally, we wish to thank the outstanding staff at the Cold Spring Harbor Laboratory Press and their compositors, particularly Hannah Turner at Techset, for their efforts in making this book a reality. Our deep appreciation goes to Project Manager, Barbara Acosta, for her sense of organization, limitless patience, and constant kindness. We also thank Acquisitions Editor, Alex Gann, whose vision and intelligence initiated the process that lead to this work.

PAOLO SASSONE-CORSI
MARGARET T. FULLER
ROBERT BRAUN

Germline Stem Cells

Allan Spradling[1], Margaret T. Fuller[2,3], Robert E. Braun[4], and Shosei Yoshida[5]

[1]Howard Hughes Medical Institute Research Laboratories, Department of Embryology, Carnegie Institution, Baltimore, Maryland 21218

[2]Department of Developmental Biology, Stanford University School of Medicine, Stanford, California 94305-5329

[3]Department of Genetics, Stanford University School of Medicine, Stanford, California 94305-5329

[4]The Jackson Laboratory, Bar Harbor, Maine 04609

[5]Division of Germ Cell Biology, National Institute for Basic Biology, 5-1 Higashiyama, Myodaiji, Okazaki, Aichi 444-8787, Japan

Correspondence: mtfuller@stanford.edu

Sperm and egg production requires a robust stem cell system that balances self-renewal with differentiation. Self-renewal at the expense of differentiation can cause tumorigenesis, whereas differentiation at the expense of self-renewal can cause germ cell depletion and infertility. In most organisms, and sometimes in both sexes, germline stem cells (GSCs) often reside in a defined anatomical niche. Factors within the niche regulate a balance between GSC self-renewal and differentiation. Asymmetric division of the germline stem cell to form daughter cells with alternative fates is common. The exception to both these tendencies is the mammalian testis where there does not appear to be an obvious anatomical niche and where GSC homeostasis is likely accomplished by a stochastic balance of self-renewal and differentiation and not by regulated asymmetric cell division. Despite these apparent differences, GSCs in all organisms share many common mechanisms, although not necessarily molecules, to guarantee survival of the germline.

In males and in females or hermaphrodites of many species, gametes are produced throughout reproductive life from adult stem cells dedicated to the germline. Some remarkably conserved themes as well as intriguing differences are beginning to emerge from studies of germline stem cells (GSCs) in the *Caenorhabditis elegans* gonad, *Drosophila* ovary, and testes from *Drosophila* and mice. GSCs commonly reside in a special microenvironment, a stem cell niche, provided by somatic support cells. GSCs in this niche both self-renew to maintain the generative population and also produce progeny germ cells that exit the niche and begin the process of differentiation. These progeny commonly execute several rounds of mitotic division before switching to the germ cell-specific cell cycle of meiosis and subsequent differentiation into mature gametes (see Kimble 2011). A strongly conserved feature is that the mitotic divisions are incomplete, so that permanent cytoplasmic bridges connect mitotic

Cite this article as *Cold Spring Harb Perspect Biol* doi: 10.1101/cshperspect.a002642

sisters descended from a single founder cell, forming a cyst of germ cells that develop in synchrony (see Greenbaum et al. 2011). Communication between germline and special somatic cells plays important roles in both stem cell maintenance and in germ cell proliferation and differentiation, as well as in the genesis of GSCs from primordial germ cells (PGCs) in the gonad. GSCs are normally unipotent, producing only differentiating gametes. However, as the only adult stem cells that contribute to the next generation, it is likely that GSC potency must be strictly controlled to prevent teratoma formation. Moreover, these controls likely differ between the sexes, reflecting the profound differences between male and female gametogenesis. In this article, we examine these intriguing characteristics of germline stem cells emerging from research on model organisms from invertebrates to mammals.

Germline stem cells, especially from *C. elegans* and *Drosophila*, are among the best understood adult stem cell types, and have provided important models for study of the regulation of adult stem cell behavior in vivo. The hallmarks of adult stem cells are a committed but relatively undifferentiated state, a long-term ability to proliferate, and an ability to produce both new stem cells (self-renewal) and differentiating progeny. Adult stem cell lineages are responsible for long-term maintenance and repair of tissues containing highly specialized, short-lived cell types, including blood, skin, and the epithelium of the intestine and colon, as well as sperm. Differentiated cells in many other tissues, including breast, lung, skeletal muscle, bladder, and prostate, are also produced from adult stem cells in response to physiological changes or damage. The mechanisms that regulate adult stem cell self-renewal and the proliferation and differentiation of stem cell progeny are key not only for harnessing the potential of adult stem cells for regenerative medicine, but also for understanding the developmental origins of cancer. Many common cancers arise in adult stem cell lineages, and there is increasing evidence that defects in the mechanisms that regulate self-renewal, proliferation, and differentiation in adult stem cell lineages can contribute to oncogenesis. Thus, the tremendous recent advances in understanding how the local microenvironment of the stem cell niche regulates self-renewal and differentiation of germline stem cells have important implications for somatic biology as well as for the germline.

GERMLINE STEM CELL LINEAGES

Germline stem cells produce progeny that proliferate through several rounds of divisions, and then enter meiosis (Fig. 1). Although this transit-amplifying (TA) strategy is widely conserved in spermatogenesis, the germline mitotic proliferation program appears very different from the perspective of the female. Although female germline stem cells produce daughter cells that initiate differentiation with rounds of mitotic division, producing cysts of interconnected mitotic sisters in both *Drosophila* and mice (Pepling and Spradling 1998, 2001), these divisions are not TA, because the number of oocytes is not amplified. The TA program likely underlies germ cell behavior in both sexes, but the strategy has been modified during evolution in the female so that most of the mitotic sisters sacrifice their contents to build the specialized oocyte. This sex difference goes along with two other main differences in germ cells between the sexes. (1) The female gamete is larger and usually contains extensive components to carry out the start of embryogenesis (see Von Stetina and Orr-Weaver 2011). (2) The meiotic divisions take place after the major events of gametogenesis in females but before the major morphogenetic events that lead to the final form of the gametes in males.

GERMLINE STEM CELL POPULATIONS MAINTAINED BY INSTRUCTIONS FROM A LOCAL NICHE

Germline stem cells reside in a special microenvironment that restricts their differentiation. A classic example occurs in *C. elegans*, where a large somatic cell at the distal tip of the gonad maintains a syncytial population of mitotically proliferating germ cell nuclei through

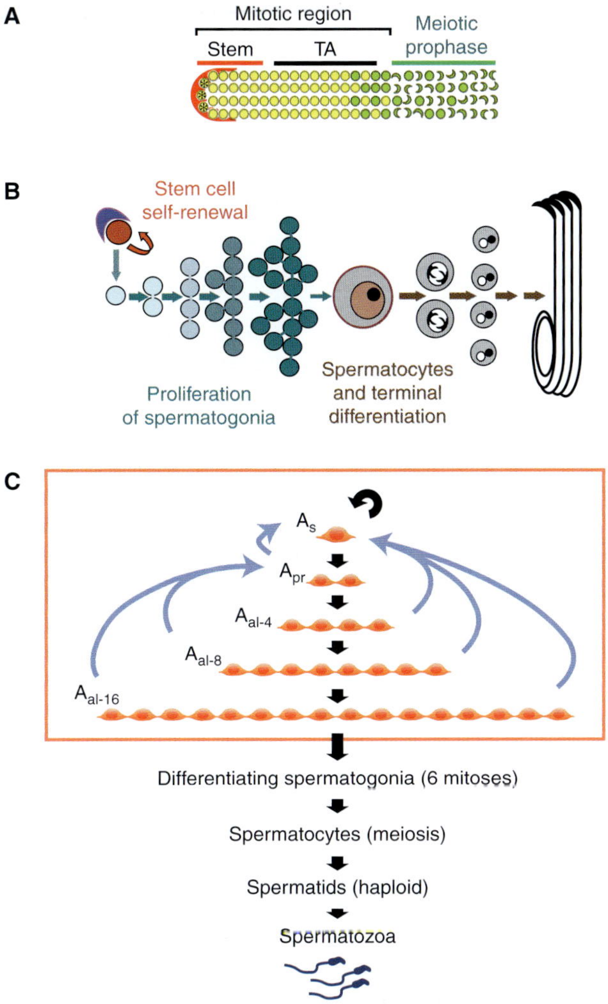

Figure 1. Germline stem cell lineages: stem cell self-renewal, transit-amplifying (TA) mitotic divisions, and the switch to meiosis and gamete differentiation. (*A*) Germline stem cell lineage in a *C. elegans* hermaphrodite undergoing oogenesis. The somatic distal tip cell (DTC, red) maintains a population of germ cell nuclei in a stem cell state. As germ cell nuclei move away from the DTC they enter a mitotic TA state during which they gradually differentiate, finally switching to the meiotic program (green). (Panel adapted from Cinquin et al. [2010]; reprinted, with permission, from the National Academy of Sciences © 2010.) (*B*) Male germline stem cell lineage in *Drosophila*. A stem cell (red) at the tip of the testis divides asymmetrically, producing a new stem cell and a gonialblast, which initiates four rounds of synchronous spermatogonial mitotic TA divisions with incomplete cytokinesis. The resulting 16 interconnected germ cells undergo premeiotic DNA replication in synchrony and switch to the spermatocyte program of cell growth, meiotic prophase, and transcription of terminal differentiation genes. (All 16 cells become spermatocytes, but only one is shown for simplicity.) (*C*) Male germline stem cell lineage in mouse. In mice, the prevailing model (shown by black arrows) assumes that the singly isolated spermatogonia (A_s) is the predominant stem cell during steady-state spermatogenesis. However, all of the cells shown in the box, the A_s, A_{pr}, and A_{al} spermatogonia have stem cell potential. At the 16 cell-cyst stage, the $A_{al\text{-}16}$ transform without cell division into differentiating spermatogonia, which subsequently undergo six rounds of mitosis before entering meiosis. The blue arrows indicate the ability of cells to detach from cysts and revert to the A_s state. Reversion occurs frequently during regeneration following insult and following germ cell transplantation, however, it has also been observed during steady-state spermatogenesis. (Panel *B* is adapted from Davies and Fuller [2008]; reprinted, with permission, from Cold Spring Harbor Press © 2008.)

close-range instruction via Notch pathway signaling (Austin and Kimble 1987). The subpopulation of the germline mitotic nuclei lying closest to the distal tip likely correspond to germline stem cells (Cinquin et al. 2010), whereas the mitotic nuclei lying further from the distal tip likely correspond to the TA population (Fig. 1A). The biology of *C. elegans* germline stem cells and their mitotic progeny is discussed in detail in Kimble (2011).

THE *DROSOPHILA* OVARIAN GERMLINE STEM CELL NICHE

Drosophila females maintain germ cell production throughout adult life by establishing a niche at the tip of each ovariole that prevents some PGCs from differentiating. The ability to form niches appears to distinguish species with female GSCs from those without rather than basic differences in germ cell differentiation. Niches in the ovary arise near the end of larval development and upload PGCs, which are produced in excess (Gilboa and Lehmann 2006), just prior to the onset of global differentiation signals. PGCs that make it into the niche are protected and become GSCs, whereas identical PGCs outside begin development (Gilboa and Lehmann 2004). In addition to sequestering a supply of undifferentiated germ cells, niches program subsequent GSC divisions to be asymmetric by providing only enough room for one of the two daughters to remain inside. In so simple a manner, GSC niches generate a lifelong supply of germ cells for egg production. The niche itself is built from three somatic cell types: terminal filament cells located at the anterior of each ovariole, cap cells, which actually contact GSCs and hold them in the niche, and escort cells, which wrap thin processes around GSCs and prevent direct GSC–GSC contact (Fig. 2A).

Genes promoting PGC/GSC differentiation are modulated by changes at the level of chromatin structure, transcription, and RNA processing, but especially at the level of translation. The conserved NANOS and PUMILIO translational repressors are essential to prevent premature germ cell differentiation (Lin and Spradling 1997; Forbes and Lehmann 1998; Wang and Lin 2004), indicating that messenger RNAs (mRNAs) capable of promoting cyst formation and oogenesis are present within PGCs. These mRNAs are controlled by proteins, including VASA, which localize them within large silencing ribonucleoprotein particles (RNPs), promote CCR4-meditated deadenylation, and which influence their binding to 40S small ribosomal subunits in conjunction with translation initiation factors such as eIF4E (Wang and Lin 2004; Fichelson et al. 2009; Liu et al. 2009; see Kugler and Lasko 2009). MicroRNAs (miRNAs) contribute to both GSC maintenance and likely provide sequence specificity to orchestrate the complex, multistep program of gametogenesis following activation (Forstemann et al. 2005; Hatfield et al. 2005; Jin and Xie 2007; Park et al. 2007; Yang et al. 2007a,b). Indeed, the perinuclear nuage, stage-specific nuclear and cytoplasmic germ granules, and chromatoid bodies that typify germ cells throughout the metazoa, are manifestations of abundant RNA processing complexes participating in germ cell regulation (Kotaja and Sassone-Corsi 2007; Lim and Kai 2007; Klattenhoff and Theurkauf 2008; Lin et al. 2008; Soper et al. 2008; see Voronina et al. 2011).

Drosophila ovarian niches control the onset of germ cell development by spatially modulating expression of a high-level differentiation inducer encoded by *bag-of-marbles* (*bam*). Cap cells and possibly anterior escort cells activate JAK-STAT signaling upon receipt of the unpaired ligand from terminal filament cells and respond by producing the BMP ligands Dpp and Gbb (López-Onieva et al. 2008; Wang et al. 2008). Consequently, germ cells located within a small zone adjacent to the cap cells are positioned to receive a strong BMP signal. Activation of BMP receptors in GSCs directly represses *bam* transcription (Chen and McKearin 2003). Only upon leaving the niche do they escape BMP-mediated repression, up-regulate *bam* expression, and become cystoblasts, i.e., cells that will found new 16-cell cysts. *bam* can initiate development equally well within larval ovarian PGCs, but premature differentiation is prevented by

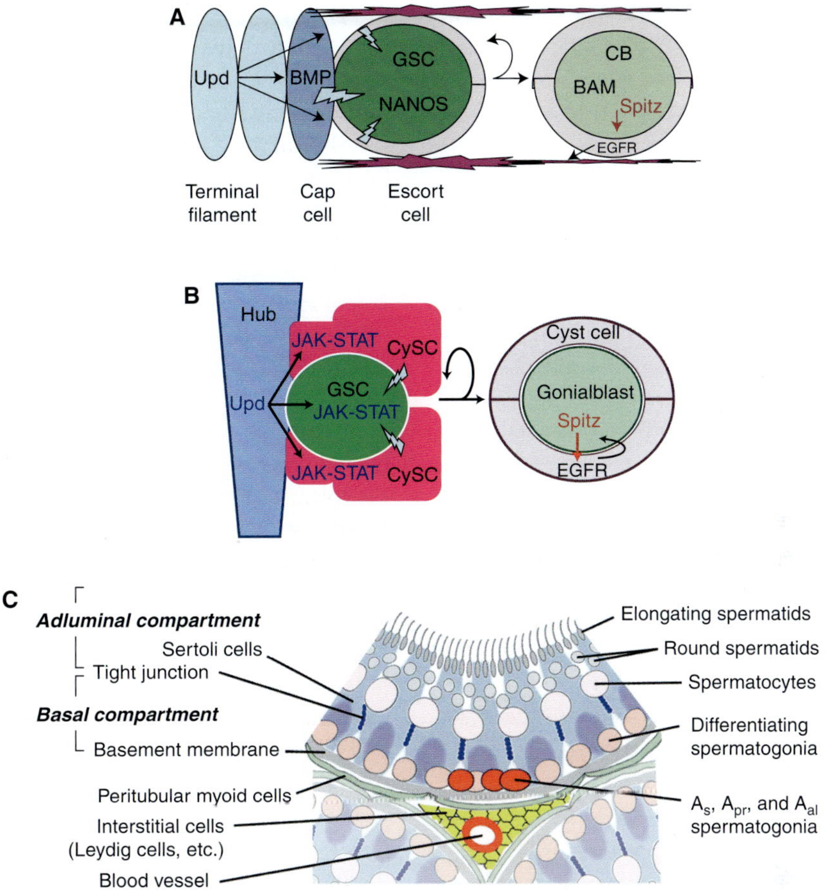

Figure 2. Germline stem cell microenvironments in *Drosophila* ovary and *Drosophila* and mouse testes. (*A*) The Drosophila ovarian stem cell niche controls cystoblast formation. *Upd* secreted by terminal filament cells (light blue) activates JAK-STAT signal transduction within cap cells (dark blue) and probably in escort cells (tan), which in response produce the BMP ligands DPP and GBB ("BMP"). Receipt of these signals in germline stem cells (GSC) activates BMP signal transduction (lightning blots), which represses bam transcription. Consequently, the NANOS translational repressor remains high. (*Right*) A GSC daughter has differentiated into a cystoblast (CB, light green) and is surrounded by escort cells. High levels of BAM present due to derepression have translationally downregulated NANOS. Signals from the CB via the ligand *Spitz* (red arrow) activate EGFR signaling in escort cells, which modulate the extracellular matrix (red lines) to block residual BMP reception. (*B*) The apical hub and flanking somatic cyst stem cells (CySCs) contribute to the germline stem cell (GSC) niche in *Drosophila* testes. (*Left*) Unpaired (Upd) secreted by hub cells (blue) activates the Janus kinase-signal transducer and activator transcription (JAK-STAT) pathway, which requires the cell to autonomously maintain the attachment of GSCs to the hub and for self-renewal of cyst stem cells (CySC). GSC self-renewal also requires as-yet-unknown signal(s) (lightning bolt) from the CySCs. (*Right*) Interactions between the progeny of somatic and germline stem cells set up the functional unit of differentiation, the cyst. A gonialblast (light green), produced by asymmetric division of a GSC, signals via the epidermal growth factor receptor (EGFR) ligand Spitz, to cyst cells (light pink) produced by asymmetric division of CySCs. Activation of EGFR triggers the somatic cyst cells to envelop the gonialblast and may also induce changes in gene expression in the cyst cells required for the germ cells to execute the TA division program. (*C*) The microenvironment of germline stem cells in mouse testes. All spermatogonia, including the GSCs, occupy the basal compartment of the seminiferous tubules between Sertoli cell tight junctions and the basement membrane. The GSCs are biased to the area adjacent to the blood vessels and interstitial cells running between the tubules. (Panel *C* is adapted from Davies and Fuller [2008]; reprinted, with permission, from Cold Spring Harbor Press © 2008.)

gonad-wide BMP signaling during these stages (Gilboa and Lehmann 2004; Kai and Spradling 2004). Interestingly, the *bam* gene is not widely conserved and shows signs of recent positive selection (Civetta et al. 2006), raising the possibility that it arose in concert with the evolution of *Drosophila* female GSCs.

A key niche requirement is to limit high-level Dpp/Gbb signaling to a one-cell-wide zone so that one daughter of each GSC division will be forced out, derepress *bam*, and become a cystoblast. The DALLY heparin sulfate glycoprotein, which accumulates at high levels only around the cap cells, binds to and stabilizes Dpp and may be necessary to effectively activate BMP signaling (Guo and Wang 2009; Hayashi et al. 2009). EGFR signaling from early germ cells to escort cells mediated by the membrane protease STET influences DALLY levels or function (Schulz et al. 2002; Liu et al. 2011). The ability of germ cells downstream from the niche to receive BMP signals is further controlled by negative feedback from the BMP-induced E3 ubiquitin ligase SMURF and the Ser/Thr kinase FUSED, which promotes turnover of the BMP receptor THICK VEINS (Tkv) (Casanueva and Ferguson 2004; Xia et al. 2010). All these mechanisms ensure that once outside the niche, *bam* transcription will be promptly and stably activated.

Bam brings about cyst formation by promoting the translation of differentiation-promoting transcripts (Li et al. 2009) and reducing E-CADHERIN (E-cad) expression to ensure that cystoblasts do not reenter the niche (Shen et al. 2009). Forming a complex with BENIGN GONIAL CELL NEOPLASM (BGCN), BAM interacts with eIF4A and other translation factors to modulate the expression of an unknown number of target mRNAs (Shen et al. 2009). The RNA-binding protein SXL, which controls sex-specific splicing in somatic cells, is also essential for cystoblast development (Chau et al. 2009). Changes in miRNA levels and activity assist in the translational changes. Within the niche, miRNAs initially help repress differentiation gene transcripts, because mutations abolishing miRNA function cause GSC loss. Following niche exit, miR-184 down-regulates the DPP receptor SAX to help maintain BAM production (Iovino et al. 2009). MEI-P26 modulates cyst growth by down-regulating repressive miRNAs, and up-regulating growth-promoting miRNAs such as *bantam* (Neumuller et al. 2008). MEI-P26 production in early cysts is itself translationally stimulated by VASA, which binds to a U-rich site in the *mei-P26* 3′ untranslated region (UTR) (Liu et al. 2009). miR-7 can repress *bam* translation and its negative control by *maelstrom* may help ensure appropriate cyst growth and stability (Pek et al. 2009).

Residence within the niche is governed by how strongly GSCs associate with cap cells via adherens junctions. This critical function is modulated in part by BAM/BGCN-mediated repression of *E-cad* mRNA translation (Shen et al. 2009). GSCs must adhere strongly enough to minimize loss, but not so tightly that the niche fills up with defective or prematurely differentiating cells. Adhesion levels within the niche respond to systemic insulin signaling and decline with age (LaFever and Drummond-Barbosa 2005; Hsu and Drummond-Barbosa 2009). Even more intriguing is the possibility that competition for niche association might maintain stem cell quality by ensuring that weakened cells are expelled (Nystul and Spradling 2007; Johnston 2009; Rhiner et al. 2009). *Bam* (or BGCN) mutant germ cells lack E-cad translational repression, up-regulate E-cad levels, and displace wild-type GSCs from the niche (Jin et al. 2008). Feedback regulation likely ensures the prompt exit of GSCs that have prematurely activated *bam*, but so far evidence for more general competition is lacking.

THE PHYLOGENETIC DISTRIBUTION OF FEMALE GERMLINE STEM CELLS

There is a widespread misconception that adult female germline stem cells are common in "lower" organisms such as invertebrates and fish, but absent in "higher" forms such as mammals. In reality, we know relatively little about the phylogenetic distribution of adult female GSCs, because like other stem cells they are difficult to document using classical techniques.

Nonetheless, available information strongly argues that species with adult female GSCs are rare across the phylogenetic spectrum. Indeed, mice and humans are typical of a wide range of invertebrate and vertebrate groups in which germ cell mitotic activity (whether from stem cell or cyst-forming divisions) is limited to pre-adult stages.

The diverse orders of insects have been investigated most widely (reviewed in Büning 1994). Oogenesis in these groups shows many common features, including terminal filaments, ovarioles, cysts (in many but not all groups), and follicle formation. Among this vast array of species, only in *Drosophila* and closely related flies have GSCs been clearly documented histologically in adult females. In other groups, ovarian mitotic germ cell divisions have not been found in adult stages. Surprisingly, even species with highly prolific females lack readily demonstrable GSCs within their adult germaria. Honeybee queens store more than one million viable sperm and individually maintain colonies containing 80,000 short-lived workers over multiple years. Yet honeybee germaria, which are highly elongated compared to *Drosophila*, lack readily identifiable GSCs. Still larger colonies numbering in the millions are sustained by some termite queens, who can live up to 50 years. Of course these queens and females from many other species may contain adult stem cells that have yet to be recognized. Nonetheless, the absence of *Drosophila*-like niches and their associated GSCs in most groups suggests that there may be evolutionary tradeoffs in maintaining adult female GSCs that have rarely proved favorable.

At least some noninsect species do maintain GSC activity into adulthood. In hydrozoans and planarians, circulating pleuripotent stem cells known as interstitial cells or neoblasts, replenish germ cells throughout life. *C. elegans* hermaphrodites maintain ongoing germ cell production. In vertebrates, adult female GSCs have been documented in a fish, medaka, where they reside in tubelike "cradles" that resemble a *Drosophila* germarium tip (Nakamura et al. 2010). Disrupting *nanos1*-gene function blocks ongoing follicle production in zebrafish (Draper et al. 2007). Outside of fish, there is currently no convincing evidence for vertebrate adult ovarian GSCs, despite contrary claims (Tilly et al. 2009). In principle, it is not difficult to recognize stem cell activity and ongoing follicle production based on histological studies. New follicles develop from a stem cell over many days, even in *Drosophila* where nurse cells speed development, ensuring that many downstream intermediates will be present whenever GSCs are active. In mice, mitotic germline divisions prior to meiosis, germline cysts, and germ cells with recombining meiotic chromosomes are all abundant during fetal development, but disappear shortly after birth (for example, see Pepling and Spradling 2001; Matzuk et al. 2011). Ultimately, demonstrating the existence of adult ovarian GSCs requires the same proof needed for identifying any stem cell. Within the intact organism, such GSCs must be lineage-labeled, shown to self-renew, and their daughter cells must be shown to develop into fertile oocytes. Demonstrating that gametes can be produced following selection in cell cultures is not definitive because cells may have undergone reprogramming (Zou et al. 2009).

Why are female GSCs apparently so rare, despite the fact that major processes of female germ cell development have been conserved during evolution? The strategy of preserving some PGCs in a niche seems simple, and male GSCs are spread widely throughout phylogenetic space. We speculate that the rarity of adult female GSCs results from the profound differences between male and female gametes. Beyond the nucleus, sperm are basically throwaway devices. What organelles remain following spermiogenesis are likely damaged by the oxidative by-products of sperm activity prior to fertilization, and either fail to enter the egg or are quickly degraded. In contrast, the oocytes of metazoans that produce relatively few progeny, such as flies and mice, must contain cellular constituents of sufficient quality to ensure that the next generation starts out with no greater load of damage than was present a generation earlier. The difficulty of this task has been discussed mostly in terms of the mitochondria and their multicopy, easily mutated DNA genome, but actually applies to every cellular component. Once set in motion, the complex processes needed to generate cytoplasm of such quality may not be easily paused. Mature oocytes may be too large to store so that follicles are instead stored at an early stage, corresponding to the primordial follicle.

Drosophila may have evolved GSCs to adapt to an ecological niche in which food and oviposition sites are limited and only sporadically available. When an opportunity arises, the ability of a *Drosophila* female to lay many eggs rapidly is crucial for reproductive success. If very young follicles were stored, they would take three days to develop to maturity, too long to be of much help in a short-lived, competitive situation. Stem cells allow adult *Drosophila* to maintain a steady stream of follicles developing until vitellogenesis, just one day from maturity. Follicles at this stage do not appear to be storable; they either mature or undergo atresia and are recycled. The net effect of using stem cells and recycling follicles, is that *Drosophila* females maintain a constant supply of developing follicles that can be completed with minimal delay to produce bursts of high-quality female gametes during the sporadic instances when they are needed.

THE *DROSOPHILA* TESTIS GERMLINE STEM CELL NICHE

Germline stem cells are also maintained by local signals from a somatic support cell niche in *Drosophila* males, and there are many themes in common with female germline stem cells in the ovary. However, sex-specific differences in how and when similar signaling pathways and control mechanisms are employed, as well as differences in which aspects have been most deeply investigated in testes versus ovaries, have contributed to intriguing differences in the current picture of how these two anatomically related systems work. For example, the translational inhibitors NANOS and PUMILIO are not required for germline stem cell maintenance in males, suggesting that male germline stem cells may not be primed by expression of translationally silenced differentiation transcripts in the same way as are female GSCs. Likewise, although *bam* also acts as a differentiation factor in males, it normally functions at a different step in the germ cell lineage than in females. Onset of *bam* expression occurs after male germ cells have entered the TA division program and *bam* function is not needed to initiate the spermatogonial divisions. Rather, in males, *bam* is required for spermatogonia to cease mitosis and enter the meiotic program (Gönczy et al. 1997) and the timing and level of accumulation of *bam* protein controls the number of cyst divisions (Insco et al. 2009). Nevertheless, forced ectopic expression of *bam* in male GSCs can induce GSC differentiation (Sheng et al. 2009a), suggesting that *bam* has the capacity to block expression of factors required for GSC maintenance, even though it is not the key component that normally does so to initiate the male germ cell differentiation program.

Oriented Spindles and a Local Niche Program the Asymmetric Outcome of Stem Cell Divisions

Approximately 6–12 *Drosophila* male GSCs lie in a rosette around a tight cluster of largely somatic support cells called the hub, located at the apical tip of each testis. *Drosophila* male GSCs normally divide asymmetrically: the mitotic spindle is set up perpendicular to the hub-GSC interface so that, upon cytokinesis, one daughter remains next to the hub and maintains stem cell identity, whereas the other daughter is displaced away and initiates differentiation, becoming a gonialblast (Gb) (Fig. 2B). This arrangement of the GSCs around the hub is set up before the end of embryogenesis and male germ cells divide with oriented spindles and begin producing differentiating daughter cells by onset of the larval stages (Kiger et al. 2001; Le Bras and Van Doren 2006; Sheng et al. 2009b). Thus, formation of the niche and conversion from PGCs to GSCs occurs much earlier in males than in females.

As in females, GSCs in the testis physically attach to the hub by localized adherens junctions, which allow GSCs to hold on to and compete for position at the hub. Male germline stem cells made homozygous mutant for E-cad are lost to differentiation. E-cad localized in the adherens junctions provides binding sites for β-catenin and APC2, a homolog of the adenomatous polyposis coli tumor suppressor, concentrating them at the GSC cortex adjacent to the hub (Yamashita et al. 2003).

The adherens junctional complexes between GSCs and the hub provide polarity cues toward which GSCs orient their centrosomes throughout interphase and their spindle during mitosis cell (Yamashita et al. 2003; Inaba et al. 2010). In G1, the single centrosome in each GSC retains a robust array of microtubules and localizes near the cell cortex where the germ cell attaches to the hub. The position of the centrosome may be set via direct or indirect interactions between the microtubules and APC2 protein localized at the adherens junction. This unusual centrosome behavior is characteristic of the GSCs but not of TA spermatogonia, and so appears to be part of the stem cell program. When the duplicated centrosomes separate, early in G2, one stays next to the hub while the other migrates to the opposite side of the cell. The stereotyped position of the centrosomes in turn orients the mitotic spindle perpendicular to the GSC-hub cell interface, ensuring the normally asymmetric outcome of GSC divisions: one daughter inherits the attachment to the hub, remains subject to local niche signals, and renews stem cell identity, whereas the other is displaced away from the niche and normally initiates differentiation (Yamashita et al. 2003, 2007; Yamashita and Fuller 2005). A checkpoint mechanism blocks progression into mitosis unless a centrosome is properly situated next to the attachment to the hub (Cheng et al. 2008), ensuring reliable spindle orientation and the programmed asymmetric outcome of stem cell divisions. Differential labeling of mother versus daughter centrosomes by transient expression of a green fluorescent protein (GFP)-tagged centriolar protein fragment revealed that the two centrosomes in male GSCs have different characters and fates. The mother centrosome normally remains adjacent to the hub and is inherited by the GSC, while the daughter centrosome migrates to the opposite side of the cell and is inherited by the Gb. As a result, male GSCs maintain a centriolar Eve, assembled many cell generations earlier (Yamashita et al. 2007).

The Testis Maintains Two Stem Cell Types, Which Produce Progeny that Work Together

The GSCs attached to the hub are flanked by somatic cyst stem cells (CySCs) (Fig. 2B). The CySCs lie with their cell bodies displaced away from the hub but maintain contact with it via thin cytoplasmic processes that extend down between the GSCs to touch the hub. CySCs and GSCs appear to compete for binding to the hub, with adhesion regulated so that GSCs normally win (Issigonis et al. 2009; Sheng et al. 2009a,b). The cytoplasmic processes of the CySCs insulate the GSCs from each other. The CySCs also divide with spindles oriented perpendicular to the interface with the hub, although spindles in CySCs are usually first set up roughly parallel with the hub surface, then reorient during anaphase so that one pole is next to the hub (Cheng et al. 2011). When CySCs divide, the daughter that maintains attachment to the hub maintains stem cell identity and the daughter displaced away from the hub normally differentiates as a cyst cell. Two cyst cells interact with a Gb produced by a nearby dividing GSC to form the functional unit of differentiation, the germline cyst (Fig. 2B, *right*). The epidermal growth factor (EGF) class ligand Spitz secreted by the germ cell activates the EGFR on the cyst cells and through the guanine exchange factor Vav differentially activates the small GTPase Rac, causing the cyst cells to send out processes and envelop the Gb (Sarkar et al. 2007). The cyst cells normally do not divide again, but accompany and codifferentiate with the germ cells they enclose throughout the process of spermatogenesis (Gönczy et al. 1992; Voog et al. 2008). In this way, the cyst cells functionally resemble Sertoli cells in mammalian testes, although individual cyst cells do not associate with germ cells at different stages. The maintenance of two stem cell types attached to the same apical hub may be an anatomical mechanism to facilitate production of daughter cells that must work together for proper differentiation.

Once enclosed by the cyst cells, the Gb embarks on the spermatogonial mitotic divisions. As in mammals, the TA spermatogonia within a cyst divide in metasynchrony and undergo incomplete cytokinesis, retaining cytoplasmic bridges that develop into ring canals (see Greenbaum et al. 2011). The number of spermatogonial divisions varies with *Drosophila* species but is four in *D. melanogaster*. After the fourth spermatogonial division, the 16 sister germ cells within the cyst together enter premeiotic DNA synthesis (see Kimble 2011) then initiate the spermatocyte program, which includes cell growth and a cell type specific gene expression program that prepares for the dramatic cell morphogenesis that follows the meiotic divisions (see White-Cooper and Davidson 2011). It is intriguing that the 16-cell stage is also an important regulatory point in the spermatogonial division program in mammals.

TA Cells Can Revert to the Stem Cell State to Replace Lost GSCs

Although TA spermatogonia are clearly differentiating and can be distinguished from stem cells based on cell division behavior, subcellular structures, and gene expression patterns, early spermatogonia are not yet committed to differentiation. Indeed, TA spermatogonia can break apart from germ cell cysts, reoccupy the niche, and revert to stem cell behavior. Male germ cells induced to begin differentiation due to temporary down-regulation of STAT activity or temporary forced early expression of *bam* will return to the niche and resume stem cell behavior once the abnormal genetic condition is reversed, as long as the spermatogonia have not advanced too far toward becoming spermatocytes (Brawley and Matunis 2004; Sheng et al. 2009a). Genetic marking experiments demonstrated that TA spermatogonia that have advanced to the four-cell cyst stage and turned on *bam* expression can, and indeed frequently do, return to the hub and resume stem cell identity and behavior, even in the absence of drastic genetic perturbations (Cheng et al. 2008). If GSCs at the hub are killed by irradiation, stem cells are replaced by massive dedifferentiation of TA spermatogonia. *Drosophila* female germ cells exhibit similar behavior: Germ cells induced to enter the cystocyte mitotic divisions by a pulse of forced expression of *bam* can revert back to the stem cell state (Kai and Spradling 2004). Dedifferentiation of TA cells may be an important mechanism to replace lost stem cells in situations where stem cells normally divide with a strictly programmed oriented spindle. As discussed below, replacement of stem cells by dedifferentiation of TA cells also appears to occur during mammalian spermatogenesis (Nakagawa et al. 2010). This ability means that transplantation assays may detect not only the most fundamental stem cells, but also possibly early TA cells that have initiated but are not yet committed to differentiation. The ability of TA germ cells to revert to the stem cell state may reflect an underlying regulatory logic where the decision to proceed from mitosis into meiosis is taken gradually (see Kimble 2011).

Cell–Cell Interactions Maintain Germline and Somatic Stem Cells in the Testis Stem Cell Niche

The niche for male GSCs in *Drosophila* is complex, involving contributions from two somatic cell types, the hub and the CySCs. A cytokinelike signaling system maintains both the GSCs and the CySCs. The somatic cells of the apical hub express the secreted ligand Upd, which binds the receptor, Domeless, activating the single *Drosophila* JAK, encoded by the *hopscotch* gene. The activated JAK then phosphorylates the transcription factor STAT in the cytoplasm, causing it to dimerize, move into the nucleus, and bind to target DNA sequences.

Action of STAT is required cell autonomously in male GSCs to maintain attachment of GSCs to the hub. Individual GSCs made homozygous mutant for STAT by recombination in a STAT/+ background are lost to differentiation (Kiger et al. 2001; Tulina and Matunis 2001). Genetic manipulations that cause loss of STAT function in all germ cells while preserving STAT function in the somatic CySC lineage result in detachment of GSCs from the hub, with CySCs moving in to fill the vacated space (Leatherman and Dinardo 2010). Strikingly, however, action of STAT within germ cells appeared neither necessary nor sufficient to maintain germ cells in the stem cell state. The germ cells displaced from the hub after loss of STAT function in germline but not in somatic cells maintained stem cell characteristics, in that they were able to produce differentiating male germ cells and support spermatogenesis for weeks (Leatherman and Dinardo 2010). Conversely, forced expression of constitutively active JAK in early male germ cells did not cause overproliferation of GSC-like cells (Leatherman and Dinardo 2008). Thus a major role of activation of STAT in GSCs by the Upd signal from the hub appears to be maintaining attachment to the hub rather than specifying stem cell self-renewal. Instead, as we shall see below, maintenance of attachment to the hub serves to hold GSCs in a privileged microenvironment provided by the somatic CySCs, whereas centrosome orientation toward the attachment site sets up an oriented spindle and programs the normally asymmetric outcome of GSC divisions (Fig. 2B).

Activation of STAT in CySCs appears to be both necessary and sufficient to maintain the somatic cells in the stem cell state (Leatherman and Dinardo 2008). CySCs made homozygous mutant for STAT in somatic cell clones failed to self-renew and were quickly lost from the tip of the testis. Tellingly, forced expression of constitutively active JAK in the somatic cyst cell lineage resulted in overproliferation of CySC-like cells. The activation of STAT in CySC is likely to specify maintenance of the stem cell state through action of the transcriptional repressors ZFH1 and CHINMO (Leatherman and Dinardo 2008; Flaherty et al. 2010). Both ZFH1 and CHINMO are expressed in CySCs and present in decreasing levels in their cyst cell progeny. Clonal analysis revealed that loss of function of either, results in loss of CySCs. Conversely, forced expression of high levels of either ZFH1 or CHINMO resulted in continued proliferation of the somatic cells, even away from the hub, and failure to turn on the cyst cell differentiation marker *Eya*, suggesting that either ZFH1 or CHINMO are sufficient to maintain CySCs in a stem cell-like state (Leatherman and Dinardo 2008; Flaherty et al. 2010). Strikingly, when CySCs were maintained in a stem cell-like state by forced expression of ZFH1, CHINMO, or constitutively active JAK in the cyst cell lineage, neighboring germ cells continued to proliferate in a stem cell or Gb-like state and rarely turned on expression of TA stage differentiation markers such as *bam* (Leatherman and Dinardo 2008; Flaherty et al. 2010).

A new picture of a complex niche governing GSC self-renewal and differentiation in the *Drosophila* testis is emerging from these recent findings (Fig. 2B). In this model, Upd signals from the hub instruct neighboring CySCs to maintain stem cell identity. The CySCs in turn provide a privileged microenvironment that maintains neighboring germ cells in a stem cell state. Meanwhile, activation of STAT in germ cells in response to the Upd signal from the hub instructs GSCs to remain attached to the hub, ensuring that they remain in the niche provided by the CySCs. Forced ectopic expression of Upd in GSCs maintains neighboring somatic cells in a proliferative CySC-like state, providing an expanded microenvironment for self-renewal of GSCs, resulting in a tumorlike mass of CySCs and GSCs. The current model is that CySCs send signal(s) that instruct GSCs to self-renew (Leatherman and Dinardo 2010) (lightning bolt in Fig. 2B). As in the ovary, transforming growth factor-β (TGF-β) signaling is a likely candidate, as the TGF-β class ligand Gbb, expressed in somatic cells including early cyst cells at the tip of the testis, is required for GSC maintenance (Shivdasani and Ingham 2003). In addition, loss of function of TGF-β pathway receptors or signal transduction components in GSCs causes loss of GSCs to differentiation. However, it remains to be determined whether TGF-β signaling is permissive for GSC self-renewal, possibly by repressing expression of *bam*, or whether it plays an instructive role in maintaining the GSC state in males.

The lesson emerging from analysis of the *C. elegans* gonad and the *Drosophila* testis and ovarian systems is that decisions between self-renewal and differentiation of GSCs are regulated by local signals from nearby somatic support cells.

THE MAMMALIAN TESTIS STEM CELL NICHE

It is safe to say that relative to *Drosophila*, we know relatively little about the regulation of mammalian GSCs and the niche in which they reside. Unlike in *Drosophila*, there does not appear to be such a strictly defined anatomical niche, and it is near certain that a structure analogous to the hub does not exist in mammals, at least in rodents and primates (including man). There is no known equivalent of the hub or the CySC (see above). Nonetheless, GSCs seem to have a tendency to reside in certain regions of the seminiferous epithelium—near areas of the interstitium that contain branching blood vessels, and GSC behavior is clearly influenced by factors secreted by adjacent somatic Sertoli cells and interstitial peritubular myoid and Leydig cells.

In the mouse, where GSCs have been best studied, it is still not currently possible to definitely identify a specific cell as being a GSC. Instead, the current state of the field refers to a group of cells as representing the GSC compartment, comprised of germ cells with different extents of stem cell potential. This may reflect the ability of early spermatogonia to dedifferentiate (discussed below) and a gradual transition from stem cell to TA state, as proposed for *C. elegans* (see Kimble 2011).

Within the GSC compartment there are many similarities between mice and *Drosophila*. At the top of the existing hierarchy is the morphologically most primitive cell, the type A single (A_s) spermatogonia (Fig. 1C). The A_s cell can divide to form two new A_s cells, or divide with incomplete cytokinesis to form a pair (A_{pr}) of cells connected by an intercellular bridge. Those A_s cells that produce an A_{pr} may be the mammalian equivalent of the Gb in the *Drosophila* hierarchy. A_{pr} cells divide to form a chain of four aligned cells ($A_{al\text{-}4}$), which can divide to form large chains of 8 ($A_{al\text{-}8}$), 16 ($A_{al\text{-}16}$), and sometimes even 32 ($A_{al\text{-}32}$) cells before relinquishing the stem cell activity. Although not exactly analogous, the A_{pr} and A_{al} spermatogonial cysts are very much like the *Drosophila* TA cells (see above). As in *Drosophila*, the mitotic sisters remain connected by cystoplasmic bridges and divide in synchrony. All of these cell types have been shown experimentally to have the capacity to behave like GSCs, in that after transplantation they can form colonies that repopulate a recipient testis with both GSCs and differentiated cells (Nakagawa et al. 2007). However, the greatest self-renewing potential likely resides within the A_s population, and self-renewing potential falls as cyst length increases (see below). In contrast to *D. melanogaster*, where male and female germ cells normally enter meiosis at the 16-cell cyst stage, in mice different lengths of A_{al} spermatogonial cysts transit into differentiating spermatogonia. Although cysts of 8 and 16 cells are the major population that undergo the switch, shorter cysts and even A_s cells also transform into differentiating spermatogonia (Nakagawa et al. 2010).

A key regulatory point in mammalian spermatogenesis is the switch from early TA spermatogonia to differentiating spermatogonial cysts (Oakberg 1956). At a specific time during the 8.6 d cycle of the murine seminiferous epithelial cycle, cohorts of A_{pr} and A_{al} spermatogonia differentiate into A_1 spermatogonia, which in turn divide to become A_2, A_3, A_4, intermediate (In), and eventually type B spermatogonia, the penultimate cell type to the preleptotene spermatocyte (Fig. 1C). The older literature tends to favor nomenclature that refers to the A_s, A_{pr}, and A_{al} spermatogonia as undifferentiated spermatogonia, and the $A_{1\text{-}4}$, In, and B spermatogonia as differentiated spermatogonia. Reasons for this include differences in cell cycle regulation (asynchronous division between different cysts of A_{pr} and $A_{al4\text{-}16}$ cells compared to relatively synchronous division of unrelated A_1 cysts, A_2 cysts, etc.) and chromatin structure. Even though there are differences between these two groups (A_s, A_{pr}, and A_{al} compared to $A_{1\text{-}4}$, In, and B) of spermatogonia, we prefer not to use this nomenclature because the cytoplasmic bridges connecting A_{pr} and larger cysts are an aspect of their differentiation, stem cell potential likely falls as cyst length increases within the A_s, A_{pr}, and A_{al} populations, and, as in *Drosophila*, spermatogonia within developing cysts (usually referred to as chains in mammals) can separate from a cyst and revert to GSCs (Hogarth and Griswold 2010).

When cells enter meiosis in mammals is a bit of a semantic argument. To the meiotic scientist, meiosis begins when preleptotene spermatocytes undergo premeiotic DNA synthesis or when double-strand breaks are initiated and chromosomes pair. Germ cells appear to commit to progress toward meiotic entry at the A_{al} to A_1 transition, a critical juncture that is under retinoic acid regulation, but only initiate meiosis after six more mitotic cell divisions.

Embryonic Induction of Primordial Germ Cells and Their Differentiation into GSCs

PGCs in mammals are an induced cell lineage that arise extragonadally around the onset of gastrulation at embryonic day E7.25 (Ginsburg et al. 1990; Lawson and Hage 1994). Embryonic position and extracellular signaling molecules, including BMP4 and BMP8b, are required for PGC formation (Lawson et al. 1999; Ying et al. 2000). Critical to becoming a PGC is the need to avoid differentiating (Seydoux and Braun 2006). Both chromatin modifying enzymes (BLIMP1/PRDM1 and PRDM14) and regulators of miRNA function (LIN28) play essential roles in PGC formation and retention of a permissive epigenetic state (Ohinata et al. 2005; Vincent et al. 2005; Yamaji et al. 2008; West et al. 2009). Perhaps not coincidentally, segregation of PGCs from embryonic signaling cues bestows on them retention of pluripotency and the ability to form teratomas (Stevens 1967). PGCs retain the ability to form teratomas until their migration into the genital ridge and until testicular cord formation in males.

The molecular mechanisms governing the transition from a PGC to a GSC are not known. However, it is likely that it happens gradually during fetal development, is influenced by signaling from somatic cells of the gonad, and does not involve all of the PGCs. Conditional ablation in fetal Sertoli cells of SIN3A, a component of a large transcriptional remodeling complex, results in down-regulation of several germ cell-expressed pluripotency and GSC genes (e.g., *Oct3/4*, *Gfra1*, *c-Ret*, promyelocytic leukemia zinc finger [PLZF]) and greatly reduces teratoma formation when fetal gonads are transferred to an ectopic site (Payne et al. 2010). Ablation of SIN3A in Sertoli cells does not lead to loss in the absolute numbers of fetal germ cells, suggesting that SIN3A-mediated signaling from fetal Sertoli cells is specifically required to retain pluripotency in germ cells. The CXCl12 ligand, which is expressed in Sertoli cells, and its receptor CXCR4, which is expressed in germ fetal germ cells, may be important mediators of SIN3A activity (Payne et al. 2010). The CXCL12/CXCR4 signaling system is also required for guidance of PGC migration and survival, and both are up-regulated in germ cell tumors in patients (Doitsidou et al. 2002; Molyneaux et al. 2003; Gilbert et al. 2009).

Properties of Postnatal GSCs

The early postnatal gonad in mice contains a heterogeneous group of germ cells in regard to both gene expression and function. The onset of postnatal germ cell proliferation and the chronology of the formation of the first mature sperm have long supported the belief that the first wave of spermatogenesis is different from steady-state spermatogenesis in adults (de Rooij 1998). Lineage tracing experiments have confirmed this notion and have demonstrated that the first sperm originate from a subset of postnatal germ cells that are the equivalent of A_1 (Yoshida et al. 2006). GSCs are derived from the remaining population of perinatal germ cells. During prepubertal growth, Sertoli cells proliferate, presumptive GSCs migrate from their perinatal embryonic luminal position to the basal lamina, and a functional niche is established to maintain self-renewal of GSCs and steady-state spermatogenesis.

Genetic studies suggest that both transcriptional and translational regulation is important for murine GSC self-renewal. Mutations in the gene encoding PLZF transcription factor result in an age-dependent loss of GSCs (Buaas et al. 2004; Costoya et al. 2004). Several other transcription factors (e.g., TAF4b, BCL6b, ETV5, and LHX1) have also been implicated in GSC self-renewal (Chen et al. 2005; Falender et al. 2005; Oatley et al. 2006). In most cases, the genes downstream from these transcription factors have not yet been identified. Less is known about the importance of translational control, but NANOS2, whose expression is highly restricted to A_s, A_{pr}, and short chains of A_{al} spermatogonia, is also required for GSC self-renewal (Sada et al. 2009). The requirement for LIN28 for PGC formation, and its expression in the GSC compartment, suggests that the let7 class of miRNAs may also function in GSCs (West et al. 2009).

As described below, glial cell-derived neurotrophic factor (GDNF) expression from Sertoli cells is critical for GSC self-renewal. GDNF interacts with GSCs via direct binding to its receptor GFRα1 and its coreceptor, c-Ret. Like all genes known to be expressed in the GSC compartment, GFRα1 expression is heterogeneous, and it tends to be expressed in most A_s, A_{pr}, and small chains of A_{al} spermatogonia and be absent in longer chains of A_{al} spermatogonia. Several signaling pathways have been suggested to function downstream from GFRα1/c-Ret in GSCs. These include the Src family kinases and the PI3-kinase-AKT pathways (Braydich-Stolle et al. 2007; Lee et al. 2007). Recent studies have suggested that mTORC1 antagonizes GFRα1/c-Ret signaling, which decreases PI3K/AKT signaling, resulting in a shift in balance toward GSC differentiation (Hobbs et al. 2010). mTORC1 levels are thought to be negatively regulated by REDD1, whose expression is under positive regulation by PLZF. In this model, loss of PLZF leads to an elevation of mTORC1 activity, increased inhibition of GFRα1/c-Ret signaling, and loss of GSC self-renewal.

Regulatory Interactions between GSCs and Their Environment

Seminiferous tubules are composed of germ cells and somatic Sertoli cells (Fig. 2B). The tubules are surrounded by a specialized extracellular matrix, the basal lamina, which in turn is wrapped with specialized peritubular myoid cells. In the interstitial space between tubules lie the steroidogenic Leydig cells, the vasculature, immune cells, and rare fibroblast-like cells. Within the tubules, extensive tight junctions between Sertoli cells create two compartments: a basal compartment demarcated by the basal lamina at the tubule circumference and an adluminal compartment that extends to the lumen of the tubule. The basal compartment segregates the spermatogonia from the more differentiated meiotic and postmeiotic germ cells. All spermatogonia (A_s, A_{pr}, A_{al}, A1-4, In, and B types) occupy the basal compartment, lie along the basal lamina of the epithelium, and are in direct contact with somatic Sertoli cells. In addition, A_s, A_{pr}, A_{al}, and more differentiated spermatogonia are intermingled with each other in the basal compartment. The basal compartment repeats the same simple structure around the circumference and along the length of seminiferous tubules. Upon differentiation into preleptotene spermatocytes the cells permanently lose contact with the basal lamina and move from the basal to the adluminal compartment of the epithelium.

Although GSCs can be found along the length of the entire seminiferous epithelium, GSCs appear to preferentially locate in regions of the epithelium in greatest contact with the interstitium, rather than juxtaposed to another tubule, and in areas that contain branching vasculature (Chiarini-Garcia et al. 2001; Yoshida et al. 2007). E-cad is expressed on GSCs in mice, suggesting adherence junctions may be important for GSC behavior. However, despite the important role of E-cad-mediated binding of GSCs to the hub in asymmetric division and self-renewal in *Drosophila* (see above), genetic ablation of E-cad in cultured mouse GSCs does not affect their ability to repopulate and establish normal spermatogenesis in transplantation studies (Kanatsu-Shinohara et al. 2008).

An important feature of murine GSCs in mice is their motility. Live imaging of EGFP-labeled A_s, A_{pr}, and A_{al} cells clearly demonstrates extensive migration along the X/Y axis of the basal lamina (Yoshida et al. 2007). Therefore, unlike in *Drosophila*, where GSCs are bound to the hub, murine GSCs can disperse themselves along the surface of the epithelium. This may be an important adaptation of mammalian GSCs to the different anatomical design of the insect versus the mammalian testis and may explain why there does not appear to be an anatomical stem cell niche in mammals. In *Drosophila*, the testis is essentially one-dimensional—GSCs reside at one end of the tubular structure and differentiate as they move down the tubule. In mammals, the seminiferous epithelium is functionally three dimensional—GSCs lie along the entire circumference of the tubule and translocate along the *Z* axis toward the lumen as they differentiate. Homeostasis of the mammalian seminiferous epithelium requires coordination of the initiation of spermatogenesis both spatially and temporally. The apparent stochastic nature of self-renewal within the GSC compartment (discussed below), requires that GSCs be motile so that they are able to respond to changes in germ cell density and relocate along the length of the tubule during both steady-state and regenerative spermatogenesis.

Sertoli cells interact with all germ cells within the epithelium and clearly play both instructive and permissive roles in spermatogenesis (refer to Matzuk et al. 2011). The development of an in vitro culture system for murine GSCs has allowed rapid screening for growth factors that promote survival or proliferation in vitro. Stem cell properties can be assayed in cultured cells (e.g., expression of Oct3/4, GFRα1, or PLZF) or in functional transplantation experiments. Several growth factors, including IGF-1 and FGF2 (also known as bFGF) have been implicated in survival or expansion of GSCs in culture (Kanatsu-Shinohara et al. 2003; Kubota et al. 2004). CSF-1, which is expressed in peritubular myoid cells and Leydig cells, and promotes GSC self-renewal in vitro (Oatley et al. 2009), may be an important extrinsic regulator of self-renewal.

Both genetic and in vitro studies unequivocally support a major role for Sertoli cell expression of GDNF in GSC maintenance. Loss of GDNF signaling in vivo results in age-dependent depletion of the GSC compartment (Meng et al. 2000) and overexpression of GDNF leads to expansion of the GSC compartment (Meng et al. 2000; Creemers et al. 2002; Yomogida et al. 2003). In vitro studies with GSC cultures suggest that GDNF may promote the self-renewal of GSCs by stimulating their survival or proliferation (Tadokoro et al. 2002; Kanatsu-Shinohara et al. 2003; Kubota et al. 2004). Not yet known is whether the local concentration of GDNF protein varies locally along the seminiferous epithelium both spatially and temporally and how self-renewal signaling from GDNF is balanced with signaling that must be required for differentiation.

Heterogeneity of GSC Potential within the GSC Compartment

A long-held presumption in mice has been that the A_s spermatogonia are the stem cells, whereas A_{pr} and A_{al} spermatogonia are irreversibly committed to differentiation (Huckins 1971; Oakberg 1971; de Rooij 1973). Support for this model stemmed from the fact that the A_s cells are the most primitive spermatogonia, they are not connected to other cells through intercellular bridges, they have a large nuclear-to-cytoplasmic ratio, and they are present in all stages of the seminiferous cycle. Three lines of evidence suggest that this long-held assumption is incorrect. First, all of these cell types have been shown experimentally to have the capacity to behave like GSCs, in that after transplantation they can form colonies that repopulate a recipient testis with both GSCs and differentiated cells (Nakagawa et al. 2010). It is likely, although not yet formally proven, that the greatest self-renewing potential resides within the A_s population, and self-renewing potential falls as cyst length increases. Second, genes that are required cell autonomously in GSCs for self-renewal, for example, *Gfra1*, *Plzf*, and *Nanos2*, are not limited in their expression to A_s cells. Every marker analyzed to date, whether present in stem cell-enriched populations used for transplantation or detected by immunostaining in whole-mount tubules, shows variable expression within the A_s, A_{pr}, and A_{al} population; this includes GFRα1, NGN3, NANOS2, E-cad, and PLZF (Buaas et al. 2004; Yoshida et al. 2004; Suzuki et al. 2009; Nakagawa et al. 2010). All of these markers are expressed in some, but not all A_s, A_{pr}, and A_{al} cells. This heterogeneity in gene expression suggests an unstable epigenetic state reflecting both self-renewal and differentiating properties within the pool of cells. Lastly, live imaging and lineage tracing during steady-state spermatogenesis have shown that, as in *Drosophila*, interconnected spermatogonia in a TA cyst occasionally fragment into single spermatogonia or shorter cysts (Nakagawa et al. 2010).

The ability of TA germ cells to revert to the stem cell state may reflect an underlying regulatory logic where the decision to proceed from mitosis into meiosis is taken gradually (see Kimble 2011). This characteristic, observed in animal species as disparate as flies and mice, may help ensure the persistence of gametogenesis, especially for animals with long reproduction periods.

Do Mouse GSCs Undergo Asymmetric Cell Division?

A prevailing idea is that stem cells are located in specialized anatomical niches and that asymmetric cell division produces one self-renewing daughter that remains in the niche and another daughter that leaves the niche and differentiates. However, mathematical analyses of single stem cell-derived cohorts in mouse testes indicate that stem cells that are located sparsely with respect to each other do not remain tethered to particular positions, and that GSCs replace each other at a surprisingly high frequency (Klein 2010). In support of this, detailed live in vivo imaging clearly demonstrates migration, expansion, and replacement of A_s, A_{pr}, and A_{al} within 8.6 days (i.e., one cycle of the seminiferous epithelium) (Yoshida et al. 2007; Klein et al. 2010). Moreover, this replacement is, as a whole, revealed to be a random event. In addition, the breakdown of TA spermatogonial cysts to generate new GSCs need not be regulated, and instead, may also be a stochastic event (Klein et al. 2010; Nakagawa et al. 2010). These findings are not easily explained by the prevailing deterministic asymmetric model and suggest a stochastic model of self-renewal and differentiation. However, it remains a formal possibility, as observed in the *Drosophila* testis, that the localized niche microenvironment that allows or instructs GSC self-renewal, influences the divisions oriented with respect to this niche so that one daughter usually remains in the niche and the other is usually displaced away.

REFERENCES

Angelo G, Van Gilst MR. 2009. Starvation protects germline stem cells and extends reproductive longevity in *C. elegans*. *Science* **326:** 954–958.

Austin J, Kimble J. 1987. glp-1 is required in the germline for regulation of the decision between mitosis and meiosis in *C. elegans*. *Cell* **51:** 589–599.

Brawley C, Matunis E. 2004. Regeneration of male germline stem cells by spermatogonial dedifferentiation in vivo. *Science* **304:** 1331–1334.

Braydich-Stolle L, Kostereva N, Dym M, Hofmann MC. 2007. Role of Src family kinases and N-Myc in spermatogonial stem cell proliferation. *Dev Biol* **304:** 34–45.

Buaas FW, Kirsh AL, Sharma M, McLean DJ, Morris JL, Griswold MD, de Rooij DG, Braun RE. 2004. Plzf is required in adult male germ cells for stem cell self-renewal. *Nat Genet* **36:** 647–652.

Büning T, Spradling AC. 2007. *The insect ovary: Ultrastructure, previtellagenic growth and evolution*. Chapman and Hall, London.

Casanueva MO, Ferguson EL. 2004. Germline stem cell number in the *Drosophila* ovary is regulated by redundant mechanisms that control Dpp signaling. *Development* **131:** 1881–1890.

Chau J, Kulnane LS, Salz HK. 2009. Sex-lethal facilitates the transition from germline stem cell to committed daughter cell in the *Drosophila* ovary. *Genetics* **182:** 121–132.

Chen D, McKearin D. 2003. Dpp signaling silences *bam* transcription directly to establish asymmetric divisions of germline stem cells. *Curr Biol* **13:** 1786–1791.

Chen C, Ouyang W, Grigura V, Zhou Q, Carnes K, Lim H, Zhao GQ, Arber S, Kurpios N, Murphy TL, et al. 2005. ERM is required for transcriptional control of the spermatogonial stem cell niche. *Nature* **436:** 1030–1034.

Cheng J, Turkel N, Hemati N, Fuller MT, Hunt AJ, Yamashita YM. 2008. Centrosome misorientation reduces stem cell division during ageing. *Nature* **456:** 599–604.

Cheng J, Tiyaboonchai A, Yamashita YM, Hunt AJ. 2011. Asymmetric division of cyst stem cells in *Drosophila* testis is ensured by anaphase spindle repositioning. *Development* **138:** 831–837.

Chiarini-Garcia H, Hornick JR, Griswold MD, Russell LD. 2001. Distribution of type A spermatogonia in the mouse is not random. *Biol Reprod* **65:** 1179–1185.

Cinquin O, Crittenden SL, Morgan DE, Kimble J. 2010. Progression from a stem cell-like state to early differentiation in the *C. elegans* germ line. *Proc Natl Acad Sci* **107:** 2048–2053.

Civetta A, Rajakumar SA, Brouwers B, Bacik JP. 2006. Rapid evolution and gene-specific patterns of selection for three genes of spermatogenesis in *Drosophila. Mol Biol Evol* **23:** 655–662.

Costoya JA, Hobbs RM, Barna M, Cattoretti G, Manova K, Sukhwani M, Orwig KE, Wolgemuth DJ, Pandolfi PP. 2004. Essential role of Plzf in maintenance of spermatogonial stem cells. *Nat Genet* **36:** 653–659.

Creemers LB, Meng X, den Ouden K, van Pelt AM, Izadyar F, Santoro M, Sariola H, de Rooij DG. 2002. Transplantation of germ cells from glial cell line-derived neurotrophic factor-overexpressing mice to host testes depleted of endogenous spermatogenesis by fractionated irradiation. *Biol Reprod* **66:** 1579–1584.

Davies EL, Fuller MT. 2008. Regulation of self-renewal and differentiation in adult stem cell lineages: Lessons from the *Drosophila* male germ line. *CSHSQB* **73:** 137–145.

Decotto E, Spradling AC. 2005. The *Drosophila* ovarian and testis stem cell niches: Similar somatic stem cells and signals. *Dev Cell* **9:** 501–510.

Deng W, Lin H. 1997. Spectrosomes and fusomes anchor mitotic spindles during asymmetric germ cell divisions and facilitate the formation of a polarized microtubule array for oocyte specification in *Drosophila. Dev Biol* **189:** 79–94.

de Rooij DG. 1973. Spermatogonial stem cell renewal in the mouse. I. Normal situation. *Cell Tissue Kinet* **6:** 281–287.

de Rooij DG. 1998. Stem cells in the testis. *Int J Exp Pathol* **79:** 767–780.

Doitsidou M, Reichman-Fried M, Stebler J, Koprunner M, Dorries J, Meyer D, Esguerra CV, Leung T, Raz E. 2002. Guidance of primordial germ cell migration by the chemokine SDF-1. *Cell* **111:** 647–659.

Draper BW, McCallum CM, Moens CB. 2007. Nanos1 is required to maintain oocyte production in adult zebrafish. *Dev Biol* **305:** 589–598.

Drummond-Barbosa D. 2008. Stem cells, their niches and the systemic environment: An aging network. *Genetics* **180:** 1787–1797.

Falender AE, Freiman RN, Geles KG, Lo KC, Hwang K, Lamb DJ, Morris PL, Tjian R, Richards JS. 2005. Maintenance of spermatogenesis requires TAF4b, a gonad-specific subunit of TFIID. *Genes Dev* **19:** 794–803.

Fichelson P, Moch C, Ivanovitch K, Martin C, Sidor CM, Lepesant JA, Bellaiche Y, Huynh JR. 2009. Live-imaging of single stem cells within their niche reveals that a U3snoRNP component segregates asymmetrically and is required for self-renewal in *Drosophila. Nat Cell Biol* **11:** 685–693.

Flaherty MS, Salis P, Evans CJ, Ekas LA, Marouf A, Zavadil J, Banerjee U, Bach EA. 2010. Chinmo is a functional effector of the JAK/STAT pathway that regulates eye development, tumor formation, and stem cell self-renewal in *Drosophila. Dev Cell* **18:** 556–568.

Flatt T, Min KJ, D'Alterio C, Villa-Cuesta E, Cumbers J, Lehmann R, Jones DL, Tatar M. 2008. *Drosophila* germ-line modulation of insulin signaling and lifespan. *Proc Natl Acad Sci* **105:** 6368–6373.

Forbes A, Lehmann R. 1998. Nanos and Pumilio have critical roles in the development and function of *Drosophila* germline stem cells. *Development* **125:** 679–690.

Forstemann K, Tomari Y, Du T, Vagin VV, Denli AM, Bratu DP, Klattenhoff C, Theurkauf WE, Zamore PD. 2005. Normal microRNA maturation and germ-line stem cell maintenance requires Loquacious, a double-stranded RNA-binding domain protein. *PLoS Biol* **3:** e236. 10.1371/journal.pbio.0030236.

Fuller MT, Spradling AC. 2007. Male and female *Drosophila* germline stem cells: Two versions of immortality. *Science* **316:** 402–404.

Gilbert DC, Chandler I, McIntyre A, Goddard NC, Gabe R, Huddart RA, Shipley J. 2009. Clinical and biological significance of CXCL12 and CXCR4 expression in adult testes and germ cell tumours of adults and adolescents. *J Pathol* **217:** 94–102.

Gilboa L, Lehmann R. 2004. Repression of primordial germ cell differentiation parallels germ line stem cell maintenance. *Curr Biol* **14:** 981–986.

Gilboa L, Lehmann R. 2006. Soma-germline interactions coordinate homeostasis and growth in the *Drosophila* gonad. *Nature* **443:** 97–100.

Ginsburg M, Snow MH, McLaren A. 1990. Primordial germ cells in the mouse embryo during gastrulation. *Development* **110:** 521–528.

Gönczy P, Viswanathan S, DiNardo S. 1992. Probing spermatogenesis in *Drosophila* with P-element enhancer detectors. *Development* **114:** 89–98.

Gönczy P, Matunis E, DiNardo S. 1997. *Bag-of-marbles* and *benign gonial cell neoplasm* act in the germline to restrict proliferation during *Drosophila* spermatogenesis. *Development* **124:** 4361–4371.

Greenbaum M, Iwamori T, Buchold G, Matzuk MM. 2011. Germ cell intercellular bridges. *Cold Spring Harb Perspect Biol* doi:10.1101/cshperspect.a005850.

Guo Z, Wang Z. 2009. The glypican Dally is required in the niche for the maintenance of germline stem cells and short-range BMP signaling in the *Drosophila* ovary. *Development* **136:** 3627–3635.

Hatfield SD, Shcherbata HR, Fischer KA, Nakahara K, Carthew RW, Ruohola-Baker H. 2005. Stem cell division is regulated by the microRNA pathway. *Nature* **435:** 974–978.

Hayashi Y, Kobayashi S, Nakato H. 2009. *Drosophila* glypicans regulate the germline stem cell niche. *J Cell Biol* **187:** 473–480.

Hobbs RM, Seandel M, Falciatori I, Rafii S, Pandolfi PP. 2010. Plzf regulates germline progenitor self-renewal by opposing mTORC1. *Cell* **142:** 468–479.

Hogarth CA, Griswold MD. 2010. The key role of vitamin A in spermatogenesis. *J Clin Invest* **120:** 956–962.

Hsu HJ, Drummond-Barbosa D. 2009. Insulin levels control female germline stem cell maintenance via the niche in *Drosophila. Proc Natl Acad Sci* **106:** 1117–1121.

Hsu HJ, LaFever L, Drummond-Barbosa D. 2008. Diet controls normal and tumorous germline stem cells via insulin-dependent and -independent mechanisms in *Drosophila. Dev Biol* **313:** 700–712.

Huckins C. 1971. The spermatogonial stem cell population in adult rats. I. Their morphology, proliferation and maturation. *Anat Rec* **169:** 533–557.

Inaba M, Yuan H, Salzmann V, Fuller MT, Yamashita YM. 2010. E-cadherin is required for centrosome and spindle orientation in *Drosophila* male germline stem cells. *PLoS One* **5:** e12473. 10.1371/journal.pone.0012473.

Insco M, Leon A, Tam C, McKearin D, Fuller MT. 2009. Accumulation of a differentiation regulator specifies transit amplifying division number in an adult stem cell lineage. *Proc Natl Acad Sci* **106:** 22311–22316.

Iovino N, Pane A, Gaul U. 2009. miR-184 has multiple roles in *Drosophila* female germline development. *Dev Cell* **17:** 123–133.

Issigonis M, Tulina N, de Cuevas M, Brawley C, Sandler L, Matunis E. 2009. JAK-STAT signal inhibition regulates competition in the *Drosophila* testis stem cell niche. *Science* **326:** 153–156.

Jin Z, Xie T. 2007. Dcr-1 maintains *Drosophila* ovarian stem cells. *Curr Biol* **17:** 539–544.

Jin Z, Kirilly D, Weng C, Kawase E, Song X, Smith S, Schwartz J, Xie T. 2008. Differentiation-defective stem cells outcompete normal stem cells for niche occupancy in the *Drosophila* ovary. *Cell Stem Cell* **2:** 39–49.

Johnston LA. 2009. Competitive interactions between cells: Death, growth, and geography. *Science* **324:** 1679–1682.

Kai T, Spradling A. 2004. Differentiating germ cells can revert into functional stem cells in *Drosophila melanogaster* ovaries. *Nature* **428:** 564–569.

Kanatsu-Shinohara M, Ogonuki N, Inoue K, Miki H, Ogura A, Toyokuni S, Shinohara T. 2003. Long-term proliferation in culture and germline transmission of mouse male germline stem cells. *Biol Reproduction* **69:** 612–616.

Kanatsu-Shinohara M, Takehashi M, Takashima S, Lee J, Morimoto H, Chuma S, Raducanu A, Nakatsuji N, Fassler R, Shinohara T. 2008. Homing of mouse spermatogonial stem cells to germline niche depends on beta1-integrin. *Cell Stem Cell* **3:** 533–542.

Kiger AA, Jones DL, Schulz C, Rogers MB, Fuller MT. 2001. Stem cell self-renewal specified by JAK-STAT activation in response to a support cell cue. *Science* **294:** 2542–2545.

Kimble J. 2011. Molecular regulation of the mitosis/meiosis decision in multicellular organisms. *Cold Spring Harb Perspect Biol* doi: 10.1101/cshperspect.a002683.

Klattenhoff C, Theurkauf W. 2008. Biogenesis and germline functions of piRNAs. *Development* **135:** 3–9.

Klein AM, Nakagawa T, Ichikawa R, Yoshida S, Simons BD. 2010. Mouse germ line stem cells undergo rapid and stochastic turnover. *Cell Stem Cell* **7:** 214–224.

Kotaja N, Sassone-Corsi P. 2007. The chromatoid body: A germ-cell-specific RNA-processing centre. *Nat Rev Mol Cell Biol* **8:** 85–90.

Kubota H, Avarbock MR, Brinster RL. 2004. Growth factors essential for self-renewal and expansion of mouse spermatogonial stem cells. *Proc Natl Acad Sci* **101:** 16489–16494.

Kugler JM, Lasko P. 2009. Localization, anchoring and translational control of oskar, gurken, bicoid and nanos mRNA during *Drosophila* oogenesis. *Fly (Austin)* **3:** 15–28.

LaFever L, Drummond-Barbosa D. 2005. Direct control of germline stem cell division and cyst growth by neural insulin in *Drosophila. Science* **309:** 1071–1073.

Lawson KA, Hage WJ. 1994. Clonal analysis of the origin of primordial germ cells in the mouse. *Ciba Found Symp* **182:** 68–84; discussion 84–91.

Lawson KA, Dunn NR, Roelen BA, Zeinstra LM, Davis AM, Wright CV, Korving JP, Hogan BL. 1999. Bmp4 is required for the generation of primordial germ cells in the mouse embryo. *Genes Dev* **13:** 424–436.

Leatherman JL, Dinardo S. 2008. Zfh-1 controls somatic stem cell self-renewal in the *Drosophila* testis and nonautonomously influences germline stem cell self-renewal. *Cell Stem Cell* **3:** 44–54.

Leatherman JL, Dinardo S. 2010. Germline self-renewal requires cyst stem cells and stat regulates niche adhesion in *Drosophila* testes. *Nat Cell Biol* **12:** 806–811.

Le Bras S, Van Doren M. 2006. Development of the male germline stem cell niche in *Drosophila. Dev Biol* **294:** 92–103.

Lee J, Kanatsu-Shinohara M, Inoue K, Ogonuki N, Miki H, Toyokuni S, Kimura T, Nakano T, Ogura A, Shinohara T. 2007. Akt mediates self-renewal division of mouse spermatogonial stem cells. *Development* **134:** 1853–1859.

Li Y, Minor NT, Park JK, McKearin DM, Maines JZ. 2009. *Bam* and *Bgcn* antagonize Nanos-dependent germ-line stem cell maintenance. *Proc Natl Acad Sci* **106:** 9304–9309.

Lim AK, Kai T. 2007. Unique germ-line organelle, nuage, functions to repress selfish genetic elements in *Drosophila melanogaster. Proc Natl Acad Sci* **104:** 6714–9719.

Lin H, Spradling AC. 1997. A novel group of pumilio mutations affects the asymmetric division of germline stem cells in the *Drosophila* ovary. *Development* **124:** 2463–2476.

Lin MD, Jiao X, Grima D, Newbury SF, Kiledjian M, Chou TB. 2008. *Drosophila* processing bodies in oogenesis. *Dev Biol* **322:** 276–288.

Liu N, Han H, Lasko P. 2009. Vasa promotes *Drosophila* germline stem cell differentiation by activating mei-P26 translation by directly interacting with a (U)-rich motif in its 3' UTR. *Genes Dev* **23:** 2742–2752.

López-Onieva L, Fernandez-Minan A, Gonzalez-Reyes A. 2008. Jak/Stat signalling in niche support cells regulates dpp transcription to contol germline stem cell maintenance in the *Drosophila* ovary. *Development* **135:** 533–540.

Meng X, Lindahl M, Hyvonen ME, Parvinen M, de Rooij DG, Hess MW, Raatikainen-Ahokas A, Sainio K, Rauvala H, Lakso M, et al. 2000. Regulation of cell fate decision of undifferentiated spermatogonia by GDNF. *Science* **287:** 1489–1493.

Molyneaux KA, Zinszner H, Kunwar PS, Schaible K, Stebler J, Sunshine MJ, O'Brien W, Raz E, Littman D, Wylie C, et al. 2003. The chemokine SDF1/CXCL12 and its receptor CXCR4 regulate mouse germ cell migration and survival. *Development* **130:** 4279–4286.

Nakagawa T, Nabeshima Y-I, Yoshida S. 2007. Functional identification of the actual and potential stem cell compartments in mouse spermatogenesis. *Dev Cell* **12:** 195–206.

Nakagawa T, Sharma M, Nabeshima Y, Braun RE, Yoshida S. 2010. Functional hierarchy and reversibility within the murine spermatogenic stem cell compartment. *Science* **328:** 62–67.

Nakamura S, Kobayashi K, Nishimura T, Higashijima S, Tanaka M. 2010. Identification of germline stem cells in the ovary of the teleost medaka. *Science* **328:** 1561–1563.

Neumuller RA, Betschinger J, Fischer A, Bushati N, Poernbacher I, Mechtler K, Cohen SM, Knoblich JA. 2008. Mei-P26 regulates microRNAs and cell growth in the *Drosophila* ovarian stem cell lineage. *Nature* **454:** 241–245.

Nystul T, Spradling AC. 2007. An epithelial niche in the *Drosophila* ovary undergoes long-range stem cell replacement. *Cell Stem Cell* **1:** 277–285.

Oakberg EF. 1956. Duration of spermatogenesis in the mouse and timing of stages of the cycle of the seminiferous epithelium. *Am J Anat* **99:** 507–516.

Oakberg EF. 1971. Spermatogonial stem-cell renewal in the mouse. *Anat Rec* **169:** 515–531.

Oatley JM, Avarbock MR, Telaranta AI, Fearon DT, Brinster RL. 2006. Identifying genes important for spermatogonial stem cell self-renewal and survival. *Proc Natl Acad Sci* **103:** 9524–9529.

Oatley JM, Oatley MJ, Avarbock MR, Tobias JW, Brinster RL. 2009. Colony stimulating factor 1 is an extrinsic stimulator of mouse spermatogonial stem cell self-renewal. *Development* **136:** 1191–1199.

Ohinata Y, Payer B, O'Carroll D, Ancelin K, Ono Y, Sano M, Barton SC, Obukhanych T, Nussenzweig M, Tarakhovsky A, et al. 2005. Blimp1 is a critical determinant of the germ cell lineage in mice. *Nature* **436:** 207–213.

Park JK, Liu X, Strauss TJ, McKearin DM, Liu Q. 2007. The miRNA pathway intrinsically controls self-renewal of *Drosophila* germline stem cells. *Curr Biol* **17:** 533–538.

Payne CJ, Gallagher SJ, Foreman O, Dannenberg JH, Depinho RA, Braun RE. 2010. Sin3a is required by sertoli cells to establish a niche for undifferentiated spermatogonia, germ cell tumors, and spermatid elongation. *Stem Cells* **28:** 1424–1434.

Pek JW, Lim AK, Kai T. 2009. Drosophila maelstrom ensures proper germline stem cell lineage differentiation by repressing microRNA-7. *Dev Cell* **17:** 417–424.

Pepling ME, Spradling AC. 1998. Female mouse germ cells form synchronously dividing cysts. *Development* **125:** 3323–3328.

Pepling ME, Spradling AC. 2001. Mouse ovarian germ cell cysts undergo programmed breakdown to form primordial follicles. *Dev Biol* **234:** 339–351.

Rhiner C, Diaz B, Portela M, Poyatos JF, Fernandez-Ruiz I, Lopez-Gay JM, Gerlitz O, Moreno E. 2009. Persistent competition among stem cells and their daughters in the *Drosophila* ovary germline niche. *Development* **136:** 995–1006.

Sada A, Suzuki A, Suzuki H, Saga Y. 2009. The RNA-binding protein NANOS2 is required to maintain murine spermatogonial stem cells. *Science* **325:** 1394–1398.

Sarkar A, Parikh N, Hearn SA, Fuller MT, Tazuke SI, Schulz C. 2007. Antagonistic roles of Rac and Rho in organizing the germ cell microenvironment. *Curr Biol* **17:** 1253–1258.

Schulz C, Wood CG, Jones DL, Tazuke SI, Fuller MT. 2002. Signaling from germ cells mediated by the rhomboid homolog stet organizes encapsulation by somatic support cells. *Development* **129:** 4523–4534.

Seydoux G, Braun RE. 2006. Pathway to totipotency: Lessons from germ cells. *Cell* **127:** 891–904.

Shen R, Weng C, Yu J, Xie T. 2009. eIF4A controls germline stem cell self-renewal by directly inhibiting BAM function in the *Drosophila* ovary. *Proc Natl Acad Sci* **106:** 11623–11628.

Sheng XR, Brawley CM, Matunis EL. 2009a. Dedifferentiating spermatogonia outcompete somatic stem cells for niche occupancy in the *Drosophila* testis. *Cell Stem Cell* **5:** 191–203.

Sheng XR, Posenau T, Gumulak-Smith JJ, Matunis E, Van Doren M, Wawersik M. 2009b. Jak-STAT regulation of male germline stem cell establishment during *Drosophila* embryogenesis. *Dev Biol* **334:** 335–344.

Shivdasani AA, Ingham PW. 2003. Regulation of stem cell maintenance and transit amplifying cell proliferation by TGF-beta signaling in *Drosophila* spermatogenesis. *Curr Biol* **13:** 2065–2072.

Song X, Call GB, Kirilly D, Xie T. 2007. Notch signaling controls germline stem cell niche formation in the *Drosophila* ovary. *Development* **134:** 1071–1080.

Soper SF, van der Heijden GW, Hardiman TC, Goodheart M, Martin SL, de Boer P, Bortvin A. 2008. Mouse maelstrom, a component of nuage, is essential for spermatogenesis and transposon repression in meiosis. *Dev Cell* **15:** 285–297.

Stevens LC. 1967. Origin of testicular teratomas from primordial germ cells in mice. *J Natl Cancer Inst* **38:** 549–552.

Suzuki H, Sada A, Yoshida S, Saga Y. 2009. The heterogeneity of spermatogonia is revealed by their topology and expression of marker proteins including the germ cell-specific proteins Nanos2 and Nanos3. *Dev Biol* **336:** 222–231.

Tadokoro Y, Yomogida K, Ohta H, Tohda A, Nishimune Y. 2002. Homeostatic regulation of germinal stem cell proliferation by the GDNF/FSH pathway. *Mech Dev* **113:** 29–39.

Tilly JL, Niikura Y, Rueda BR. 2009. The current status of evidence for and against postnatal oogenesis in mammals: A case of ovarian optimism versus pessimism? *Biol Reprod* **80:** 2–12.

Tulina N, Matunis E. 2001. Control of stem cell self-renewal in *Drosophila* spermatogenesis by JAK-STAT signaling. *Science* **294:** 2546–2549.

Vincent SD, Dunn NR, Sciammas R, Shapiro-Shalef M, Davis MM, Calame K, Bikoff EK, Robertson EJ. 2005. The zinc finger transcriptional repressor Blimp1/Prdm1 is dispensable for early axis formation but is required for specification of primordial germ cells in the mouse. *Development* **132:** 1315–1325.

Von Stetina JR, Orr-Weaver TL. 2011. Developmental control of oocyte maturation and egg activation in metazoan models. *Cold Spring Harb Perspect Biol* doi:10.1101/cshperspect.a005553.

White 1981; Morgan et al. 2010). In adults, ~35–70 mitotic germ cells at the distal-most end are maintained in an essentially equivalent stem cell–like state; more proximally, transit-amplifying germ cells remain in the mitotic cell cycle but are organized in a distal-to-proximal gradient of increasing maturation (Fig. 2A) (Cinquin et al. 2010). Therefore, the meiotic program is triggered before germ cells leave the mitotic cell cycle and overtly enter meiotic S-phase. Even more proximally, all germ cells have entered meiotic S-phase or progressed into early stages of meiotic prophase I (Fig. 2A). When late larvae are starved, they metabolize most of their germline tissue, with only ~35 GSCs remaining to replenish the entire tissue and generate progeny after refeeding during adulthood (Angelo and Van Gilst 2009). The molecular regulation of GSCs and the mitosis/meiosis decision have been analyzed in well-fed animals, but this recent result shows unequivocally that nutritional cues have a major impact. Consistent with this idea, insulin signaling can affect the extent of larval GSC proliferation (Michaelson et al. 2010).

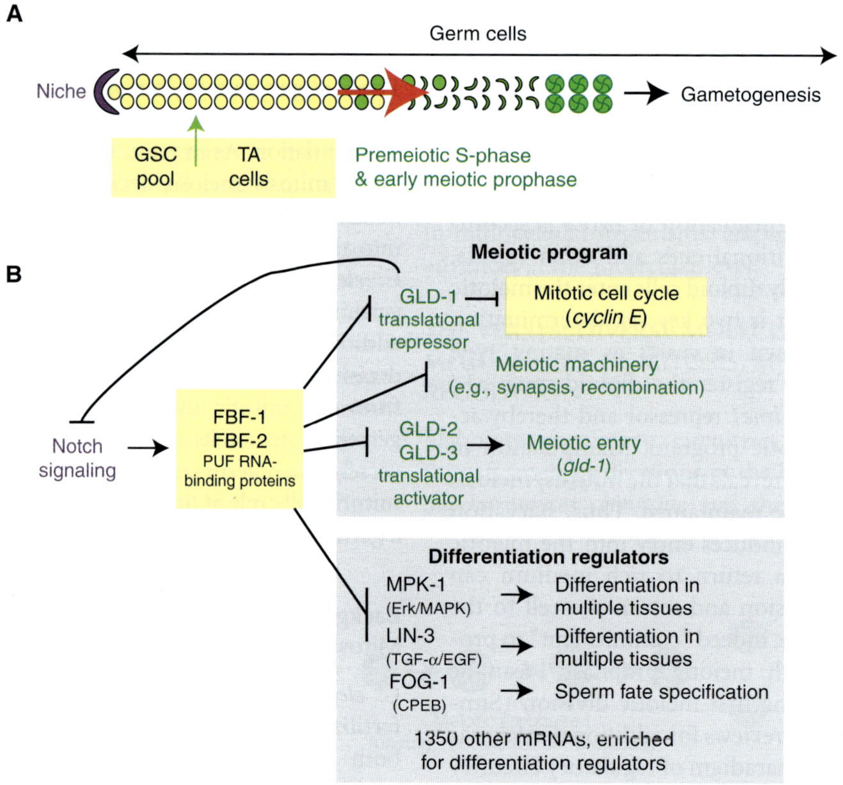

Figure 2. The *C. elegans* mitosis/meiosis decision. (*A*) The nematode germline possesses a GSC pool within its somatic niche. Germ cells in the mitotic cell cycle (yellow) extend beyond the niche and include transit-amplifying (TA) germ cells, which have been triggered to begin maturation toward meiotic entry (green arrow); germ cells in the meiotic cell cycle (green) are more proximal. Overt entry into the meiotic cell cycle (red arrow) occurs asynchronously. Conventions are as in Figure 1. (*B*) Molecular regulation of the *C. elegans* mitosis/meiosis decision includes Notch signaling from the niche and FBF maintenance of GSCs, including repression of the meiotic program. Arrows indicate positive regulation; barred lines indicate negative regulation. Solid arrows and lines indicate direct molecular regulation: Notch signaling directly activates *fbf-2* transcription and FBF directly represses mRNAs of the meiotic program as well as key differentiation regulators.

Cite this article as *Cold Spring Harb Perspect Biol* doi: 10.1101/cshperspect.a002683

Regulators that Promote Germline Self-Renewal Directly Repress the Meiotic Program

Notch signaling and two PUF (for Pumilio and FBF) RNA-binding proteins maintain germline self-renewal and also promote continued mitotic divisions at the expense of meiotic entry (Fig. 2B) (e.g., Kimble and Crittenden 2007). A Notch ligand from the niche keeps GSCs in the mitotic cell cycle and in a stem cell–like state. When signaled, the GLP-1/Notch receptor triggers transcriptional activation of target genes (Petcherski and Kimble 2000). Two GLP-1/Notch target genes relevant to the mitosis/meiosis decision are *fbf-2* (Lamont et al. 2004) and *lip-1* (Lee et al. 2006). FBF-2 is discussed below. LIP-1 is the homolog of a dual specificity phosphatase that normally delays meiotic entry, probably by inactivating MPK-1, the single *C. elegans* homolog of ERK/MAP kinase (Berset et al. 2001; Lee et al. 2006). Other targets of Notch transcriptional activation must exist, because an *fbf-2 lip-1* double-mutant maintains GSCs and supports a virtually normal mitosis/meiosis decision. Importantly, both Notch signaling and FBF control GSCs and the mitosis/meiosis decision in both sexes during adulthood—when XX germlines are oogenic and XO germlines are spermatogenic.

Two nearly identical PUF proteins, FBF-1 and FBF-2 (collectively known as FBF), function within germ cells to regulate both self-renewal and the mitosis/meiosis decision (Fig. 2B) (e.g., Kimble and Crittenden 2007). FBF prevents meiotic entry by directly repressing the expression of a battery of mRNAs critical to the meiotic program, including regulators of meiotic entry (*gld* mRNAs, see below) as well as components of the meiotic machinery (e.g., synaptonemal complex formation, recombination) (Crittenden et al. 2002; Eckmann et al. 2004; Merritt et al. 2008; Kershner and Kimble 2010; Merritt and Seydoux 2010). In addition to its direct control of the meiotic program, FBF also represses *mpk-1* mRNA, *lin-3* mRNA, and *fog-1* mRNA, which encode the *C. elegans* homologs of ERK/MAP kinase, TGF-α/EGF, and CPEB, respectively (Fig. 2B) (Thompson et al. 2005, 2006; Lee et al. 2007a; Kershner and Kimble 2010). MPK-1 and LIN-3 regulate multiple cell fates in somatic tissues (Sundaram 2006); MPK-1 also affects germ cell fate specification (Lee et al. 2007b); FOG-1 is one of the terminal regulators of sperm fate specification (Barton and Kimble 1990). Indeed, FBF has at least 1350 likely targets, many of which encode potent regulators of differentiation (Kershner and Kimble 2010). Therefore, FBF appears to maintain GSCs in a naïve stem cell–like state by repressing a broad spectrum of differentiation regulators, including those in the meiotic program.

Regulators that Promote Meiotic Entry Oppose Germline Self-Renewal and the Mitotic Cell Cycle

Three GLD proteins drive germ cells into the meiotic cell cycle (Fig. 2B) (Kadyk and Kimble 1998; Eckmann et al. 2004; Hansen et al. 2004). These GLD proteins form two branches of control, with GLD-1 in one branch and GLD-2/GLD-3 in the other branch. Importantly, the two GLD branches control meiotic entry in both XX and XO animals (Kadyk and Kimble 1998); therefore, regulation of meiotic entry appears to be gender-neutral.

GLD-1 is a translational repressor (Lee and Schedl 2010) that controls regulators of both self-renewal and the mitotic cell cycle. Its direct targets include mRNAs encoding the GLP-1/Notch receptor and cyclin E (Marin and Evans 2003; Biedermann et al. 2009). GLD-1 also maintains commitment to progression through the meiotic cell cycle; germ cells revert to mitotic divisions and form a germline tumor in *gld-1* single mutants (Francis et al. 1995; Subramaniam and Seydoux 2003; Biedermann et al. 2009); indeed within that tumor, germ cells can *trans*-differentiate into a variety of somatic cell types (Ciosk et al. 2006). Therefore, GLD-1 promotes meiotic entry and also maintains germ cells in a totipotent state during meiotic progression.

The GLD-2/GLD-3 heterodimer is a cytoplasmic poly(A) polymerase and translational activator (Wang et al. 2002; Suh et al. 2006).

One direct target of GLD-2/GLD-3 activation is the *gld-1* mRNA, a regulatory link that provides positive feed-forward to drive meiotic entry (Suh et al. 2006); other direct targets are not yet known but must exist because GLD-2/GLD-3 promotes meiotic entry in *gld-1* null mutants. Like the GLD-1 regulator, GLD-2 and GLD-3 also govern meiotic progression, but their roles are distinct (Kadyk and Kimble 1998; Eckmann et al. 2002).

Shifting the Balance between Mitotic and Meiotic Modes of the Network

The FBF and GLD proteins form major hubs in the network of mitosis/meiosis control and the nature of those hubs confers plasticity on the decision (Kimble and Crittenden 2007). Feedback within the network is provided by GLD-1 repression of Notch signaling (Marin and Evans 2003) and FBF negative auto- and cross-regulation (Lamont et al. 2004). When either FBF-1 or FBF-2 is removed, the other FBF is up-regulated and GSCs are maintained, but when both are removed, GSCs are lost. Similarly, when only one GLD branch is removed, meiotic entry occurs, albeit abnormally, but when both branches are removed a germline tumor results. Intriguingly, the individual FBFs affect the balance between mitotic and meiotic modes differently, probably because of the gene-specific regulation of the two proteins (Crittenden et al. 2002; Lamont et al. 2004). In addition to the FBFs and GLDs, other regulators also shift the number of mitotically dividing germ cells (e.g., LIP-1, see above). This redundancy and modulation renders the mitosis/meiosis decision both robust and plastic, features that permit its regulatory network to respond to varying inputs (e.g., gender, aging, stress, pathogenesis, natural selection).

Many other regulatory proteins influence GSC mitotic divisions and the mitosis/meiosis balance (Kimble and Crittenden 2005; van den Heuvel 2005). The canonical cell cycle machinery controls GSC mitotic divisions, as expected; in addition, many RNA regulatory proteins affect the mitosis/meiosis balance, underscoring the importance of posttranscriptional controls of *C. elegans* germ cell cycles (e.g., Belfiore et al. 2004; Ciosk et al. 2004; Eckmann et al. 2004; Maine et al. 2004; Hansen et al. 2005; Thompson et al. 2005; Hasegawa et al. 2006; Ariz et al. 2009; Kerins et al. 2010). In addition, micro RNAs have been implicated in GSC proliferation controls (Cox et al. 1998; Smardon et al. 2000). Elucidation of the complete molecular network controlling the *C. elegans* mitosis/meiosis decision is clearly a work in progress.

DROSOPHILA: CYST FORMATION MEETS THE MITOSIS/MEIOSIS DECISION

The *Drosophila* germline tissue is organized with stem cells at one end and differentiating gametes at the other (Fig. 3A, male germline). Although male and female germlines differ in many respects, some striking similarities do exist. In both sexes, GSCs divide asymmetrically to produce one GSC daughter and one daughter destined to differentiate; the latter is called a "cystoblast" in females and a "gonialblast" in males. The cystoblast/gonialblast then generates a 16-cell cyst in both sexes. In males, meiotic S-phase occurs in the germ cells of the 16-cell cyst and their meiotic divisions generate 64 sperm. In females, two of the germ cells in the 16-cell cyst enter the meiotic cell cycle and one of those goes on to generate a single egg; the 15 others become nurse cells. Regulation of the GSC–cystoblast/gonialblast decision has been elucidated in exquisite detail in both males and females (see Davies and Fuller 2008; Zhang and Xie 2009). Importantly, two key regulators of that early decision also control meiotic entry in the germ cells of the 16-cell cyst and appear to do so in both sexes.

The *bag-of-marbles (bam)* and *benign gonial cell neoplasm (bgcn)* proteins function in both males and females to promote "postcyst" differentiation, including meiotic entry (Fig. 3B) (McKearin and Spradling 1990; Gönczy et al. 1997; Insco et al. 2009). Germ cells lacking either *bam* or *bgcn* undergo extra rounds of mitotic division and fail to enter the meiotic cell cycle. Moreover, GSCs exposed to ectopic Bam protein are induced to differentiate (Ohlstein and McKearin 1997). The Bam molecular

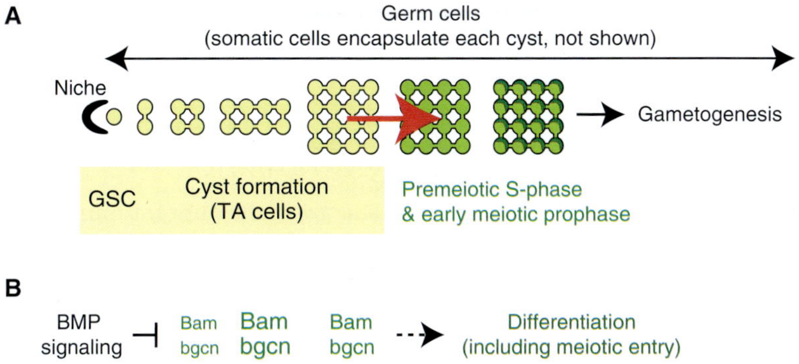

Figure 3. The *Drosophila* mitosis/meiosis decision. (*A*) The *Drosophila* germline possesses asymmetrically dividing GSCs within its somatic niche; transit-amplifying divisions form a 16-cell cyst and meiotic entry (red arrow) occurs in that 16-cell cyst. In males, all 16 germ cells enter the meiotic cell cycle, as depicted here; in females, only two of the 16 germ cells enter the meiotic cell cycle (not shown). Conventions and acronyms are as in Figures 1 and 2. (*B*) Molecular regulation of the *Drosophila* mitosis/meiosis decision relies on extrinsic BMP signaling and intrinsic Bam/bgcn RNA regulation, with abundant Bam controlling the position of meiotic entry. Solid arrows and lines indicate direct molecular regulation; dashed line indicates control that may be either direct or indirect.

function was not apparent from its amino acid sequence (McKearin and Spradling 1990), but bgcn is related to DExH proteins, suggesting a role in RNA regulation (Ohlstein et al. 2000). Indeed Bam and bgcn form a complex that binds and antagonizes both Pumilio and eIF4A proteins and that also regulates the *Nanos* mRNA via its 3′UTR (Li et al. 2009; Shen et al. 2009; Kim et al. 2010). Therefore, Bam/bgcn functions posttranscriptionally.

The transition from mitotic germ cell divisions within the cyst into meiotic S-phase normally occurs at the 16-cell stage. In *Drosophila* testes, Bam is normally expressed in 4-, 8-, and 16-cell cysts; its levels peak in 8-cell cysts, continue at a lower level in early 16-cell cysts that are in meiotic S-phase, and drop precipitously in later 16-cell cysts that have completed DNA replication (Insco et al. 2009). Manipulations of Bam dosage dramatically affect the timing of meiotic entry. In *bam*/+ heterozygotes, Bam peaks later than normal and the switch into meiosis occurs late, resulting in 32-cell and >32-cell cysts; by contrast, a PEST-deleted mutant BAM peaks earlier than normal and the switch into meiosis occurs early, resulting in 8-cell cysts that go on to differentiate. Therefore, in *Drosophila* testes, the amount of Bam protein is crucial for determining the time at which cyst cells stop dividing mitotically and instead begin meiotic S-phase. Indeed, normal cyst formation relies on the Bam gradient extending between GSC and the site of meiotic entry (Fig. 3B) (Insco et al. 2009).

BMP signaling down-regulates Bam expression in both male and female GSCs, and that down-regulation is crucial for GSC maintenance (Fig. 3B) (Chen and McKearin 2003; Kawase et al. 2004). Taking these results together, the *Drosophila* mitosis/meiosis decision appears to be integrated with controls of GSC self-renewal maintenance, much as it is in *C. elegans*. In addition to the few regulators mentioned here, which interface clearly with the mitosis/meiosis decision, many other regulators affect *Drosophila* GSC self-renewal and differentiation and therefore are implicated in the mitosis/meiosis decision.

MOUSE: SEX DETERMINATION MEETS THE MITOSIS/MEIOSIS DECISION

The murine mitosis/meiosis decision shows a dramatic sex-specific timing of meiotic entry that historically has been analyzed as a sex determination decision. The sex-specific difference occurs soon after the primordial germ cells (PGCs) migrate into the fetal gonadal primordium, when all female PGCs enter the meiotic cell cycle and all male PGCs arrest at G_1/G_0 of the mitotic cell cycle (Fig. 4A, left). The female germ cells quickly progress through meiotic prophase I and arrest at metaphase of the first meiotic division, still within the fetal ovary; then after puberty, those female germ cells develop in cyclical fashion to produce mature oocytes. By contrast, male germ cells at puberty take on the dual task of maintaining spermatogonial stem cells and generating a continuous stream of sperm (Fig. 4A, right). This striking sex-specific meiotic entry led to a major question in the field: Are mitosis/meiosis and sperm/oocyte two decisions or only a single decision (Kimble and Page 2007)? An answer to that question has now emerged through groundbreaking studies that have outlined central players in the murine mitosis/meiosis pathway.

Retinoic Acid and STRA8 Promote Meiotic Entry

Success in the hunt for murine regulators of the mitosis/meiosis decision began with two key discoveries. A target gene of retinoic acid signaling, STRA8 (stimulated by retinoic acid 8), was expressed sex-specifically in female fetal gonads at the time of female-specific meiotic entry (Oulad-Abdelghani et al. 1996; Koubova et al. 2006), and the *Cyp26b1* inhibitor of retinoic acid signaling was expressed sex-specifically in male fetal gonads at the same time (Bowles et al. 2006). Both findings led investigators to the idea that retinoic acid (RA) might induce meiotic entry in the fetal ovary, whereas

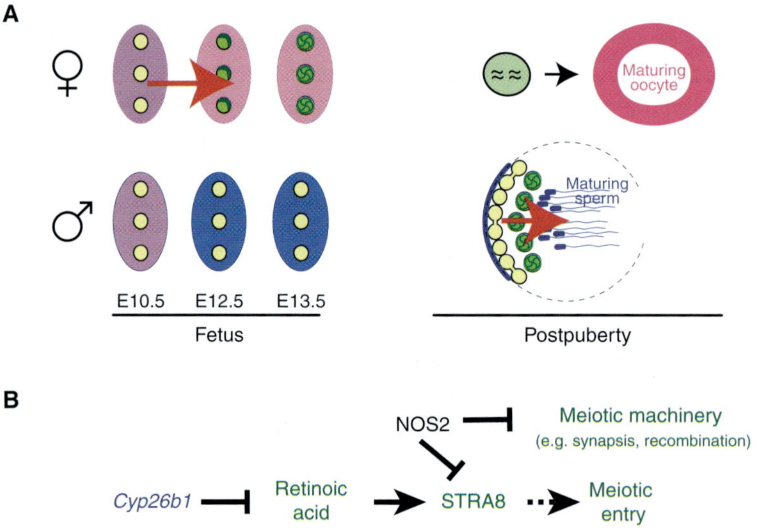

Figure 4. The mouse mitosis/meiosis decision. (*A*) Meiotic entry (red arrow) occurs during fetal development in female mice and after puberty in males. PGCs (yellow) enter the fetal somatic gonad, which is sexually bipotential (purple). Sex determination of the fetal somatic gonad results in a female (rose) or male (blue) somatic gonad; in these fetal developing gonads, female PGCs enter the meiotic cell cycle (green), whereas male PGCs arrest in the mitotic cell cycle (yellow). After puberty, female germ cells produce mature oocytes, whereas the male maintain spermatogonial stem cells (yellow), transit-amplifying cells (yellow) and meiotic cells (green) that produce mature sperm. No individual cross-section of the seminiferous tubule (dotted line) has all stages because of the wave of maturation along its length. (*B*) Molecular regulation of the murine mitosis/meiosis decision. Retinoic acid activates expression of STRA8, which governs meiotic entry by an unknown mechanism. NOS2 has been implicated in repression of the *stra8* mRNA as well as mRNAs encoding the meiotic machinery. In the male fetal gonad, Cyp26b1 inhibits retinoic acid and blocks meiotic entry. Solid arrows and lines indicate direct molecular regulation; dashed line indicates control that may be either direct or indirect.

degradation of retinoic acid might prevent meiotic entry in the fetal testis (Fig. 4B). Subsequent experiments supported that model (Bowles et al. 2006; Koubova et al. 2006). For example, RA addition to a fetal male gonad induced early markers of meiosis, including STRA8, and suppressed the pluripotency marker Oct4, whereas RA antagonists had the opposite effect. Moreover, germ cells in the fetal testis entered the meiotic cell cycle in a *Cyp26b1* mutant. Therefore, RA is capable of driving both female and male fetal germ cells into meiosis.

STRA8 is expressed specifically in meiotic germ cells in the fetal ovary and in the juvenile testis, suggesting that STRA8 might act downstream from RA to induce meiotic entry in both sexes (Fig. 4B) (Oulad-Abdelghani et al. 1996; Baltus et al. 2006; Anderson et al. 2008; Mark et al. 2008). Indeed, in *stra8* null mutants, germ cells fail to enter the meiotic cell cycle, both in fetal female gonads and juvenile male gonads (Baltus et al. 2006; Anderson et al. 2008). An alternate view, that STRA8 instead functions in early meiotic prophase, comes from a parallel analysis of a *stra8* mutant (Mark et al. 2008). The likely resolution is that STRA8 governs both initiation into the meiotic cell cycle and early meiotic progression, with genetic modifiers generating the different phenotypes. The STRA8 "novel" amino acid sequence includes a putative helix–loop–helix domain (Baltus et al. 2006), and its subcellular localization encompasses both nucleus and cytoplasm (Oulad-Abdelghani et al. 1996; Tedesco et al. 2009). The idea that STRA8 may be a transcription factor has garnered some experimental support (Tedesco et al. 2009) but that conclusion is not yet confirmed. Most importantly, STRA8 is required for meiotic entry in both male and female germ cells, putting to rest the hypothesis that the mitosis/meiosis decision might be equivalent to the sperm/oocyte decision.

NOS2 and DAZ Also Influence the Fetal Mitosis/Meiosis Decision

Two conserved RNA binding proteins also influence the fetal mitosis/meiosis decision. The Nanos-related NOS2 protein prevents meiotic entry in fetal male germ cells and can be induced to do the same in fetal female germ cells when ectopically expressed (Fig. 4B) (Suzuki and Saga 2008). NOS2 is normally expressed in fetal male germ cells, just as *Cy26b1* begins to decrease. In the absence of NOS2, male germ cells fail to maintain their G_0/G_1 cell cycle arrest and appear to embark on the meiotic cell cycle, but eventually die. However, in a double mutant that removes both NOS2 and the BAX2 proapoptotic regulator, fetal male germ cells expressed STRA8 aberrantly and entered the meiotic cell cycle (Suzuki and Saga 2008). Nanos proteins are translational repressors (Gray and Wickens 1998), and NOS2 itself interacts with the CCR4-Not deadenylation complex (Suzuki et al. 2010). Importantly, immunoprecipitation of a FLAG-tagged NOS2 brings down *stra8* mRNA along with other transcripts implicated in meiosis (e.g., sycp3) (Suzuki et al. 2010). The likely scenario is that NOS2 represses both a regulator of meiotic entry (*stra8*) plus critical components of the meiotic machinery (Fig. 4B). In addition to NOS2, the DAZ (deleted in azoospermia) protein renders fetal germ cells of both sexes responsive to RA signaling (Lin et al. 2008). Therefore, both NOS2 and DAZ are capable of influencing meiotic entry in murine germ cells of either sex.

PLANTS: GERMLINE/SOMA DECISION MEETS THE MITOSIS/MEIOSIS DECISION

An embryonic germline lineage does not exist in plants. Instead, individual germ cells are induced from somatic tissues in the developing flower (Walbot and Evans 2003). Those germ cells, termed generically "sporocytes" in the two sexes or "megaspore mother cell" and "pollen mother cell" in females and males, respectively, typically enter into the meiotic cell cycle soon after their induction from the somatic hypodermal tissue. Therefore, meiotic entry is typically linked to germ cell specification in plants. However, in some mutant strains, germ cells can be specified and generate gametes without meiotic entry, a process termed apomixis (e.g., Ravi et al. 2008). The germline/soma and mitosis/meiosis decisions are therefore not inextricably linked.

Multiple regulators of germ cell specification and meiotic entry have been identified in *Arabidopsis*, maize, and rice (Ma 2005; Hamant et al. 2006; Mercier and Grelon 2008; Yang et al. 2010). The analyses of these regulators have provided important clues about controls of meiotic entry in plants. First, extrinsic signaling induces germ cell differentiation, including meiotic entry. In both the *Arabidopsis* dicot and rice monocot, a leucine-rich repeat receptor-like protein kinase controls germ cell induction (reviewed in Ma 2005), suggesting an ancient signaling mechanism among plants. Second, intrinsic regulators control meiotic entry or early meiotic prophase in both ovules and anthers (Motamayor et al. 2000; Mercier et al. 2003; Pawlowski et al. 2009). These include the *Arabidopsis* SWITCH1 and maize AMEIOTIC1 genes, which encode homologous plant-specific proteins that localize to the nucleus. SWITCH1 is required for the normal pattern of histone modifications, suggesting a role in chromatin modification (Boateng et al. 2008). Similarly, a germ cell specific Argonaute controls both early meiotic prophase and histone modifications in rice (Nonomura et al. 2007). Therefore, chromatin modification has been implicated as a likely molecular mechanism that influences meiotic entry in all three model plants.

EMERGING THEMES

Recent advances in understanding molecular controls of the mitosis/meiosis decision reveal several unifying themes. Some of those themes extend to yeast and plants, whereas others appear specific to metazoans. Rather than repeat references here, readers are referred for the most part to individual sections above for citations.

1. Extrinsic signals control the mitosis/meiosis decision. Extrinsic cues control the mitosis/meiosis decision in *S. cerevisiae* and *S. pombe* (nutritional environment), *C. elegans* (Notch signaling), *Drosophila* (BMP signaling), and mice (retinoic acid signaling). In addition, signaling is responsible for germ cell induction in plants, which immediately precedes meiotic entry.
2. Intrinsic regulators of the mitosis/meiosis decision control the early meiotic program, often by regulation of mRNAs. Many of the key intrinsic regulators of the animal mitosis/meiosis decision regulate mRNAs: *C. elegans* FBF directly represses a battery of mRNAs encoding the meiotic program, *Drosophila* Bam/bgcn regulates expression of the *nanos* mRNA, and murine NOS2 RNA-binding protein appears to control the expression of mRNAs encoding STRA8 and a synaptonemal complex protein. The prevalence of RNA regulation is striking and extends to *S. pombe*, in which Mei2 and meiRNA antagonize the sequestration of meiotic mRNAs. By contrast, intrinsic regulators of the decision in plants have been implicated in chromatin regulation, which is reminiscent of the transcriptional activation of the early meiotic program in *S. cerevisiae*. The common thread is use of intrinsic regulators to govern the meiotic program.
3. Key regulators of meiotic entry are gender-neutral. The mitosis/meiosis decision can differ in the two sexes in a variety of ways, including timing and position of meiotic entry. Yet the same central regulators control meiotic entry in both sexes in *C. elegans*, mice, plants, and probably *Drosophila*. Because the mitosis/meiosis decision in *C. elegans* is not dramatically different in the two sexes, it was not surprising that gender-neutral regulators controlled the decision. However, in mice, where the decision is dramatically different in the two sexes, the discovery of regulators that control meiotic entry in both sexes was a real surprise and major breakthrough. The emerging theme is that gender-neutral regulators of the mitosis/meiosis decision can be modulated to drive meiotic entry in a sex-specific manner.
4. A close relationship exists between GSC self-renewal and mitosis/meiosis controls in animals. GSC loss occurs aberrantly in *C. elegans* and *Drosophila* on removal of self-renewal regulators, and it occurs naturally in female

mammals during fetal development. In all three cases, GSC loss is accompanied by the meiotic entry and differentiation of all germ cells. In *C. elegans* and *Drosophila*, molecular regulators of GSC self-renewal and meiotic entry are directly antagonistic. In mice, pluripotency factors (e.g., Oct4) are similarly antagonistic with RA signaling and STRA8, although the mechanism of that interaction is not yet known (Bowles and Koopman 2007).

5. Commitment to the meiotic cell cycle must be actively maintained. *S. cerevisiae* and *C. elegans* cells can enter the meiotic cell cycle and progress into pachytene, but then return to mitotic divisions, suggesting that commitment to the meiotic cell cycle must be actively maintained. In *Drosophila*, spermatogonial cells can dedifferentiate into GSCs (Sheng et al. 2009), suggesting a similar need for active maintenance of meiotic progression in flies. An irreversible commitment to meiosis only occurs at the meiosis I division in *S. cerevisiae* and may be similarly late in *C. elegans*.

OPEN QUESTIONS

The major advances made in understanding the mitosis/meiosis decision over the past decade have answered some key questions and opened the door to others.

1. What are the terminal regulators of the mitosis/meiosis decision in *Drosophila*, mice and plants? Terminal regulators act at the end of a pathway to execute the function of the pathway. The terminal regulators of the mitosis/meiosis pathway include the IME1 transcription factor in *S. cerevisiae* and the Mei2/MeiRNA regulators in *S. pombe*. In *C. elegans*, FBF responds to Notch signaling to directly repress the meiotic program, and GLD-1 directly represses cyclin E mRNA and the mitotic cell cycle. What are the terminal regulators in *Drosophila*, mice, and plants? Bam/bgcn, STRA8, and SWI1/AM1 are the best current candidates, but pathways must be filled out and direct links established with the meiotic program to be sure.

2. Are regulators of the mitosis/meiosis decision conserved? Mitosis/meiosis regulators are not conserved among all eukaryotes (yeast to human), but the pathways in multicellular organisms are fragmentary and conservation of at least some key regulators remains a possibility. One mitosis/meiosis regulator that may be broadly conserved among metazoans is FBF, a *C. elegans* PUF RNA-binding protein. FBF regulates GSC maintenance in nematodes and directly represses the meiotic program as one part of this role. The Pumilio PUF RNA-binding protein is similarly required for GSC maintenance in *Drosophila* ovaries (Lin and Spradling 1997; Forbes and Lehmann 1998) and the Pum1 and Pum2 PUF RNA-binding proteins are enriched in GSCs in mice and humans (Moore et al. 2003). However, a role for these fly and mammalian PUF counterparts in the mitosis/meiosis decision remains an open question.

3. How does mitosis/meiosis regulation interface with controls of other germ cell fates? Research during the past few years has clarified relationships between the mitosis/meiosis decision and the decisions of self-renewal versus differentiation and male versus female, at least in nematodes and mice, respectively. An open question is how universal the answers are. For example, do GSC regulators control both the meiotic program and general differentiation regulators in all metazoans or indeed in all multicellular organisms? And what about regulatory relationships with other decisions (e.g., proliferation versus self-renewal and maintenance, germline survival versus apoptosis) and other more global events (e.g., aging, starvation)?

4. Do nutritional cues affect the mitosis/meiosis decision in animal germ cells? Nutritional cues are central to the mitosis/meiosis decision in *S. cerevisiae* and *S. pombe*, but do they play a similar role in other creatures? In male mice, vitamin A depletion (VAD) yields germlines with only GSCs or early differentiating premeiotic germ cells (van Pelt and de Rooij 1990); an attractive hypothesis

is that VAD blocks meiotic entry (Bowles and Koopman 2007; Clagett-Dame and Knutson 2011). In nematodes, starvation has a similar effect, which is dependent on a nuclear hormone receptor, NHR-49, that controls fatty acid metabolism in response to starvation (Van Gilst et al. 2005; Angelo and Van Gilst 2009). In flies, nutrition affects the rate of mitotic proliferation (Hsu et al. 2008), but its effect on the mitotic/meiotic decision has not yet been reported. The broader role of nutrition on the mitosis/meiosis decision in multicellular creatures is virtually virgin territory for future exploration.

5. What is the clinical impact of understanding the mitosis/meiosis decision? Understanding the mitosis/meiosis decision has clear implications for human fertility and germ cell tumors. For example, in adult men, the mitotic and meiotic cell cycles are normally kept in a tightly controlled balance between spermatogonial stem cells (SSCs) and maturing spermatocytes. If that balance is tipped aberrantly toward meiotic entry, SSCs are depleted, and if tipped toward mitotic divisions, uncontrolled germ cell proliferation and perhaps cancer result. An in depth understanding of the molecular mechanisms that modulate the balance will therefore be crucial for developing therapies that impact human reproduction and health. One possible example might be the development of a male contraceptive that is based on the inhibition of molecular events specific to the mitosis/meiosis decision. Vitamin A depletion renders males sterile and its effects can be reversed—but Vitamin A depletion affects many tissues deleteriously. The generation of a drug that targets a germ cell-specific component of the pathway (e.g., STRA8) might be equally effective but have few side effects.

CONCLUSIONS AND PERSPECTIVES

Our understanding of the mitosis/meiosis decision in germ cells of multicellular organisms has been transformed during the past 5 years. One conceptual breakthrough is that mitosis/meiosis regulators are gender-neutral—they control meiotic entry in the germ cells of both sexes. This basic idea holds true in nematodes, mice, plants, and probably flies. The dramatic sex-specific timing of the mammalian mitosis/meiosis decision, which historically was equated with the sperm/oocyte decision, actually relies on sex-specific regulation of the gender-neutral mitosis/meiosis regulators. A second conceptual breakthrough is that regulators of germline stem cells control entry into the meiotic cell cycle as part of their larger program to maintain the stem cells in a naïve, undifferentiated state. Plants do not have germline stem cells, per se, but instead possess totipotent somatic cells that may play a similar but plant-specific role. The relationship between stem cell and mitosis/meiosis regulators may therefore be specialized for animal germ cells.

The decision of a germ cell to progress through either the mitotic or the meiotic cell cycle is an ancient one with important implications for reproduction and health in humans and improved methods of agriculture in plants. Although breakthroughs have been made, our molecular understanding of the mitosis/meiosis decision remains in its infancy. The next decade will expand our knowledge of the mitosis/meiosis molecular circuitry in multiple organisms and clarify the relationship of that circuitry with GSC controls, totipotency controls, and the sperm/oocyte decision.

ACKNOWLEDGMENTS

I thank Kathy Barton, Sarah Crittenden, Matt Evans, Clint Morgan, and Dave Zarkower for comments, and Anne Helsley and Laura Vanderploeg for assistance in preparation of the manuscript and figures. J.K. is an investigator of the Howard Hughes Medical Institute and is supported by the NIH.

REFERENCES

Anderson EL, Baltus AE, Roepers-Gajadien HL, Hassold TJ, de Rooij DG, van Pelt AMM, Page DC. 2008. *Stra8* and its inducer, retinoic acid, regulate meiotic initiation in both spermatogenesis and oogenesis in mice. *Proc Natl Acad Sci* **105:** 14976–14980.

Angelo G, Van Gilst M. 2009. Starvation protects germline stem cells and extends reproductive longevity in *C. elegans*. *Science* **326:** 954–958.

Ariz M, Mainpal R, Subramaniam K. 2009. *C. elegans* RNA-binding proteins PUF-8 and MEX-3 function redundantly to promote germline stem cell mitosis. *Dev Biol* **326:** 295–304.

Baltus AE, Menke DB, Hu YC, Goodheart ML, Carpenter AE, de Rooij DG, Page DC. 2006. In germ cells of mouse embryonic ovaries, the decision to enter meiosis precedes premeiotic DNA replication. *Nat Genet* **38:** 1430–1434.

Barton MK, Kimble J. 1990. *fog-1*, a regulatory gene required for specification of spermatogenesis in the germ line of *Caenorhabditis elegans*. *Genetics* **125:** 29–39.

Belfiore M, Pugnale P, Saudan Z, Puoti A. 2004. Roles of the *C. elegans* cyclophilin-like protein MOG-6 in MEP-1 binding and germline fates. *Development* **131:** 2935–2945.

Berset T, Fröhli Hoier E, Battu G, Canevascini S, Hajnal A. 2001. Notch inhibition of RAS signaling through MAP kinase phosphatase LIP-1 during *C. elegans* vulval development. *Science* **291:** 1055–1058.

Biedermann B, Wright J, Senften M, Kalchhauser I, Sarathy G, Lee MH, Ciosk R. 2009. Translational repression of cyclin E prevents precocious mitosis and embryonic gene activation during *C. elegans* meiosis. *Dev Cell* **17:** 355–364.

Boateng KA, Yang X, Dong F, Owen HA, Makaroff CA. 2008. SWI1 is required for meiotic chromosome remodeling events. *Mol Plant* **1:** 620–633.

Bowles J, Koopman P. 2007. Retinoic acid, meiosis and germ cell fate in mammals. *Development* **134:** 3401–3411.

Bowles J, Knight D, Smith C, Wilhelm D, Richman J, Mamiya S, Yashiro K, Chawengsaksophak K, Wilson MJ, Rossant J, et al. 2006. Retinoid signaling determines germ cell fate in mice. *Science* **312:** 596–600.

Chen D, McKearin DM. 2003. A discrete transcriptional silencer in the *bam* gene determines asymmetric division of the *Drosophila* germline stem cell. *Development* **130:** 1159–1170.

Cinquin O, Crittenden SL, Morgan DE, Kimble J. 2010. Progression from a stem cell–like state to early differentiation in the *C. elegans* germ line. *Proc Natl Acad Sci* **107:** 2048–2053.

Ciosk R, DePalma M, Priess JR. 2004. ATX-2, the *C. elegans* ortholog of ataxin 2, functions in translational regulation in the germline. *Development* **131:** 4831–4841.

Ciosk R, DePalma M, Priess JR. 2006. Translational regulators maintain totipotency in the *Caenorhabditis elegans* germline. *Science* **311:** 851–853.

Clagett-Dame M, Knutson D. 2011. Vitamin A in reproduction and development. *Nutrients* 3: 395–428.

Cox DN, Chao A, Baker J, Chang L, Qiao D, Lin H. 1998. A novel class of evolutionarily conserved genes defined by *piwi* are essential for stem cell self-renewal. *Genes Dev* **12:** 3715–3727.

Crittenden SL, Bernstein DS, Bachorik JL, Thompson BE, Gallegos M, Petcherski AG, Moulder G, Barstead R, Wickens M, Kimble J. 2002. A conserved RNA-binding protein controls germline stem cells in *Caenorhabditis elegans*. *Nature* **417:** 660–663.

Crittenden SL, Leonhard KA, Byrd DT, Kimble J. 2006. Cellular analyses of the mitotic region in the *Caenorhabditis elegans* adult germ line. *Mol Biol Cell* **17:** 3051–3061.

Davies EL, Fuller MT. 2008. Regulation of self-renewal and differentiation in adult stem cell lineages: Lessons from the *Drosophila* male germ line. *Cold Spring Harbor Symp Quant Biol* **73:** 137–145.

Eckmann CR, Kraemer B, Wickens M, Kimble J. 2002. GLD-3, a Bicaudal-C homolog that inhibits FBF to control germline sex determination in *C. elegans*. *Dev Cell* **3:** 697–710.

Eckmann CR, Crittenden SL, Suh N, Kimble J. 2004. GLD-3 and control of the mitosis/meiosis decision in the germline of *Caenorhabditis elegans*. *Genetics* **168:** 147–160.

Forbes A, Lehmann R. 1998. Nanos and Pumilio have critical roles in the development and function of *Drosophila* germline stem cells. *Development* **125:** 679–690.

Forsburg SL. 2002. Only connect: Linking meiotic DNA replication to chromosome dynamics. *Mol Cell* **9:** 703–711.

Francis R, Barton MK, Kimble J, Schedl T. 1995. *gld-1*, a tumor suppressor gene required for oocyte development in *Caenorhabditis elegans*. *Genetics* **139:** 579–606.

Gönczy P, Matunis E, DiNardo S. 1997. *bag-of-marbles* and *benign gonial cell neoplasm* act in the germline to restrict proliferation during *Drosophila* spermatogenesis. *Development* **124:** 4361–4371.

Gray NK, Wickens M. 1998. Control of translation initiation in animals. *Annu Rev Cell Dev Biol* **14:** 399–458.

Hamant O, Ma H, Cande WZ. 2006. Genetics of meiotic prophase I in plants. *Annu Rev Plant Biol* **57:** 267–302.

Hansen D, Wilson Berry L, Dang T, Schedl T. 2004. Control of the proliferation versus meiotic development decision in the *C. elegans* germline through regulation of GLD-1 protein accumulation. *Development* **131:** 93–104.

Hansen M, Hsu A-L, Dillin A, Kenyon C. 2005. New genes tied to endocrine, metabolic, and dietary regulation of lifespan from a *Caenorhabditis elegans* genomic RNAi screen. *PLoS Genet* **1:** 119–128.

Harigaya Y, Yamamoto M. 2007. Molecular mechanisms underlying the mitosis/meiosis decision. *Chromosome Res* **15:** 523–537.

Harigaya Y, Tanaka H, Yamanaka S, Tanaka K, Watanabe Y, Tsutsumi C, Chikashige Y, Hiraoka Y, Yamashita A, Yamamoto M. 2006. Selective elimination of messenger RNA prevents an incidence of untimely meiosis. *Nature* **442:** 45–50.

Hasegawa E, Karashima T, Sumiyoshi E, Yamamoto M. 2006. *C. elegans* CPB-3 interacts with DAZ-1 and functions in multiple steps of germline development. *Dev Biol* **295:** 689–699.

Hsu HJ, LaFever L, Drummond-Barbosa D. 2008. Diet controls normal and tumorous germline stem cells via insulin-dependent and -independent mechanisms in *Drosophila*. *Dev Biol* **313:** 700–712.

Insco ML, Leon A, Tam CH, McKearin DM, Fuller MT. 2009. Accumulation of a differentiation regulator specifies transit amplifying division number in an adult stem cell lineage. *Proc Natl Acad Sci* **106:** 22311–22316.

Kadyk LC, Kimble J. 1998. Genetic regulation of entry into meiosis in *Caenorhabditis elegans*. *Development* **125:** 1803–1813.

Kassir Y, Adir N, Boger-Nadjar E, Guttmann Raviv N, Rubin-Bejerano I, Sagee S, Shenhar G. 2003. Transcriptional regulation of meiosis in budding yeast. *Int Rev Cytol* **224:** 111–171.

Kawase E, Wong MD, Ding BC, Xie T. 2004. Gbb/Bmp signaling is essential for maintaining germline stem cells and for repressing *bam* transcription in the *Drosophila* testis. *Development* **131:** 1365–1375.

Kerins JA, Hanazawa M, Dorsett M, Schedl T. 2010. PRP-17 and the pre-mRNA splicing pathway are preferentially required for the proliferation versus meiotic development decision and germline sex determination in *Caenorhabditis elegans. Dev Dyn* **239:** 1555–1572.

Kershner AM, Kimble J. 2010. Genome-wide analysis of mRNA targets for *Caenorhabditis elegans* FBF, a conserved stem cell regulator. *Proc Natl Acad Sci* **107:** 3936–3941.

Kim JY, Lee YC, Kim C. 2010. Direct inhibition of Pumilo activity by Bam and Bgcn in *Drosophila* germ line stem cell differentiation. *J Biol Chem* **285:** 4741–4746.

Kimble J, Crittenden SL. 2005. Germline proliferation and its control. In *WormBook* (ed. The *C. elegans* Research Community), 10.1895/wormbook.1.13.1. www.wormbook.org.

Kimble J, Crittenden SL. 2007. Controls of germline stem cells, entry into meiosis, and the sperm/oocyte decision in *Caenorhabditis elegans. Annu Rev Cell Dev Biol* **23:** 405–433.

Kimble J, Page DC. 2007. The mysteries of sexual identity. The germ cell's perspective. *Science* **316:** 400–401.

Kimble JE, White JG. 1981. On the control of germ cell development in *Caenorhabditis elegans. Dev Biol* **81:** 208–219.

Koubova J, Menke DB, Zhou Q, Capel B, Griswold MD, Page DC. 2006. Retinoic acid regulates sex-specific timing of meiotic initiation in mice. *Proc Natl Acad Sci* **103:** 2474–2479.

Lamont LB, Crittenden SL, Bernstein D, Wickens M, Kimble J. 2004. FBF-1 and FBF-2 regulate the size of the mitotic region in the *C. elegans* germline. *Dev Cell* **7:** 697–707.

Lee M-H, Schedl T. 2010. *C. elegans* STAR proteins, GLD-1 and ASD-2, regulate specific RNA targets to control development. In *Posttranscriptional regulation by STAR proteins: Control of RNA metabolism in development and disease* (ed. Volk T, Atrzt K), pp. 106–122. Landes Bioscience and Springer SBM, New York.

Lee M-H, Hook B, Lamont LB, Wickens M, Kimble J. 2006. LIP-1 phosphatase controls the extent of germline proliferation in *Caenorhabditis elegans. EMBO J* **25:** 88–96.

Lee M-H, Hook B, Pan G, Kershner AM, Merritt C, Seydoux G, Thomson JA, Wickens M, Kimble J. 2007a. Conserved regulation of MAP kinase expression by PUF RNA-binding proteins. *PLoS Genet* **3:** e233.

Lee M-H, Ohmachi M, Arur S, Nayak S, Francis R, Church D, Lambie E, Schedl T. 2007b. Multiple functions and dynamic activation of MPK-1 extracellular signal-regulated kinase signaling in *Caenorhabditis elegans* germline development. *Genetics* **177:** 2039–2062.

Li Y, Minor NT, Park JK, McKearin DM, Maines JZ. 2009. Bam and Bgcn antagonize *Nanos*-dependent germ-line stem cell maintenance. *Proc Natl Acad Sci* **106:** 9304–9309.

Lin H, Spradling AC. 1997. A novel group of *pumilio* mutations affects the asymmetric division of germline stem cells in the *Drosophila* ovary. *Development* **124:** 2463–2476.

Lin Y, Gill ME, Koubova J, Page DC. 2008. Germ cell-intrinsic and -extrinsic factors govern meiotic initiation in mouse embryos. *Science* **322:** 1685–1687.

Ma H. 2005. Molecular genetic analyses of microsporogenesis and microgametogenesis in flowering plants. *Annu Rev Plant Biol* **56:** 393–434.

Maine EM, Hansen D, Springer D, Vought VE. 2004. *Caenorhabditis elegans atx-2* promotes germline proliferation and the oocyte fate. *Genetics* **168:** 817–830.

Marin VA, Evans TC. 2003. Translational repression of a *C. elegans* Notch mRNA by the STAR/KH domain protein GLD-1. *Development* **130:** 2623–2632.

Mark M, Jacobs H, Oulad-Abdelghani M, Dennefeld C, Féret B, Vernet N, Codreanu C-A, Chambon P, Ghyselinck NB. 2008. STRA8-deficient spermatocytes initiate, but fail to complete, meiosis and undergo premature chromosome condensation. *J Cell Sci* **121:** 3233–3242.

McKearin DM, Spradling AC. 1990. *bag-of-marbles*: A *Drosophila* gene required to initiate both male and female gametogenesis. *Genes Dev* **4:** 2242–2251.

Mercier R, Grelon M. 2008. Meiosis in plants: Ten years of gene discovery. *Cytogenet Genome Res* **120:** 281–290.

Mercier R, Armstrong SJ, Horlow C, Jackson NP, Makaroff CA, Vezon D, Pelletier G, Jones GH, Franklin FC. 2003. The meiotic protein SWI1 is required for axial element formation and recombination initiation in *Arabidopsis. Development* **130:** 3309–3318.

Merritt C, Seydoux G. 2010. Transgenic solutions for the germline. In *WormBook* (ed. The *C. elegans* Research Community), 10.1895/wormbook.1.148.1. www.wormbook.org.

Merritt C, Rasoloson D, Ko D, Seydoux G. 2008. 3′ UTRs are the primary regulators of gene expression in the *C. elegans* germline. *Curr Biol* **18:** 1476–1482.

Michaelson D, Korta DZ, Capua Y, Hubbard EJ. 2010. Insulin signaling promotes germline proliferation in *C. elegans. Development* **137:** 671–680.

Mitchell AP. 1994. Control of meiotic gene expression in *Saccharomyces cerevisiae. Microbiol Rev* **58:** 56–70.

Moore FL, Jaruzelska J, Fox MS, Urano J, Firpo MT, Turek PJ, Dorfman DM, Reijo Pera RA. 2003. Human Pumilio-2 is expressed in embryonic stem cells and germ cells and interacts with DAZ (Deleted in AZoospermia) and DAZ-like proteins. *Proc Natl Acad Sci* **100:** 538–543.

Morgan DE, Crittenden SL, Kimble J. 2010. The *C. elegans* adult male germline: Stem cells and sexual dimorphism. *Dev Biol* **346:** 204–214.

Motamayor JC, Vezon D, Bajon C, Sauvanet A, Grandjean O, Marchand M, Bechtold N, Pelletier G, Horlow C. 2000. *Switch* (*swi1*), and *Arabidopsis thaliana* mutant affected in the female meiotic switch. *Sex Plant Reprod* **12:** 209–218.

Nonomura K, Morohoshi A, Nakano M, Eiguchi M, Miyao A, Hirochika H, Kurata N. 2007. A germ cell specific gene of the ARGONAUTE family is essential for the progression of premeiotic mitosis and meiosis during sporogenesis in rice. *Plant Cell* **19:** 2583–2594.

Ohlstein B, McKearin D. 1997. Ectopic expression of the *Drosophila* Bam protein eliminates oogenic germline stem cells. *Development* **124:** 3651–3662.

Ohlstein B, Lavoie CA, Vef O, Gateff E, McKearin DM. 2000. The *Drosophila* cystoblast differentiation factor, *benign gonial cell neoplasm*, is related to DExH-box proteins and interacts genetically with *bag-of-marbles*. *Genetics* **155:** 1809–1819.

Oulad-Abdelghani M, Bouillet P, Decimo D, Gansmuller A, Heyberger S, Dolle P, Bronner S, Lutz Y, Chambon P. 1996. Characterization of a premeiotic germ cell–specific cytoplasmic protein encoded by Stra8, a novel retinoic acid–responsive gene. *J Cell Biol* **135:** 469–477.

Pawlowski WP, Wang CJ, Golubovskaya IN, Szymaniak JM, Shi L, Hamant O, Zhu T, Harper L, Sheridan WF, Cande WZ. 2009. Maize AMEIOTIC1 is essential for multiple early meiotic processes and likely required for the initiation of meiosis. *Proc Natl Acad Sci* **106:** 3603–3608.

Petcherski AG, Kimble J. 2000. LAG-3 is a putative transcriptional activator in the *C. elegans* Notch pathway. *Nature* **405:** 364–368.

Ravi M, Marimuthu MP, Siddiqi I. 2008. Gamete formation without meiosis in *Arabidopsis*. *Nature* **451:** 1121–1124.

Shen R, Weng C, Yu J, Xie T. 2009. eIF4A controls germline stem cell self-renewal by directly inhibiting BAM function in the *Drosophila* ovary. *Proc Natl Acad Sci* **106:** 11623–11628.

Sheng XR, Brawley CM, Matunis EL. 2009. Dedifferentiating spermatogonia outcompete somatic stem cells for niche occupancy in the *Drosophila* testis. *Cell Stem Cell* **5:** 191–203.

Simchen G. 2009. Commitment to meiosis: What determines the mode of division in budding yeast? *BioEssays* **31:** 169–177.

Smardon A, Spoerke JM, Stacey SC, Klein ME, Mackin N, Maine EM. 2000. EGO-1 is related to RNA-directed RNA polymerase and functions in germ-line development and RNA interference in *C. elegans*. *Curr Biol* **10:** 169–178.

Subramaniam K, Seydoux G. 2003. Dedifferentiation of primary spermatocytes into germ cell tumors in *C. elegans* lacking the Pumilio-like protein PUF-8. *Curr Biol* **13:** 134–139.

Suh N, Jedamzik B, Eckmann CR, Wickens M, Kimble J. 2006. The GLD-2 poly(A) polymerase activates *gld-1* mRNA in the *Caenorhabditis elegans* germ line. *Proc Natl Acad Sci* **103:** 15108–15112.

Sundaram MV. 2006. RTK/Ras/MAPK signaling. In *WormBook* (ed. The *C. elegans* Research Community), 10.1895/wormbook.1.80.1. www.wormbook.org.

Suzuki A, Saga Y. 2008. Nanos2 suppresses meiosis and promotes male germ cell differentiation. *Genes Dev* **22:** 430–435.

Suzuki H, Saba R, Sada A, Saga Y. 2010. The *Nanos3*-3′UTR is required for germ cell specific NANOS3 expression in mouse embryos. *PLoS ONE* **5:** e9300.

Tedesco M, La Sala G, Barbagallo F, De Felici M, Farini D. 2009. STRA8 shuttles between nucleus and cytoplasm and displays transcriptional activity. *J Biol Chem* **284:** 35781–35793.

Thompson BE, Bernstein DS, Bachorik JL, Petcherski AG, Wickens M, Kimble J. 2005. Dose-dependent control of proliferation and sperm specification by FOG-1/CPEB. *Development* **132:** 3471–3481.

Thompson BE, Lamont LB, Kimble J. 2006. Germ-line induction of the *Caenorhabditis elegans* vulva. *Proc Natl Acad Sci* **103:** 620–625.

van den Heuvel S. 2005. Cell-cycle regulation. In *WormBook* (ed. The *C. elegans* Research Community), 10.1895/wormbook.1.28.1. www.wormbook.org.

Van Gilst MR, Hadjivassiliou H, Yamamoto KR. 2005. A *Caenorhabditis elegans* nutrient response system partially dependent on nuclear receptor NHR-49. *Proc Natl Acad Sci* **102:** 13496–13501.

van Pelt AM, de Rooij DG. 1990. Synchronization of the seminiferous epithelium after vitamin A replacement in vitamin A–deficient mice. *Biol Reprod* **43:** 363–367.

Vershon AK, Pierce M. 2000. Transcriptional regulation of meiosis in yeast. *Curr Opin Cell Biol* **12:** 334–339.

Walbot V, Evans MM. 2003. Unique features of the plant life cycle and their consequences. *Nat Rev Genet* **4:** 369–379.

Wang L, Eckmann CR, Kadyk LC, Wickens M, Kimble J. 2002. A regulatory cytoplasmic poly(A) polymerase in *Caenorhabditis elegans*. *Nature* **419:** 312–316.

Watanabe Y, Yamamoto M. 1994. *S. pombe mei2*$^+$ encodes an RNA-binding protein essential for premeiotic DNA synthesis and meiosis I, which cooperates with a novel RNA species meiRNA. *Cell* **78:** 487–498.

Watanabe Y, Yokobayashi S, Yamamoto M, Nurse P. 2001. Pre-meiotic S phase is linked to reductional chromosome segregation and recombination. *Nature* **409:** 359–363.

Yamamoto M. 1996. The molecular control mechanisms of meiosis in fission yeast. *Trends Biochem Sci* **21:** 18–22.

Yang WC, Shi DQ, Chen YH. 2010. Female gametophyte development in flowering plants. *Annu Rev Plant Biol* **61:** 89–108.

Zhang N, Xie T. 2009. Extrinsic and intrinsic control of germline stem cell regulation in the *Drosophila* ovary. In *Regulatory networks in stem cells* (ed. Rajasekhar VK, Vemuri MC), pp. 155–164. Humana Press, New York.

Function of the Sex Chromosomes in Mammalian Fertility

Edith Heard[1] and James Turner[2]

[1]Mammalian Developmental Epigenetics Group, Institut Curie, CNRS UMR3215 INSERM U934, 75248 Paris Cedex 05, France

[2]Division Stem Cell Biology and Developmental Genetics, MRC National Institute for Medical Research, Mill Hill, London NW7 1AA, United Kingdom

Correspondence: edith.heard@curie.fr

The sex chromosomes play a highly specialized role in germ cell development in mammals, being enriched in genes expressed in the testis and ovary. Sex chromosome abnormalities (e.g., Klinefelter [XXY] and Turner [XO] syndrome) constitute the largest class of chromosome abnormalities and the commonest genetic cause of infertility in humans. Understanding how sex-gene expression is regulated is therefore critical to our understanding of human reproduction. Here, we describe how the expression of sex-linked genes varies during germ cell development; in females, the inactive X chromosome is reactivated before meiosis, whereas in males the X and Y chromosomes are inactivated at this stage. We discuss the epigenetics of sex chromosome inactivation and how this process has influenced the gene content of the mammalian X and Y chromosomes. We also present working models for how perturbations in sex chromosome inactivation or reactivation result in subfertility in the major classes of sex chromosome abnormalities.

The evolution of sex chromosomes from a pair of autosomes is clearly advantageous to species in enabling sexual reproduction and the genetic fitness that results, but also necessitates the coevolution of strategies to deal with unique situations that arise. These include the lack of homology between the X and Y chromosomes in the male germline in which autosomal homologs must pair and the imbalance in X-linked gene dosage between males and females. In mammals, this sex chromosome imbalance is dealt with by inactivating one of the two X chromosomes during early female development. In the female germline, the inactive X chromosome is reactivated before meiosis. In the male germline, the X and Y chromosomes undergo meiotic inactivation that is linked to their unique unpaired status. This is a conserved process and has important implications for the evolution of sex chromosome gene content. In this article, we will cover these different aspects of sex chromosome behavior and function in mammals, including the role of sex chromosomes in sex determination and the behavior of the sex chromosomes during gametogenetesis, particularly in the male germline.

Cite this article as *Cold Spring Harb Perspect Biol* doi: 10.1101/cshperspect.a002675

OVERVIEW OF SEX CHROMOSOME INFERTILITY

Infertility, defined as failure to conceive after twelve months of regular intercourse, affects one in seven couples in the Western world and evidence suggests that its incidence is rising (De Kretzer 1997). Genetic abnormalities account for 10% and 15% of cases of infertility in females (XX) and males (XY), respectively (Ferlin et al. 2006). A striking feature of these abnormalities is that they frequently affect the sex chromosomes; in males, for instance, between 10% and 15% of cases of azoospermia (i.e., no sperm in the ejaculate) are caused by Y chromosome deletions (Foresta et al. 2001) and a further 10% are caused by Klinefelter syndrome (XXY) (Ferlin et al. 2006). These statistics alone provide compelling evidence that the sex chromosomes play a highly specialized role in normal germ cell development.

Sex chromosome abnormalities can be divided into two classes. The first comprises gross abnormalities that are readily identified by standard karyotypic analysis (Table 1). This includes the common sex chromosome aneuploidies—for example, Turner syndrome (XO; prevalence 0.04%), Klinefelter syndrome (XXY; prevalence 0.1%), and Double Y syndrome (XYY; prevalence 0.1%)—as well as various types of X-autosome and Y-autosome translocations. Women with Turner syndrome normally have no germ cells (Speroff et al. 1991) and analysis of XO mice has found that this is because of oocyte losses at the pachytene substage of meiosis (Burgoyne and Baker 1981, 1985; Speed 1986). Klinefelter men and mice are usually sterile with a block occurring before meiosis, during the early spermatogonial divisions (Ferguson-Smith 1959; Mroz et al. 1999; Lue et al. 2001). For Double Y syndrome men, the reproductive outlook varies; the extra Y chromosome is often lost during the early spermatogonial stages and this generates XY (euploid) cell lines that can complete spermatogenesis and support fertility (Chandley 1997; Shi and Martin 2000). In XYY men and mice in which XY cell lines do not arise, infertility results, with germ cell losses occurring during pachytene and the meiotic divisions (Speed et al. 1991; Solari and Rey Valzacchi 1997; Mahadevaiah et al. 2000; Rodriguez et al. 2000).

DNA polymorphism analysis to identify the parental origin of human aneuploid conditions has shown that most cases of sex chromosome aneuploidy result from meiotic errors arising in the paternal germline (Hall et al. 2006). This stands in stark contrast to the autosomal aneuploidies (e.g., Down syndrome), which arise predominantly from maternal meiotic

Table 1. Summary of the three common sex chromosome abnormalities

Syndrome	Karyotype	Prevalence	Somatic features	Reproductive impact
Turner	XO[a]	0.04%	Growth retardation Congenital heart disease Horseshoe kidney Visual impairment	Humans: sterile Mice: fertile but reduced oocyte pool and reproductive lifespan
Klinefelter	XXY[b]	0.1%	Variable Tall stature Gynecomastia Mild developmental and behavioral problems	Humans and mice sterile unless spontaneous X chromosome loss
Double Y	XYY	0.1%	Tall stature	Humans: commonly fertile because of high incidence of Y chromosome loss Mice: sterile because of low probability of Y chromosome loss

[a]Turner females are likely to be XX/XO or XY/XO mosaics; furthermore, some Turner females are partially monosomic.
[b]Some Klinefelter men have more than two X chromosomes.

errors (reviewed in Hassold and Hunt 2001). Double Y syndrome by definition always involves a paternal error as the embryo inherits two Y chromosomes; one study found that 84% of cases result from failure of the two Y sister chromatids to separate at the second meiotic division (Robinson and Jacobs 1999). For Klinefelter syndrome, approximately one-half of cases result from failure of the paternal X and Y chromosomes to disjoin at the first meiotic division (Thomas and Hassold 2003). Finally, in the case of Turner syndrome, ~75% of affected females inherit their single X chromosome from their mother, indicating that loss of a paternal sex chromosome is the causative factor (Jacobs et al. 1997). The different types of paternal and maternal meiotic errors that give rise to these conditions are summarized in Figure 1.

The second class of sex chromosome abnormalities is the Y chromosome deletions, most of which occur in the long arm of the Y chromosome at Yq11 (Tiepolo and Zuffardi 1976). These are difficult to detect cytogenetically and as a result are usually diagnosed by PCR-based methods. Three common Y deletions that cause azoospermia are the *AZF* (*azoo*spermia *f*actor) deletions *AZFa*, *AZFb*, and *AZFc* (Vogt et al. 1996; Vogt 2005), named according to their subchromosomal position and associated spermatogenic phenotype. *AZFa* deletions affect proximal Yq11 and result in a complete absence of germs cells, whereas *AZFb* deletions affect the middle of Yq11 and cause a block during meiosis. *AZFc* deletions, which are the most common, arising in roughly 1 in 4000 males (Kuroda-Kowaguchi et al. 2001), cause arrest in late spermatogenesis with small numbers of aberrantly shaped sperm being produced. A fourth type of Y chromosome deletion, which removes distal Yq sequences and also results in the formation of a mirror image or isodicentric Y chromosome with duplicated proximal Yp and centromeric sequences, is present in 8% of azoospermic men (Jacobs and Ross 1966; Hsu 1994; Lange et al. 2009). Y chromosomal deletions are also associated with infertility in mice; Yp deletions cause a block during early spermatogenesis, whereas Yq deletions give rise to sperm head abnormalities (Mazeyrat et al. 2001; Mitchell 2004).

The discovery of these Y chromosome deletions and their deleterious effects on spermatogenesis created a paradigm shift in our understanding of sex chromosome genetics, because they countered the long-held view that Y chromosomes are "genetic deserts" containing few functional genes (Marshall Graves 2002). This belief was based on the fact that the Y chromosome has no homologous partner with which to undergo recombination-related repair (Charlesworth and Charlesworth 2000). Without such a partner, mutations affecting the Y chromosome would essentially be irreparable and this would lead to the stepwise decay of Y-linked genes over evolutionary time, and possibly even to the eventual extinction of the Y chromosome (Marshall Graves 2002). However, analysis of the sequence of the human Y chromosome has revealed that it contains at least 78 protein-coding genes (Skaletsky et al. 2003). The Y chromosome also contains a large number of direct repeats and inverted repeats, or palindromes. Palindromes, which comprise two arms of highly similar (>99.9%) nucleotide sequence separated by a small spacer region, contain many of the critical spermatogenesis genes and these are therefore present as multiple copies (Skaletsky et al. 2003). The human Y chromosome contains eight palindromes ranging in size from 30 kb to 2.9 MB and together, they occupy one quarter of the euchromatic DNA content. These palindromes provide a counterpoint to Y chromosome decay, because mutations occurring in genes in one arm of the palindrome can be corrected using the nonmutated copy within the other arm. This is achieved using gene conversion; the nonreciprocal transfer of information from one DNA duplex to another (Szostak et al. 1983). Gene conversion events occur between palindromes at a high frequency, estimated at one gene conversion per 600 nucleotides per newborn male (Rozen et al. 2003).

An alternative outcome of recombination is reciprocal exchange or crossover formation, which can also occur between direct repeats

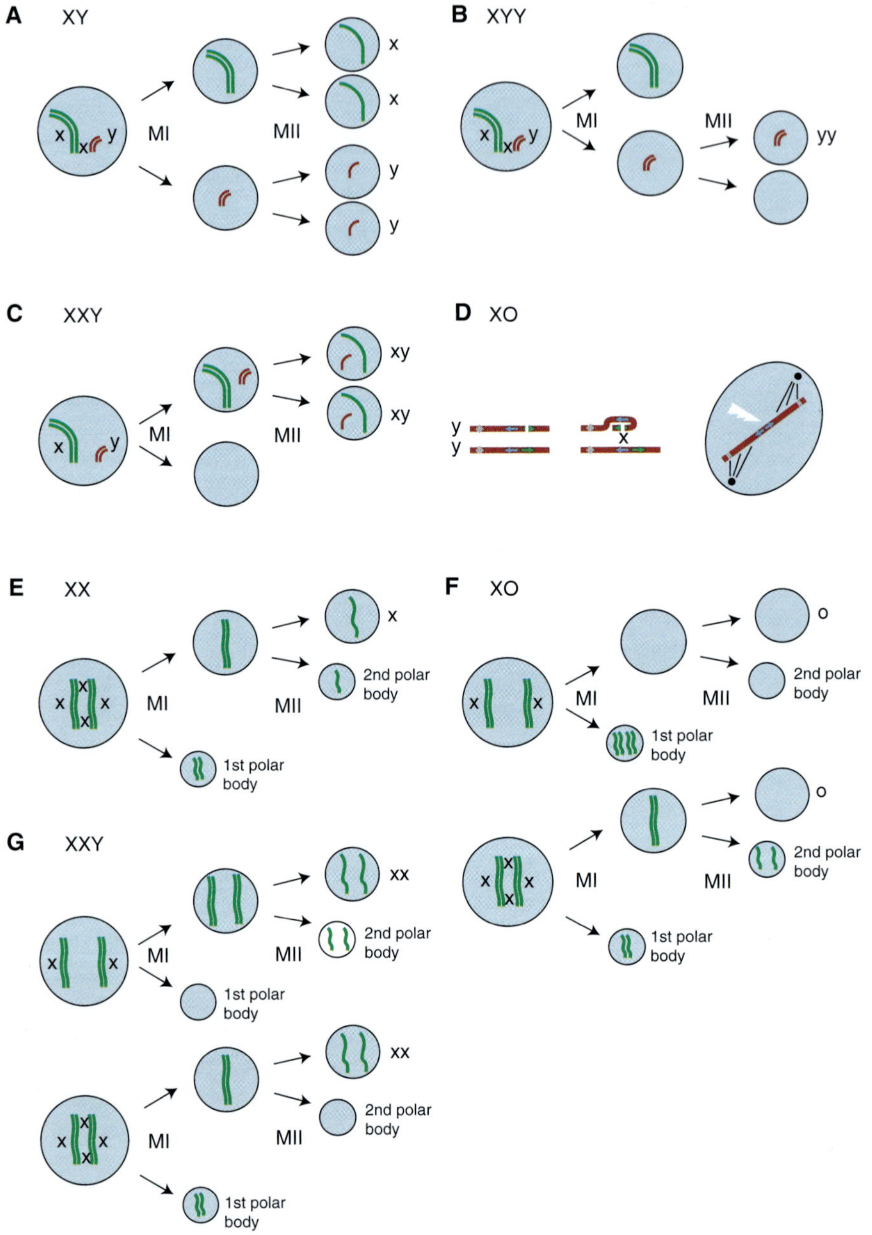

Figure 1. The mechanisms by which errors in meiosis give rise to chromosome abnormalities. (*A*) During normal male meiosis, the X and Y chromosomes synapse and crossover within the distal pseudoautosomal region; this crossover is essential in ensuring normal homologous segregation during the first meiotic division (MI). The resulting secondary spermatocytes then enter the second meiotic division (MII), during which sister chromatids segregate, thereby generating four haploid products each containing a single sex chromosome. (*B*) Nondisjunction of the two Y sisters to segregate at MII results in YY spermatids; these give rise to XYY sons at fertilization. (*C*) Nondisjunction of X and Y homologs to segregate at MI, either because of failed establishment or maintenance of crossing over, followed by normal sister separation at MII results in XY spermatids; these give rise to XXY sons at fertilization. (*D*) Formation of a DNA double-strand break in one of two Y sister chromatids followed by crossing over between palindromes of sisters results in an isodicentric Y chromosome. (*See facing page for legend.*)

and between palindromes. These recombination events result in loss of genetic information and as such are the principle mechanism by which *AZF* deletions and isodicentric Y chromosomes originate (Fig. 1) (Blanco et al. 2000; Kamp et al. 2000; Sun et al. 2000; Kuroda-Kawaguchi et al. 2001; Repping et al. 2002, 2003; Lange et al. 2009). In the case of isodicentric Y chromosomes, impaired fertility is not the only deleterious outcome of this form of recombination. Isodicentric Y chromosomes carry two centromeres that can form a bipolar spindle attachment during cell division. This results in breakage of the Y chromosome during anaphase and subsequently to Y chromosome loss (Fig. 1) (Lange et al. 2009). If this loss occurs in the male germline, sperm will be produced that have no sex chromosome and this can lead to formation of XO embryos at fertilization. Thus, palindrome–palindrome recombination could give rise to a significant proportion of cases of Turner syndrome (Fig. 1) (Lange et al. 2009).

The mechanism by which Y chromosome deletions cause infertility is straightforward; these deletions remove Y-linked genes that are essential for germ cells to transit through defined stages of spermatogenesis. However, the way in which sex chromosome aneuploidy causes infertility has, until recently, remained mysterious. In this case a major factor appears to be linked to the activity of the X chromosome and its unique dynamics of reactivation in the female germline and inactivation in the male germline. These processes are described in the following sections.

SEX CHROMOSOME EXPRESSION DURING MALE AND FEMALE GAMETOGENESIS

Our knowledge of the dynamics and behavior of the sex chromosomes during human gametogenesis is limited, for obvious reasons. In mice on the other hand, extensive studies have been performed on germ cell development. In both males and females, primordial germ cells (PGCs) originate from the proximal epiblast of the postimplantation mouse embryo at around E7.25 (Saitou 2009). In males, the single X chromosome is active in newly formed PGCs; however, in females one of the Xs is initially inactive in PGCs (Fig. 2) (Xi; Chuva de Sousa Lopes et al. 2008). Indeed, random X chromosome inactivation (XCI) initiates well before this stage, in epiblast precursor cells at around E5.5 (McMahon and Monk 1983). During XCI, up-regulation of *Xist* RNA leads to the onset of gene silencing across the length of the X chromosome. *Xist* RNA induces changes in chromosome organization and results in recruitment of histone modifying enzymes, principally the PRC1 and PRC2 complexes, which catalyze mono-ubiquitylation of histone H2A at lysine 119 (H2AK119u1) (de Napoles et al. 2004; Fang et al. 2004) and trimethylation of histone H3 at lysine 27 (H3K27me3) (Silva et al. 2003; Plath et al. 2003), respectively. The inactive X is further modified by DNA methylation at promoters. One X chromosome is thus globally inactive by the time PGCs first start to appear during female embryogenesis, which implies that all of the above epigenetic changes must be reversed during the reprogramming

Figure 1. (*Continued*) This forms a bipolar attachment during cell division and is subsequently broken and lost. In principle, this form of recombination could take place during spermatogonial mitosis, meiosis, or even in the early embryo; all of which will give rise to XO daughters. (*E*) During normal female meiosis, synapsis and crossover formation are followed by dictyate arrest. The first meiotic division at puberty generates a polar body and a secondary oocyte; the second at fertilization generates a second polar body and a single haploid product. (*F*) Nondisjunction of either homologs at MI or sisters at MII has similar effect, with haploid products with no sex chromosome ("O" gametes) generated. These generate XO daughters at fertilization. (*G*) Similar meiotic errors give rise to XX haploid products and thereafter XXY sons; the outcome depends on whether the nondisjoined chromosomes end up in the oocyte or the polar body. Although not detailed here, some cases of XXY and XYY arise during embryogenesis as a result of mitotic nondisjunction of the maternally inherited X and paternally inherited Y, respectively.

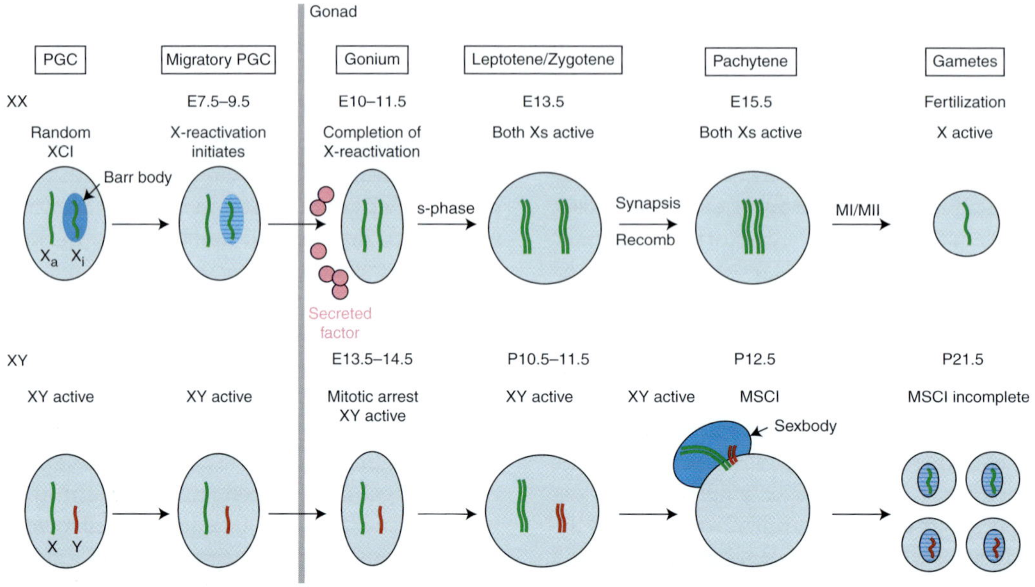

Figure 2. Sex chromosome activity during female and male germ cell development. Female PGCs have one inactive and one active X chromosome. X chromosome reactivation occurs during migration of PGCs to the genital ridge between E7.5 and E9.5; full reactivation is achieved after colonization of the genital ridge between E10 and E11.5 in response to an unknown secreted factor/s. Female germ cells then enter meiosis. Meiotic DNA replication (S-phase) is followed by leptotene (E13.5) when DNA DSBs are formed. Chromosomes initiate synapsis during zygotene (E14) and the completion of synapsis heralds the onset of pachytene (E15.5). Throughout and after meiosis the X chromosomes remain active. In males, the X and Y are active in male throughout early germ cell development. In contrast to females, male germ cells undergo mitotic arrest from E13.5 to E14.5 until shortly after birth, when the spermatogonial divisions begin. The X and Y remain active until pachytene, when meiotic sex chromosome inactivation (MSCI) takes place. After meiosis, mean expression from the X chromosome remains low compared to autosomes, but at this stage genes are not silenced as completely as they are during pachytene.

that occurs in the female germline (Chuva de Sousa Lopez et al. 2008).

Following their specification, PGCs divide and migrate through the hindgut endoderm, eventually colonizing the gonad between E10 and E11.5 (Fig. 2) (Saitou 2009). During this time, the Xi in female PGCs is reactivated. X chromosome reactivation was originally believed to occur after germ cells enter the gonad, just before the initiation of meiosis (Kratzer and Chapman 1981; Monk and McLaren 1981; Tam et al. 1994; Nesterova et al. 2002). However, more recent analyses have found that reactivation initiates much earlier during PGC development, with *Xist* RNA, PRC2, and H3K27me3 being lost from Xi during the migratory phase (E7.5–9.5) (de Napoles et al. 2007; Sugimoto and Abe 2007; Chuva de Sousa Lopes et al. 2008). Intriguingly, Xi reactivation is dependent on factors secreted by the XX urogenital ridge rather than being an intrinsic property of female germ cells (Chuva de Sousa Lopes et al. 2008). The nature of the secreted factors that enable Xi reactivation are unknown however. Following entry into the gonad, male and female germ cells exhibit a striking sexual dimorphism in their behavior; germ cells in the male form prospermatogonia and enter mitotic arrest between E13.5 and E14.5 (reviewed by McLaren 2003), whereas female germ cells enter meiosis. Throughout meiosis, both X chromosomes within the oocyte remain active (Fig. 2) (Epstein 1969; Gartler et al. 1975; Andina 1978).

Shortly after birth, prospermatogonia in the male gonad resume mitosis and initiate

 Cite this article as *Cold Spring Harb Perspect Biol* doi: 10.1101/cshperspect.a002675

spermatogenesis (Russell et al. 1990). Spermatogenesis comprises three broad phases; spermatogonial proliferation, meiosis, and spermiogenesis, together lasting 35 and 64 days in mouse and man, respectively (Heller and Clermont 1963; Russell et al. 1990). During the first phase, spermatogonia undergo approximately ten mitotic divisions, forming A, Intermediate and then finally B type-spermatogonia. Analyses of transcription across the X chromosome by microarray analysis (Khil et al. 2004; Namekawa et al. 2006) and RNA fluorescence in-situ hybridization (RNA FISH; Turner et al. 2006; Mueller et al. 2008) have shown that the X and Y are active during these stages, with the collective mean expression level of genes on the X chromosome being approximately equal to that of the autosomes (Fig. 2) (Mueller et al. 2008). In spite of this similarity in expression *levels*, the X chromosome is more densely populated with spermatogonially expressed genes when compared to the autosomes. This is consistent with the evolutionary concept of "sexually antagonistic" genes (Fisher 1931), which are predicted to become enriched on the X chromosome because of its hemizygous state in males (see later for more detailed description; Wang et al. 2001; Khil et al. 2004).

After the spermatogonial divisions germ cells become spermatocytes and enter meiosis. Meiosis is a highly specialized cell division that serves to convert diploid cells into haploid daughters and to allow the exchange of genetic information between maternal and paternal homologs (Zickler and Kleckner 1999). During this phase, the X and Y chromosomes become transcriptionally silenced, a process called meiotic sex chromosome inactivation (MSCI) (McKee and Handel 1993; Turner 2007), and form the heterochromatic "sex body" (Fig. 2) (Solari 1974).

When exactly during meiosis does MSCI take place? Early cytological studies found that the X and Y chromosomes are heterochromatic from as early as meiotic S-phase (Odartchenko and Pavillard 1970) and semiquantitative RTPCR experiments concluded that repression of specific X-linked genes initiated in leptotene and/or zygotene (Singer-Sam et al. 1990; McCarrey et al. 1992). Analyzing gene expression during these very early meiotic stages is inherently difficult because transcription levels throughout the genome, as assayed by tritiated uridine uptake (Monesi 1965) and Cot1 RNA FISH (Turner et al. 2005) are very low. This is possibly the result of the high degree of chromosome compaction that accompanies DNA double-strand break (DSB) formation and repair (Hunter et al. 2001). However, the limited data available for gene-specific RNA FISH for X- and Y-linked genes indicate that silencing is initiated at early pachytene (Turner et al. 2006), when modifications necessary for MSCI (e.g., γH2AX) appear on the sex chromosomes (Mahadevaiah et al. 2001; Fernandez-Capetillo et al. 2003).

Following the two meiotic divisions, newly formed haploid round spermatids undergo a protracted period of differentiation into mature spermatozoa, accompanied by the replacement of histones with transition proteins and, thereafter, protamines. For a number of years, the transcriptional state of the X and Y chromosomes during this period remained unclear; anecdotal evidence focusing on specific sex-linked genes reported expression in purified round spermatids populations (Hendriksen et al. 1995; Hendriksen 1999), and this fostered the view that MSCI was restricted to meiosis. However, this dogma was challenged when it was noted that repressive chromatin marks, such as histone H3 dimethylated at lysine 9 (H3K9me2) and the heterochromatin-associated CBX1 protein remain associated with the X and Y during the meiotic divisions and into early round spermatid development (Khalil et al. 2004).

Shortly thereafter, four studies provided further evidence that the repressive effects induced by MSCI are retained postmeiosis (Fig. 2) (Greaves et al. 2006; Namekawa et al. 2006; Turner et al. 2006; Mueller et al. 2008). The X chromosome was found to be conspicuously heterochromatic and transcription poor, based on Cot1 RNA FISH and RNA polymerase II staining throughout round and elongating spermatid differentiation (Greaves et al. 2006; Namekawa et al. 2006; Turner et al. 2006).

Microarray analysis showed that the mean expression level for X-linked genes, which predictably dropped between spermatogonia and pachytene as a result of MSCI, remained low in round spermatids (Namekawa et al. 2006; Mueller et al. 2008). However, an important detail became apparent when the mean X-gene expression levels in spermatids and pachytene were scrutinized more closely. Although low, the mean X-expression level in round spermatids was found to be slightly higher than in pachytene (Mueller et al. 2008). This provided preliminary evidence that silencing of the X chromosome in spermatids might not be as complete as during pachytene. This was subsequently borne out by RNA FISH data; although X-linked genes were found to be robustly silenced in pachytene cells, the same genes showed variable levels of reactivation in spermatids; some remained silent whereas others reactivated (Mueller et al. 2008).

In summary, in female germ cells the inactive X chromosome is reactivated before meiosis and thereafter remains active, whereas in males the X chromosome is inactivated during meiosis and then retains a significant but not complete level of silencing during spermiogenesis (Fig. 2).

FUNCTION AND MECHANISM OF MEIOTIC SEX CHROMOSOME INACTIVATION

In female mammals, XCI serves to equalize the levels of X-linked products with that in males (Heard and Disteche 2006). In contrast, the function of MSCI remains mysterious, although it is assumed to be an essential process because of its conservation across the animal kingdom (McKee and Handel 1993). Insight into the processes that drive MSCI in mammals came from an unlikely source, the fungus *Neurospora crassa*. Shiu and colleagues (2001) were studying the effects of different types of mutation of the *Asm-1* (*Ascospore maturation-1*) gene on the development of the haploid products of meiosis, called ascospores. They noted that heterozygous loss of function mutations in *Asm-1* resulted in inviability in those ascospores that inherited this mutation, whereas those ascospores inheriting the wild-type allele were unaffected. When heterozygous for a complete deletion of *Asm-1* all the resulting ascospores, including those with the wild type *Asm-1* allele, were inviable. Thus, deletion of one *Asm-1* allele had rendered the second allele susceptible to silencing. This led the investigators to propose that physical "pairing" between genes on different homologs during meiosis was required for them to be expressed.

Might a similar "meiotic silencing" process be operating in mammals? During normal male meiosis, the autosomes pair/synapse fully whereas the single X and Y chromosomes are largely nonhomologous and therefore remain unsynapsed. This asynapsis could trigger gene silencing across the X chromosome, resulting in MSCI. Two key observations showed that MSCI was indeed driven by the unsynapsed state of the sex chromosomes. First, when either the X or the Y chromosomes were provided with a pairing partner, this prevented them being silenced (Turner et al. 2006). For instance, in XYY males, the two Y chromosomes form a completely synapsed bivalent, and as a result continue to express Y-linked genes during pachytene, when MSCI should normally have been completed. This escape from MSCI may be directly responsible for the pachytene arrest observed in XYY mice and men (Turner 2006). Second, in male mice carrying X-autosome or autosome–autosome translocations, both of which prevent synapsis between autosomal homologs, the resulting unsynapsed autosomal regions became transcriptionally silenced, just like the X and Y chromosomes in normal male pachytene cells (Baarends et al. 2005; Turner et al. 2005). Thus, any unpaired meiotic chromosome was silenced if it failed to synapse. Studies in other organisms, principally *Caenorhabditis elegans*, has shown that asynapsis of the X chromosome is a universal driving force for MSCI (Bean et al. 2004).

Interestingly, meiotic silencing is not restricted to the male germline. In organisms in which the female is the heterogametic sex, for instance birds, the partly synapsed ZW sex bivalent is silenced during oogenesis, whereas the fully synapsed ZZ bivalent remains active during spermatogenesis (Schoenmakers et al. 2009). In female mice, meiotic silencing is also seen when chromosome synapsis is impaired. In XX females the two X chromosomes are fully synapsed and transcriptionally active throughout pachytene, but in XO females, the single X chromosome becomes transcriptionally silenced during pachytene (Baarends et al. 2005; Turner et al. 2005). Similarly, in human oocytes, unsynapsed autosomes, such as the additional copy of chromosome 21 in Down syndrome, are silenced (Garcia-Cruz et al. 2009).

A critical step in the initiation of MSCI is phosphorylation of histone H2AX at serine-139, generating γH2AX (Rogakou et al. 1999). Indeed, *H2AX*-null male mice are sterile, with a block midway through pachytene, and MSCI fails to initiate (Fernandez-Capetillo et al. 2003). H2AX is abundantly expressed in the testis and the appearance of γH2AX throughout the XY chromatin at early pachytene is temporally linked to the formation of the sex body (Mahadevaiah et al. 2001) as well as to silencing of X- and Y-linked genes as determined by RNA FISH (Turner et al. 2006). In mitotic cells, H2AX phosphorylation is achieved through the combined kinase activities of *Atm*, *Atr*, and *DNA-PK* (Fillingham et al. 2006). In mammals, both ATR (Turner et al. 2004) and ATM (Hamer et al. 2004) has been localized to the XY bivalent during the initiation of MSCI, but MSCI initiates normally in mice with mutations in *Atm* (Bellani et al. 2005; Turner et al. 2005), suggesting either that ATR acts alone in H2AX phosphorylation or that these kinases can compensate for one another in MSCI. Further evidence implicating ATR in MSCI has been provided by studies of mice with a mutation in *Brca1* (Turner et al. 2004). BRCA1 also localizes to the XY bivalent at the onset of MSCI, and in *Brca1* mutants ATR localization to the X and Y chromosomes is impaired. This impairment correlates with defective H2AX phosphorylation and MSCI failure.

Other epigenetic modifications more classically associated with chromatin silencing also become enriched on the sex chromosomes during the initiation of MSCI, most notably H3K9me2 (Peters et al. 2001; Khalil et al. 2004) and histone H3 trimethylated at lysine 9 (H3K9me3) (Greaves et al. 2006), as well as SUMO-1 modification of an as-yet unidentified chromatin target (Rogers et al. 2004; Vigodner and Morris 2005). Whether these modifications are essential for MSCI is unclear; SUMO-1 enrichment precedes H2AX phosphorylation and is therefore a very early event in the inactivation process (Vigodner 2009). Other histone modifications and variants—for example, H2AK119u1 (Baarends et al. 1999), hypoacetylated histones H3 and H4 (Khalil et al. 2004), H2AFY (Hoyer-Fender et al. 2000), H2AZ (Greaves et al. 2006), CBX1 (Motzkus et al. 1999; Metzler-Guillemain et al. 2003), and CBX3 (Metzler-Guillemain et al. 2003)—appear after H2AX phosphorylation and therefore may be involved in the maintenance of silencing. Recently, a study has found that the XY bivalent is dramatically remodeled following the initiation of MSCI; with H3.1/2 nucleosomes being evicted and replaced by H3.3 (van der Heijden et al. 2007). This exchange correlates with the transient disappearance and reappearance of H3-linked inactivating histone marks and is thought to be linked to the requirement to reactivate selected X-linked genes in the postmeiotic period (van der Heijden et al. 2007).

The mechanisms underlying the maintenance of X chromosome repression during spermiogenesis are poorly understood. A new study implicates a Y-linked gene in this process (Cocquet et al. 2009). The long arm of the mouse Y chromosome (Yq) comprises at least four gene families each present as multiple copies (Toure et al. 2004, 2005). One of these is *Sly* (*Sycp3-like Y-linked*), which encodes a protein related to the synaptonemal complex protein SYCP3 (Reynard et al. 2009). Deletions of Yq cause abnormal sperm head morphogenesis and infertility, and this is associated with increased mRNA levels of X-linked and Yp-linked genes in developing spermatids (Ellis et al. 2005). This elevation in mRNA levels is caused by increased nascent transcription of X and Y genes and therefore reflects a chromatin-mediated defect in the maintenance of XY silencing (Reynard and Turner 2009). Accordingly, repressive histone marks, including H3K9me3 and CBX1, are reduced or absent on the spermatid X and Y chromosomes in Yq mutants (Reynard and Turner 2009). Interestingly, SLY protein localizes to both the X and the Y chromosomes during spermatid development, and mice depleted in SLY through small interfering RNA-based approaches show exactly the same sex chromatin defects as males with Yq deletions (Cocquet et al. 2009). Thus, SLY is a major regulator of X chromosome repression in the mouse germline. The *Sly* gene family is found only mice; it would therefore be interesting to study to what extent X chromosome silencing is maintained during postmeiotic development in other mammals.

MEIOTIC SEX CHROMOSOME INACTIVATION AND X CHROMOSOME GENE CONTENT

Recent studies have shown that MSCI has had a profound effect on the gene content of the X chromosome. One would predict that MSCI and repression of the X chromosome during spermiogenesis would drive meiotic and postmeiotic genes off the X chromosome, and this is indeed the case. Microarray-based analysis in *Drosophila* (Parisi et al. 2003; Ranz et al. 2003), *C. elegans* (Reinke et al. 2004), and mice (Khil et al. 2004) has shown that the X chromosome is depleted in spermatogenesis genes. This has been accompanied in *Drosophila* and mice by an excess of retrotransposition of genes off the X chromosome and onto autosomes (Betran et al. 2002; Emerson et al. 2004; Wang 2004; Khil et al. 2005). The resulting retroposed copies are expressed specifically during and after MSCI, suggesting that they compensate for the silencing of their X-linked counterparts (Wang 2004).

At first, these findings seem at variance with the predictions of Rice (1984), who argued that genes involved in spermatogenesis should actually be enriched on the X chromosome. This argument was based on the concept of "sexually antagonistic" alleles (Fisher 1931); those that enhance reproductive fitness in one sex but diminish it in the other. Rice (1984) argued that a recessive mutation in an X-linked gene that conferred a reproductive advantage in males but was deleterious to females would spread rapidly in a population, because it would be immediately apparent phenotypically because of its hemizygous state. Eventually, females would arise that are homozygous for the mutation. This would impair their reproductive capacity, and this in turn would lead to the evolution of modifiers that restrict the expression of the gene to the testis. Perhaps paradoxically, the X chromosome is also expected to be enriched in genes involved in oogenesis (Rice 1984), because the X chromosome spends two-thirds of its evolutionary time in females, thereby increasing the chance that a mutation conferring a female reproductive advantage will become fixed. An enrichment of female-biased genes on the X chromosome has indeed been found both in *Drosophila* (Ranz et al. 2003) and mice (Khil et al. 2004), although not in humans (Lercher et al. 2003).

So how can the apparent depletion of spermatogenesis genes on the mouse X chromosome be reconciled with Rice's prediction? The answer, it appears, lies in the specific substage of spermatogenesis being scrutinized. Although the X chromosome appears to be depleted in testis-expressed genes when the testis is considered as a whole entity, studies on specific spermatogenic subtypes painted a different picture. For instance, in a screen designed to identify spermatogonial-specific genes, Wang et al. (2001) found that one-third of the 36 genes identified mapped to the X chromosome, suggesting that this chromosome plays a predominant role in spermatogonial development. A similar finding was made later by Khil et al. (2004); in their analysis spermatogonially expressed genes were overrepresented on the X chromosome, but genes expressed during meiosis and spermiogenesis were underrepresented. Because these later cell types outnumbered spermatogonia when the testis was considered as a whole, the overall impression was that the X chromosome was depleted in spermatogenesis genes.

 Cite this article as *Cold Spring Harb Perspect Biol* doi: 10.1101/cshperspect.a002675

So is this the complete picture? Probably not; the conclusion of Khil et al. (2004) that spermatid-expressed genes are underrepresented on the X chromosome was based on analysis of single-copy genes. A recent study has found that the X chromosome is enriched in multiple-copy genes and that most of these genes are expressed exclusively in spermatids (Mueller et al. 2008). Together, these genes represent 33 gene families, totaling 273 genes or 18% of the total protein-coding X-gene complement. Mueller et al. (2008) noted that in contrast to single-copy genes, which showed reactivation in only a small proportion of spermatids, multiple-copy genes were expressed in high numbers of spermatids, and the expression levels of multiple-copy genes were similar to those of autosomal, spermatid-expressed genes. They concluded that gene amplification antagonizes the repressive chromatin environment on the spermatid X chromosome.

Furthermore, most studies on X-gene expression have focused on protein-coding genes. Recent reports have now examined the expression of miRNAs during spermatogenesis (Ro et al. 2007; Song et al. 2009). Interestingly testis-expressed miRNAs also preferentially map to the X chromosome (Ro et al. 2007) and, more strikingly, around 80% of these X-linked miRNAs are expressed during pachytene (Song et al. 2009). This finding is unprecedented, as it represents the first example of a class of genes that are able to evade MSCI. These genes are proposed to play roles in either meiosis or in MSCI itself (Song et al. 2009).

X CHROMOSOME REACTIVATION AND MEIOTIC SILENCING AS CAUSES OF INFERTILITY

The studies outlined on X chromosome reactivation in the female germline and MSCI in the male germline provide a framework for understanding how sex chromosome aneuploidy causes infertility (Fig. 3). X chromosome reactivation occurs in the female during germ cell development, but it also occurs in males with two X chromosomes, notably XXY/Klinefelter mice (Mroz et al. 1999). However, the exact timing of X-reactivation in males is unclear. Although two active X chromosomes are clearly tolerated by female germ cells, the situation may be different in males, because in this case germ cells will express a double dosage of spermatogonial-specific X-linked genes, in addition to "housekeeping" X-linked genes. Any toxic effect caused by this expression would be expected to manifest itself during spermatogonial development, and this fits well with the early germ cell arrest seen in XXY men and mice (Fig. 3) (Ferguson-Smith 1959; Mroz et al. 1999; Lue et al. 2001).

In sex chromosome aneuploid models that germ cell arrest occurs later, during pachytene (e.g., Turner syndrome, X- or Y-autosome translocations), meiotic silencing could be a contributing factor. In X- or Y-autosome translocations, and even autosome–autosome translocations, the disruption to autosomal synapsis results in silencing of all genes located within the unsynapsed autosomal segments via meiotic silencing. Clearly, if one or more of these genes are essential for progression through pachytene, then this will result in germ cell arrest. Pachytene is a stage during which abundant autosomal transcription is known to occur (Monesi 1965; Turner et al. 2005). The meiotic silencing hypothesis fits well with the observation that small amounts of autosomal asynapsis do not invoke arrest whereas larger unsynapsed autosomal segments do (Ashley 2000); the larger the unsynapsed chromosome region, the higher the probability that it will contain a meiosis-critical gene. The model is also particularly compelling when considering XO oocytes; in this case the entire X chromosome will be silenced, rendering the oocyte nullizygous for X-linked gene products (Fig. 3). The timing of oocyte loss in XO females fits well with that of the initiation of meiotic silencing (i.e., during pachytene [Burgoyne and Baker 1981, 1985]).

A converse argument can be used to explain meiotic losses in XYY mice and men. In XYY mice, a significant proportion of early pachytene cells exhibit complete YY synapsis and this allows escape of Y genes from MSCI (Turner et al. 2006). Interestingly, these cells

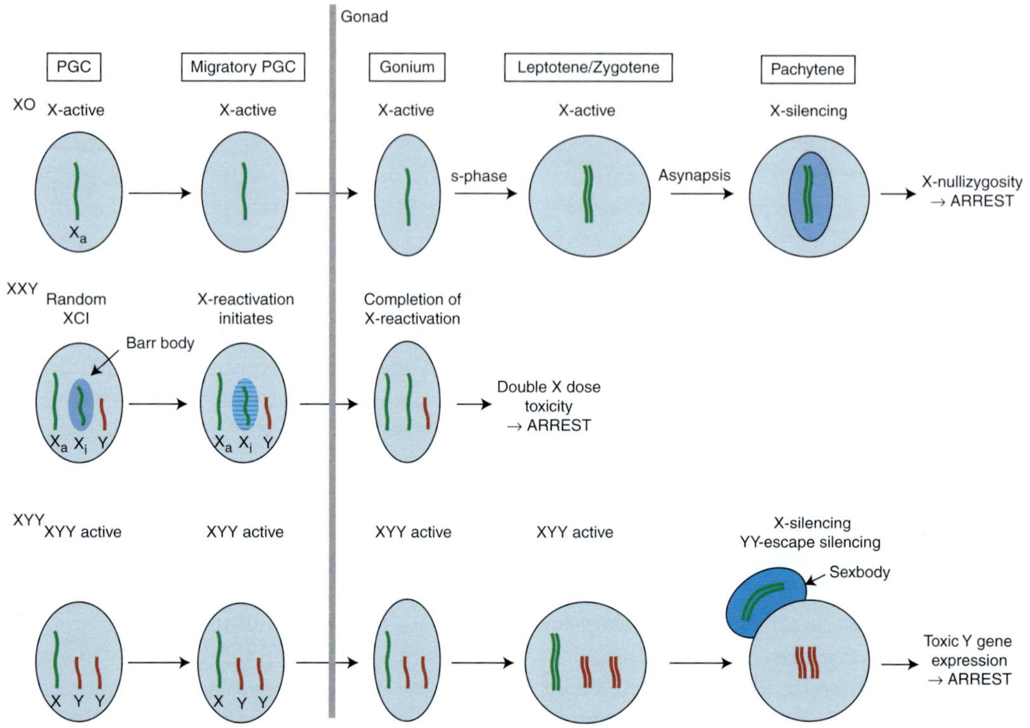

Figure 3. A model for germ cell arrest in sex chromosome aneuploid conditions. In XO females, XCI does not take place during early development, the X chromosome is therefore active until pachytene. At this stage, the unsynapsed state of the X chromosome triggers meiotic silencing; the resulting inactivation of X-linked genes leads to pachytene arrest. In XXY males, the presence of two X chromosomes initiates random XCI in the epiblast; PGCs therefore have one inactive X. The exact timing of X chromosome reactivation during germ cell development is unclear, but may follow similar kinetics to that in XX females. The resulting double dose of X-linked gene expression causes spermatogonial arrest. In XYY males, all three sex chromosomes are active until pachytene. At this stage, the X chromosome undergoes MSCI because it is unsynapsed. However, the two Y chromosomes undergo synapsis and therefore escape silencing; the resulting sustained expression of Y-linked genes is toxic and leads to pachytene arrest.

are eliminated during mid-pachytene, whereas those harboring completely unsynapsed, and hence fully silenced X and Y chromosomes, are able to progress to late pachytene (Turner et al. 2006). These findings are consistent with a model in which escape from MSCI causes germ cell arrest, presumably because of the toxic effects of misexpression of one or more genes on the Y chromosome (Fig. 3). This model has recently been corroborated: Preventing Y–Y synapsis in XYY males circumvents this mid-pachytene block (Royo et al. 2010). Furthermore, forced expression of Y-linked genes during mid-pachytene in XY males causes a meiotic block indistinguishable from that seen in XYY males (Royo et al. 2010). Independent evidence that MSCI is essential for transition through pachytene comes from *Brca1* and *H2AX* null males; in both cases there is a meiotic block and the stage of germ cell arrest phenocopies that seen in XYY males (i.e., mid-pachytene [Turner et al. 2004; Fernandez-Capetillo et al. 2003]). Thus illegitimate YY synapsis drives meiotic failure in Double Y syndrome mice and men.

The sex chromosome abnormalities so far discussed are an unusually common cause of infertility in men. Most of the remaining cases are classed as idiopathic; however, even here as yet uncharacterized genetic defects affecting the sex chromosomes may be responsible. In particular, mutation screens have focused predominantly on the Y chromosome, because of its male specificity and enrichment for spermatogenesis genes. However, we now know that the X chromosome is also enriched in spermatogenesis genes, with many having roles in spermatogonial differentiation (predominantly single-copy genes) and others in spermiogenesis (predominantly multiple-copy genes). Mutations in spermatogonial-specific X-linked genes could predispose to infertility associated with early germ cell arrest, whereas copy number loss or gain in the highly amplified spermatid-expressed genes could cause late spermatogenic phenotypes.

DISCUSSION

In this article we have highlighted the very dynamic behavior of the sex chromosomes in the mammalian germline, as well as the impact that an incorrect dosage of X- or Y-linked genes can have on fertility. The sex chromosomes have a unique genetic constitution because of the specific evolutionary pressures that they are under. They also undergo unique processes, including meiotic XY inactivation in the male germline and X chromosome inactivation in the female soma. Numerous questions and hypotheses remain to be explored in this area. For example, what is the exact role of meiotic inactivation? Can sterility in XO females be rescued in if meiotic inactivation is abolished? Do men with early spermatogenic blocks harbor mutations in single copy X linked genes? Does dampening the expression of pachytene-toxic Y genes rescue germ cell losses in XYY males? Answers to such questions should be of profound importance to our understanding of infertility.

REFERENCES

Andina RJ. 1978. A study of X chromosome regulation during oogenesis in the mouse. *Exp Cell Res* **111:** 211–218.

Ashley T. 2000. An integration of old and new perspectives of mammalian meiotic sterility. In *Results and problems in cell differentiation*, Vol. 2 (ed. McElreavey K), pp. 131–173. Springer-Verlag, Berlin.

Baarends WM, Hoogerbrugge JW, Roest HP, Ooms M, Vreeburg J, Hoeijmakers JH, Grootegoed JA. 1999. Histone ubiquitination and chromatin remodeling in mouse spermatogenesis. *Dev Biol* **207:** 322–333.

Baarends WM, Wassenaar E, van der Laan R, Hoogerbrugge J, Sleddens-Linkels E, Hoeijmakers JH, de Boer P, Grootegoed JA. 2005. Silencing of unpaired chromatin and histone H2A ubiquitination in mammalian meiosis. *Mol Cell Biol* **25:** 1041–1053.

Baudat F, de Massy B. 2007. Regulating double-stranded DNA break repair towards crossover or non-crossover during mammalian meiosis. *Chromosome Res* **15:** 565–577.

Bean CJ, Schaner CE, Kelly WG. 2004. Meiotic pairing and imprinted X chromatin assembly in *Caenorhabditis elegans*. *Nat Genet* **36:** 100–105.

Bellani MA, Romanienko PJ, Cairatti DA, Camerini-Otero RD. 2005. SPO11 is required for sex-body formation, and *Spo11* heterozygosity rescues the prophase arrest of $Atm^{-/-}$ spermatocytes. *J Cell Sci* **118:** 3233–3245.

Betran E, Thornton K, Long M. 2002. Retroposed new genes out of the X in *Drosophila*. *Genome Res* **12:** 1854–1859.

Blanco P, Shlumukova M, Sargent CA, Jobling MA, Affara N, Hurles ME. 2000. Divergent outcomes of intrachromosomal recombination on the human Y chromosome: Male infertility and recurrent polymorphism. *J Med Genet* **37:** 752–758.

Burgoyne PS, Baker TG. 1981. Oocyte depletion in XO mice and their XX sibs from 12 to 200 days post partum. *J Reprod Fertil* **61:** 207–212.

Burgoyne PS, Baker TG. 1985. Perinatal oocyte loss in XO mice and its implications for the aetiology of gonadal dysgenesis in XO women. *J Reprod Fertil* **75:** 633–645.

Cabrero J, Teruel M, Carmona FD, Jiménez R, Camacho JP. 2002. Histone H3 lysine 9 acetylation pattern suggests that X and B chromosomes are silenced during entire male meiosis in a grasshopper. *Cytogenet Genome Res* **119:** 135–142.

Chandley AC. 1997. Karyotype and oligospermia. In *Genetics of human male fertility* (ed. Barratt C, et al.), pp. 111–122. Editions EDK, Paris.

Charlesworth B, Charlesworth D. 2000. The degeneration of Y chromosomes. *Philos Trans R Soc Lond B Biol Sci* **355:** 1563–1572.

Chen B, Chu T, Harms E, Gergen JP, Strickland S. 1998. Mapping of *Drosophila* mutations using site-specific male recombination. *Genetics* **149:** 157–163.

Chuva de Sousa Lopes SM, Hayashi K, Shovlin TC, Mifsud W, Surani MA, McLaren A. 2008. X chromosome activity in mouse XX primordial germ cells. *PLoS Genet* **4:** e30.

Cocquet J, Ellis PJ, Yamauchi Y, Mahadevaiah SK, Affara NA, Ward MA, Burgoyne PS. 2009. The multicopy gene Sly represses the sex chromosomes in the male mouse germline after meiosis. *PLoS Biol* **7:** e1000244.

Cocquet J, Ellis PJ, Yamauchi Y, Riel JM, Karacs TP, Rattigan A, Ojarikre OA, Affara NA, Ward MA, Burgoyne PS. 2010. The multicopy gene Sly represses the sex chromosomes in the male mouse after meiosis. *PLoS Biology* **7:** e1000244.

De Kretser DM. 1997. Male infertility. *Lancet* **349:** 787–790.

complex, flagellar formation, and finally cytoplasmic exclusion. Transcription is repressed during spermatid elongation as the histone octamer-based chromatin organization is replaced by the incorporation of transition proteins and protamines. In mouse and in humans, however, a small fraction of the genome retains a nucleosomal organization (Gatewood et al. 1990). There is much speculation and interest as to whether inheritance of these histones confers epigenetic information that plays a role in directing gene expression following fertilization (van der Heijden et al. 2008; Albert and Peters 2009).

Study of transcription regulation in spermatogenesis is hampered by the lack of an in vitro differentiation model. Most insight has therefore come from model organisms such as *Drosophila* and mouse in which the power of genetics has proved essential for identifying the factors and mechanisms involved. This article will discuss several of these mechanisms that are unique to germ cells in both of these model organisms.

SPECIALIZED TRANSCRIPTION MECHANISMS IN MALE GERM CELLS OF *DROSOPHILA MELANOGASTER*

The Gene Expression Program of *Drosophila* Primary Spermatocytes

Drosophila primary spermatocytes transcribe genes they themselves require; additionally they transcribe the vast majority of genes required at later stages in spermatids. Although there is significant postmeiotic transcription in mammalian round spermatids (Schultz et al. 2003; Potrzebowski et al. 2008), until recently it was believed to be essentially absent from *Drosophila* spermatogenesis (Barreau et al. 2008). For example, protamines that repackage chromatin in spermatids are only detected late in elongation, however, protamine transcripts accumulate in primary spermatocytes (Jayaramaiah Raja and Renkawitz-Pohl 2005). Like other organs, testes express a little over 50% of the annotated *Drosophila* protein coding genes, however, unlike other organs, many of these genes are testis-specific or highly testis enriched (Parisi et al. 2004; Chintapalli et al. 2007). Indeed, transcription of about half the genes for known sperm proteins are testis-specific or highly enriched (Dorus et al. 2006). This large set of testis-expressed genes are activated in primary spermatocytes by the concerted actions of two protein complexes.

The Meiotic Arrest Genes Control and Coordinate Transcription in Primary Spermatocytes

There is remarkably little interdependence of cellular events within spermatid differentiation, as defects in one aspect of spermatid formation typically do not cause differentiation arrest; rather they result in abnormal progression through later stages. For example, spermatocytes that fail in the meiotic divisions continue to spermatid differentiation as 4N cells (Gonzalez et al. 1988; Alphey et al. 1992). An exception to this rule is supplied by the "meiotic arrest" mutants in which testes accumulate arrested mature primary spermatocytes (Lin et al. 1996; Table 1).

The meiotic arrest genes function in primary spermatocytes to ensure the coordinate progression of meiosis and spermatid differentiation. Reasoning that failure to initiate meiotic divisions in meiotic arrest spermatocytes was likely to result from failure to produce or activate key cell cycle proteins, the expression of the meiotic regulatory proteins CyclinB and Twine was examined. Accumulation of Twine protein was disrupted in all the mutants, and CyclinB failed to accumulate in *aly* mutants. In *aly* mutants, the spermatocytes failed to express the mRNAs for these two genes, whereas *can* (and *mia* and *sa*) mutants expressed *twine* mRNA but not protein. Because meiosis and spermatid differentiation are independent, the expression of eight known spermiogenesis genes in meiotic arrest mutant testes was examined and found to be severely abrogated (in *can*, *mia*, and *sa* mutants) or totally undetectable (*aly* mutant) (White-Cooper et al. 1998). Thus, the meiotic arrest loci were defined as being critical for transcription of certain spermiogenic genes in primary spermatocytes, and

Table 1. Classification of the meiotic arrest genes in *Drosophila*

	Gene name	Homology/domains	Reference(s)
aly-class	*aly (always early)*	DIRP	White-Cooper et al. 2000
	achi+vis (achintya and vismay)	TALE class homeodomain	Ayyar et al. 2003; Wang and Mann 2003
	comr (cookie monster)	Winged helix	Jiang and White-Cooper 2003
	topi (matotopetli)	11 Zn fingers	Perezgasga et al. 2004
	tomb (tombola)	CXC	Jiang et al. 2007
can-class	*can (cannonball)*	TAF5 WD40 repeats	Hiller et al. 2001
	mia (meiosis I arrest)	TAF6 histone fold	Hiller et al. 2004
	mip40 (Myb interacting protein 40kD)	lin-37	Beall et al. 2007
	nht (no hitter)	TAF4 histone fold	Hiller et al. 2004
	rye (ryan express)	TAF12 histone fold	Hiller et al. 2004
	sa (spermatocyte arrest)	TAF8 histone fold	Hiller et al. 2004

The gene names are shown along with distinguishing structural domains and similarity to other factors.

were subdivided into *aly*-class and *can*-class on the basis of their effects on particular target genes.

To date eleven meiotic arrest mutant loci have been described, on the basis of this classification five are *aly*-class, whereas six are *can*-class (see Table 1). Transcription of all eight known spermiogenesis genes tested in the original 1998 study was dramatically reduced in the meiotic arrest mutants. To assess what proportion of spermiogenesis genes are controlled by the meiotic arrest transcriptional module, a large scale RNA in situ hybridization screen has been performed (www.fly-ted.org) (Barreau et al. 2008). Typically, genes that are transcribed in primary spermatocytes and whose transcripts persist in spermatids consistent with a postmeiotic role are not expressed in mutant testes. As a conservative estimate, based also on unpublished microarray data, over 1500, and probably more than 2000, genes require meiotic arrest gene function for their transcription in *Drosophila* primary spermatocytes. Therefore, the meiotic arrest genes are critical regulators of gene expression in these cells.

tMAC: A Testis-Specific DREAM Complex

Four of the five known *aly*-class genes are expressed exclusively or almost exclusively in primary spermatocytes, consistent with their critical function in these cells (Jiang and White-Cooper 2003; Perezgasga et al. 2004; Jiang et al. 2007; White-Cooper 2009). One *aly*-class locus is complex, with both copies of a recent gene duplication (*achi* and *vis*) being mutated to give the male sterile phenotype; *vis* is strongly testis-enriched in adults; its duplication partner, *achi* is expressed ubiquitously at low levels (Ayyar et al. 2003; Wang and Mann 2003). *aly* contains a DIRP domain (domain in Rb-related pathway), whereas *topi*, *comr*, *tomb*, and *achi+vis* proteins all have predicted DNA binding domains, of the Zn-finger, winged helix, CXC, and homeodomain classes, respectively. Thus, these proteins probably act to directly control transcription by interacting with DNA.

Most animal genomes have only one DIRP protein and one CXC-domain protein, however, in *Drosophila* these genes have duplicated and *aly* and *tomb* are testis-specific paralogs of the broadly expressed genes *mip130* and *mip120*. Mip130 and Mip120 proteins copurify in a complex (MMB/dREAM) from somatic tissues with Mip40, Myb, CAF1, Rbf, E2F2, Dp, and Lin-52 (Korenjak et al. 2004; Lewis et al. 2004). An orthologous complex has also been purified from *Caenorhabditis elegans* (DRM) (Harrison et al. 2006) and from humans (DREAM/LINC) (Litovchick et al. 2007; Schmit et al. 2007).

Given the evolutionary conservation of the protein families it is not surprising that Aly and Tomb proteins copurify with Mip40 from testes in the tMAC complex (testis meiotic arrest complex) that also contains Topi, Comr, and CAF1 (Beall et al. 2007). Direct protein interactions have been shown for Topi:Comr, for Tomb:Aly and for Tomb:Comr (Perezgasga et al. 2004; Jiang et al. 2007). Notably, Vis (and Achi) was not found in tMAC, neither were Myb, Rbf, Dp, or E2F2. Because Achi/Vis, Aly, and Comr have been shown to coimmunoprecipitate from testes (Wang and Mann 2003), it is likely that tMAC comes in at least two forms, one with Achi/Vis, but lacking Mip40, the other with Mip40, but lacking Achi/Vis. Curiously, whereas *mip40* mutants are meiotic arrest male sterile, they do not have an *aly*-class phenotype, indicating that Mip40 functions differently from the other components of the complex (Beall et al. 2007).

Analysis of protein localizations in various mutant backgrounds also sheds light on the interdependence of the proteins. Aly and Comr proteins must both be present for either to localize to the nucleus (Jiang and White-Cooper 2003), and for Tomb to be stable, whereas Tomb is essential to allow Aly and Comr to interact with chromatin (Jiang et al. 2007). Topi and Achi/Vis seem to be more peripheral to the core complex, although all testis-specific members of tMAC must be present for the complex to efficiently interact with chromatin (Ayyar et al. 2003; Perezgasga et al. 2004).

The Meiotic Arrest Genes—Transcriptional Activators or Repressors of Repressors?

Theoretically, activation of a set of genes in a developmentally orchestrated manner can be achieved by several different mechanisms. The most straightforward is direct activation by a transcription factor (complex) of all the target genes. More complex are transcription factor cascades, in which factors are sequentially activated as differentiation proceeds. Finally, the genes could have a default setting of ON, and be repressed in all cells in which their transcription is not desired. Activation would then involve repressing this repressor. The finding that transcription of any one meiotic arrest locus is independent of all the others argues against a potential transcription factor cascade mechanism (White-Cooper et al. 2000; Hiller et al. 2004).

Although the human homolog of Achi/Vis, TGIF, is a transcriptional repressor, Achi in *Drosophila* testes is an activator (Wang et al. 2008). dREAM (Beall et al. 2004; Korenjak et al. 2004; Lewis et al. 2004), and orthologous complexes in other species have predominantly been linked to a transcriptional repression rather than activation (Cui et al. 2006), although some activating functions are reported (Georlette et al. 2007). LINC does activate expression of certain cell cycle genes (Pilkinton et al. 2007), and the activation versus repression role of LINC seems to be related to the DNA binding factors with which it is associated (Osterloh et al. 2007; Schmit et al. 2007). It is likely that dREAM, which contains Myb, Rbf, and Dp is predominantly repressive, whereas tMAC, using Comr, Topi, and Achi/Vis to interact with promoters, is predominantly activatory. Consistent with this, all the tMAC subunits associate with euchromatin in primary spermatocytes, although their binding to specific target promoters has not been reported. The *C. elegans* SynMuvB pathway genes also include known chromatin modifiers, for example, subunits of the NuRD nucleosome remodeling and histone deacetylase complex (Fay and Yochem 2007). It is possible that tMAC interacts with such a complex in primary spermatocytes, and activates gene expression by regulating chromatin architecture and accessibility in primary spermatocytes (White-Cooper et al. 2000).

tTAFs as Repressors of a Repressor

An alternative mechanism is observed for the members of the *can*-class factors that encode paralogs of the subunits of the general transcription factor TFIID. The TFIID complex is formed by the TATA-binding protein (TBP) and 13–14 TBP-associated factors (TAFs) (Tora 2002; Matangkasombut et al. 2004; Cler et al. 2009). The composition and organization of TFIID is highly conserved through evolution. TFIID has a modular structure with a core domain formed by TAF5 and a set of histone fold-containing TAFs, in which a second module comprising TAF1, TAF7, and TBP associates to form TFIID (Fig. 1) (Gangloff et al. 2001; Papai et al. 2009). In addition to the TFIID TAFs that are widely expressed in embryonic and adult tissues, male germ cells express paralogs of TBP and several TAFs that play critical, but distinct roles in spermatogenesis in *Drosophila* and mouse.

The *can*-class meiotic arrest genes *can*, *mia*, *nht*, *rye*, and *sa*, encode testis-specific paralogs of ubiquitously expressed TAFs that form the TFIID core (Hiller et al. 2001, 2004). These tTAFs share the same structural domains (histone-folds, WD40 repeats) and form heterodimers similar to those described for the corresponding core TAFs (see Table 1, and Kolthur-Seetharam et al. 2008). A simple hypothesis is that these factors along with non-testis-specific components, and testis-specific splice isoforms of TAFs (e.g., TAF1-2; Metcalf and Wassarman 2007) assemble to form a testis-specific TFIID complex. Alternatively, as a subset of TAFs are also components of other transcriptional regulatory complexes, for example, in the Polycomb Repression Complex 1 (PRC1) with Polycomb group (PcG) proteins (Saurin et al. 2001) or the GCN5/PCAF histone acetyltransferases in the SAGA complex (Timmers and Tora 2005), the tTAFs could also function through their incorporation in these complexes.

The idea that the tTAFs form a testis-specific TFIID complex that binds directly to the promoters of target genes would predict that this complex be localized to euchromatin in primary spermatocytes. Although some tTAF association with the chromosomal masses in these cells is observed, the strongest labeling is in a subcompartment of the nucleolus (Chen et al. 2005; Metcalf and Wassarman 2007). The action of tMAC is important for localization of tTAFs to the nucleolus, as in *aly* mutant testes Sa protein localizes in a peri-nucleolar ring (Metcalf and Wassarman 2007). Only tTAFs and the testis-specific isoform TAF1-2 have this nucleolar localization, other TFIID subunits are found associated with euchromatin.

Polycomb and other PRC1 components localize exclusively to chromatin in spermatogonia, but significantly relocalize to the nucleolus in primary spermatocytes (Chen et al. 2005). This nucleolar localization of PRC1 depends on the action of the tTAFs, as Pc remains exclusively on euchromatin in *sa* mutant testes. Analysis by chromatin immuno precipitation (ChIP) revealed that tTAFs are normally associated with the promoters of target spermiogenesis genes, and that Pc preferentially accumulates at these regions in tTAF mutant testes. Together these observations suggest that tTAFs in primary spermatocytes act to remove the PRC1 repressor complex from target promoters and sequester it in the nucleolus (i.e., they repress the repressors), but they may also directly function as transcriptional activators, replacing the ubiquitous TFIID at target promoters.

SPECIALIZED TRANSCRIPTION MECHANISMS IN MURINE MALE GERM CELLS

This next section will describe several examples of specialized transcription mechanism in murine male germ cells, and highlight the observation that, although mouse germ cells also express paralogs of TAFs and other general transcription factors, their mechanism of action is quite distinct from that in *Drosophila*.

CREM, ACT, and KIF17b Cooperate in an Integrated Regulatory Mechanism Unique to Haploid Cells

Cyclic AMP response element (CRE) binding protein (CREB) and cyclic AMP response element modulator (CREM) are highly related proteins that regulate transcription in response to various stress, metabolic and developmental signals (Hummler et al. 1994; Sassone-Corsi 1995). CREB and CREM bind to the consensus palindromic CRE 5′-TGACGTCA-3′ or half-CRE 5′-TGACG-3′ and 5′-CGTCA-3′ elements present in the promoters of target genes. CREM exists in multiple isoforms that act either as repressors or activators (Foulkes and Sassone-Corsi 1992). In male germ cells of

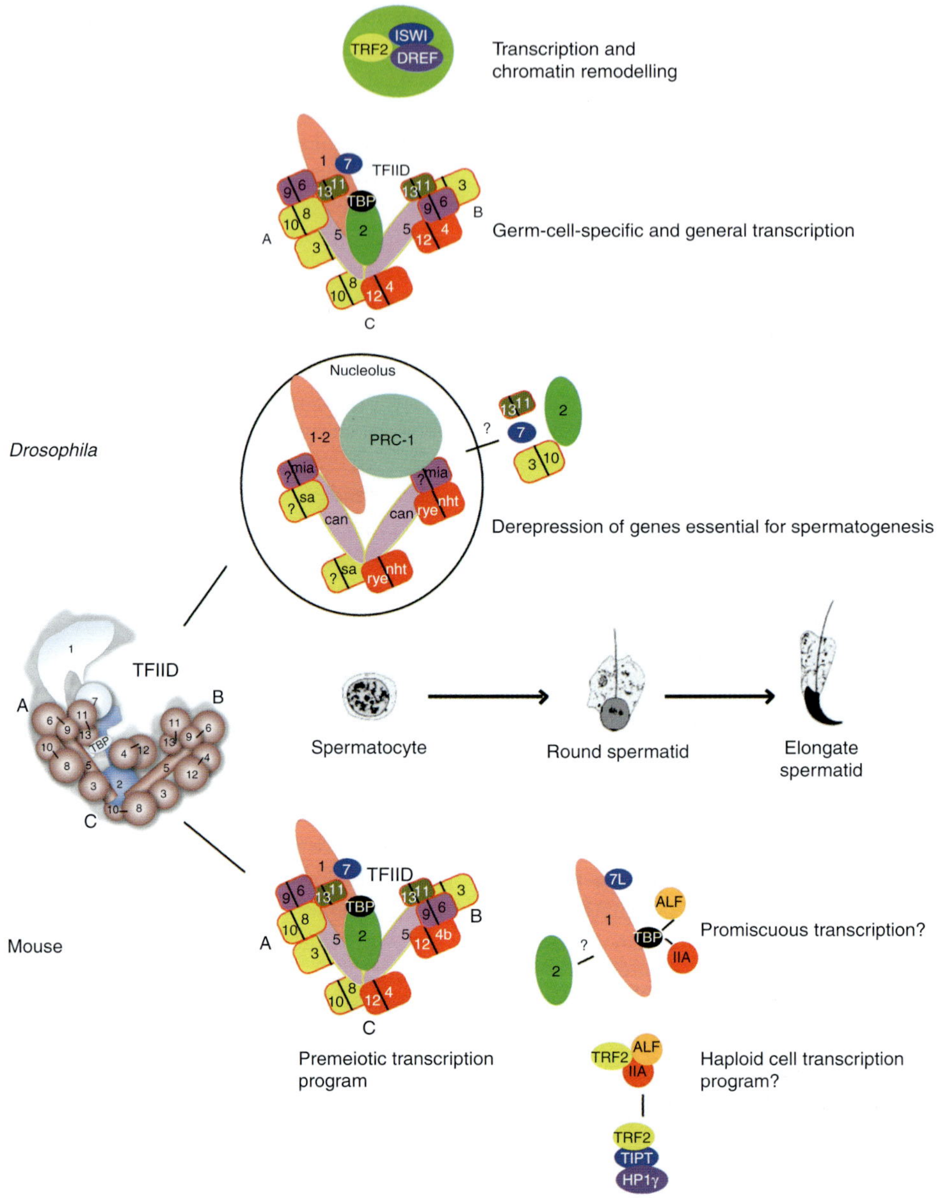

Figure 1. Schematic description of TFIID complexes that are involved in transcription in *Drosophila* and murine male germ cells. The structure of TFIID as determined from immunoelectron microscopy (EM) experiments is shown on the *left* of the figure. The TBP and TAF subunits are represented by colored balls superimposed on the reconstructed EM model. *Drosophila* spermatocytes express paralogs of several TAFs that are shown organized in the form of a partial TFIID complex. The heterodimerization partners of *mia* and *sa* remain to be determined along with the presence of TAFs 2, 7, 11, and 13 in this complex. These spermatocytes likely also contain a TFIID complex comprising somatically expressed TAFs and the TRF2 complex. Murine spermatocytes predominantly contain a normal TFIID complex containing TAF4 and/or TAF4b. In haploid cells, TAF4 is strongly down-regulated leading to loss of full TFIID and accumulation of the TBP-TAF1-TAF7L submodule. TBP can interact with either TFIIA or ALF to mediate elevated levels of mRNA transcription in these cells. TRF2 also interacts with either TFIIA or ALF and may mediate transcription of a subset of essential genes or it may interact with chromatin via TIPT and HP1γ.

prepubertal animals, low expression of the repressor isoforms are observed. At puberty, follicle stimulating hormone (FSH), by an as yet undefined mechanism modulates the usage of alternative polyadenylation sites such that several destabilizer signals in the 3′ untranslated region of the CREM τ activator isoform mRNA are eliminated leading to increased stability and the accumulation of the CREM τ protein to high levels in postmeiotic round spermatids (Foulkes et al. 1992, 1993).

CREM knockout leads to early apoptosis of haploid cells, around step 4 of their development, and male sterility showing that CREM plays an essential role in spermiogenesis (Blendy et al. 1996; Nantel et al. 1996; Nantel and Sassone-Corsi 1996). Amongst the direct CREM targets are the genes encoding protamines, transition proteins, and outer dense fiber protein whose expression initiates in round spermatids, but whose gene products are required only later during the elongation stage. Diminished expression of these genes may account for the reduced sperm count and abnormalities seen in heterozygous CREM mutant mice, but do not readily explain the apoptosis. ChIP from adult testis coupled to high throughput sequencing shows that CREM occupies the promoters of more than 6000 target genes in haploid cells (Martianov et al. 2010). Comparison with transcriptome data shows that only a small subset of these genes are deregulated following CREM inactivation. The deregulated genes include the anti-apoptotic factor *Bcl6b* and genes involved in signal transduction and metabolic processes (Beissbarth et al. 2003; Martianov et al. 2010). It is rather the deregulation of these genes that accounts for the apoptotic phenotype seen on CREM inactivation.

In somatic tissues, a variety of stimuli and several different kinases converge to phosphorylate a serine residue (S133 or S117) in the activation domains of CREB and CREM, respectively (Johannessen et al. 2004). Serine phosphorylation results in recruitment of the p300 and CREB binding protein (CBP) coactivators (Chrivia et al. 1993; Nakajima et al. 1997). Alternatively, CREB can also activate transcription of a subset of target genes through a phosphorylation-independent interaction with the TORC (transducers of regulated CREB activity) family of coactivators via the CREB bZIP domain that enhances interaction with the TFIID subunit TAF4 (Fig. 2) (Conkright et al. 2003; Xu et al. 2007). In haploid cells, CREM functions by yet another mechanism that bypasses the requirement for phosphorylation through interaction with the LIM-only domain protein ACT (activator of CREM in testis) encoded by the *Fhl5* gene (Fimia et al. 1999) specifically expressed in haploid cells. ACT functions as a coactivator in yeast and in mammalian cells in which it stimulates CREM transcriptional activity. Inactivation of ACT in mice does not phenocopy loss of CREM as there is no block in spermatogenesis and males are fertile (Kotaja et al. 2004). ACT knockout does, however, lead to a severe reduction in the number of mature sperm and the appearance of frequent malformations of the sperm head and flagellum. These observations suggest only a subset of CREM target genes are regulated using ACT as a coactivator, or that ACT is required for a general fine-tuning of CREM target transcript levels.

The activity of the CREM-ACT pathway is controlled by KIF17b, a testis-specific form of the kinesin protein KIF17, that interacts with ACT (Macho et al. 2002). KIF17b colocalizes with ACT in the nucleus of haploid cells, but at the onset of the elongation phase, KIF17b and ACT cotranslocate to the cytoplasm. In ex vivo models, KIF17b-mediated export of ACT from the nucleus represses CREM activation of target genes. Intracellular translocation of KIF17b is dependent neither on its motor domain nor on microtubules, revealing a novel microtubule-independent function for kinesins (Kotaja et al. 2005). In cell culture assays, cyclic AMP-dependent protein kinase A (PKA) phosphorylates KIF17b to regulate its subcellular localization. Inhibition of the PKA pathway affects nuclear-cytoplasmic shuttling and promotes cytoplasmic accumulation of KIF17b. If such a mechanism were to operate in germ cells, it would provide a mechanism in which cyclic AMP signaling could modulate

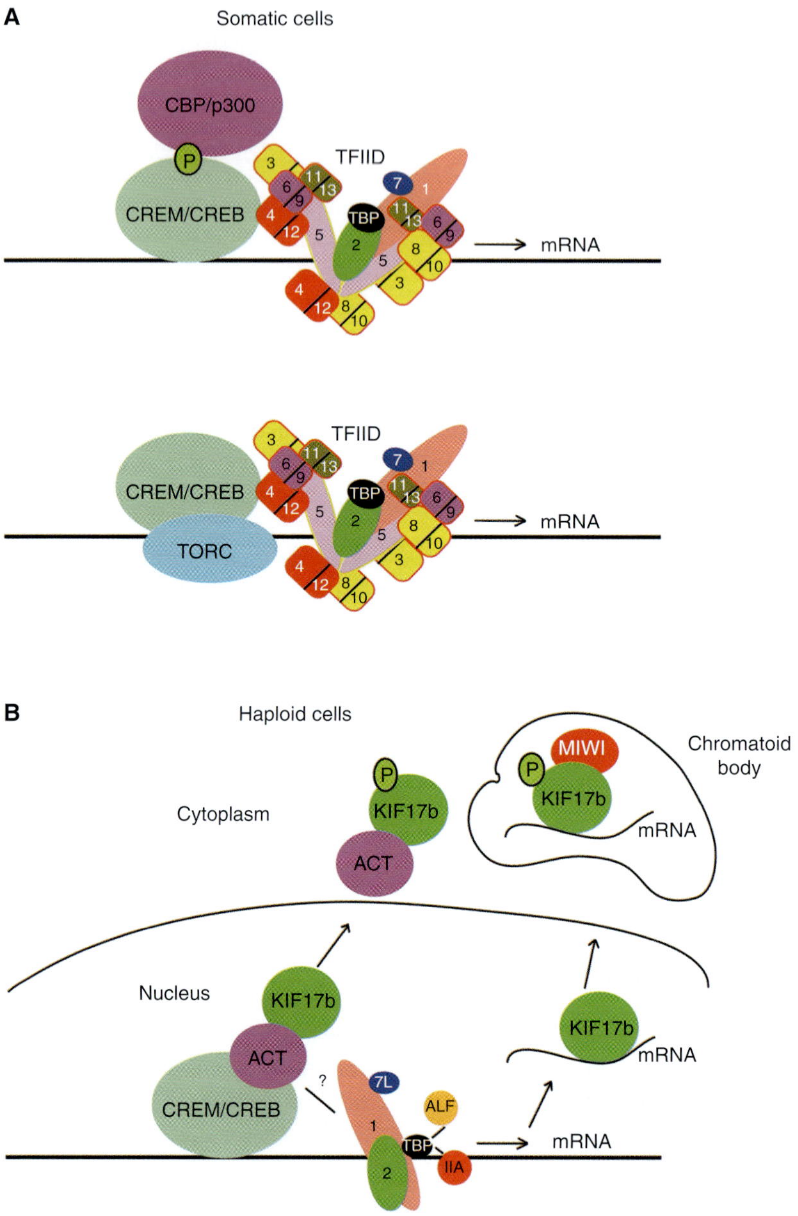

Figure 2. Schematic model of the CREM-ACT-KIF17b regulatory pathway. (*A*) In somatic cells, CREB and CREM activate transcription following phosphorylation of a conserved serine residue and recruitment of the CBP and p300 coactivators. CREB and CREM can also interact with TAF4 that is required as a coactivator for their activity. Alternatively, CREB can also activate transcription independent of CBP/p300 via the TORC coactivators. (*B*) In germ cells, CREM interacts with ACT. At the elongation phase ACT is exported to the cytoplasm by KIF17B to repress CREM activation of target genes. KIF17B also transports mRNAs to the chromatoid body where it interacts with the MIWI protein. Phosphorylation of KIF17B by PKA increases its cytoplasmic localization.

 Cite this article as *Cold Spring Harb Perspect Biol* doi: 10.1101/cshperspect.a002626

CREM-mediated transcription in male germ cells through control of KIF17b rather than direct phosphorylation of the CREM activation domain.

KIF17b is a multifunctional protein that is involved not only in the transport of CREM, but also of a set of CREM-regulated mRNAs (Chennathukuzhi et al. 2003). KIF17b physically interacts with a testis-specific member of the PIWI/Argonaute family, MIWI, a component of the chromatoid body implicated in RNA metabolism (Kotaja et al. 2006; Kotaja and Sassone-Corsi 2007). This interaction may play a role in the transportation of mRNAs to the chromatoid body. Thus, KIF17b links the processes of transcription and transport of specific mRNAs in male germ cells.

Together the above studies describe a unique and integrated mechanism by which CREM, ACT, and KIF17b act together to regulate transcription in developing male germ cells (summarized in Fig. 2).

TRF2 and TRF3: Two TBP-Related Proteins with Specific Roles in Male and Female Germ Cell Development

As described above, the tTAFs play an important role in *Drosophila* spermatogenesis. In mouse, several testis-expressed paralogs of TFIID components have been identified, such as TAF4b that plays an as yet undefined role in maintaining spermatogonia self-renewal (Falender et al. 2005), and TAF7L (see below). One of the most striking examples is, however, that of the TBP-related factors TRF2 and TRF3.

TBP has a bipartite structure with a poorly conserved amino-terminal region and a highly conserved carboxy-terminal domain that folds as a molecular saddle-shaped structure and binds DNA via the concave surface (Burley 1996). In vertebrates, two TBP-related proteins, TRF2 and TRF3 (also designated as TLF and TBP2, respectively), share this conserved carboxy-terminal domain (Dantonel et al. 1999; Rabenstein et al. 1999; Teichmann et al. 1999; Persengiev et al. 2003; Bartfai et al. 2004; Jallow et al. 2004). In mouse, TRF2 is highly expressed in male germ cells, whereas TRF3 expression is restricted to the ovary (Martianov et al. 2001; Xiao et al. 2006).

Genetic experiments in mice show a specific function of these factors in germ cells. TRF2-null mice are viable and females are fertile (Martianov et al. 2001; Zhang et al. 2001). In contrast, male animals are sterile with an almost complete arrest of spermiogenesis because of apoptosis of step 7 round spermatids. TRF2 mutant round spermatids are characterized by the fragmentation of the chromocenter, a nuclear structure comprising the centromeric heterochromatin from each chromosome (Martianov et al. 2002). In normal spermatids, a single dense nuclear structure is observed, whereas in TRF2 mutant cells, 3–4 large foci are found. The chromocenter is proposed to act as a chromatin organizer prior to the elongation phase (Zalensky et al. 1995). The TRF2 mutant spermatids undergo apoptosis at the beginning of the elongation phase suggesting that chromatin disorganization activates a checkpoint at this step.

At present, it is not clear whether TRF2 acts as a transcription factor or has a more direct structural role on the chromatin and evidence exists for both possibilities. TRF2, like TBP, interacts with TFIIA (Teichmann et al. 1999) and with a paralog of the TFIIA α/β subunit, ALF (TFIIA-like factor) (Upadhyaya et al. 1999; Ozer et al. 2000), that is specifically expressed in late pachytene spermatocytes and in haploid round spermatids (Catena et al. 2005). Like TFIIA, ALF is cleaved into α and β subunits that associate with TFIIAγ. Immunoprecipitation experiments show that male germ cells contain not only TFIIA$\alpha/\beta/\gamma$ and ALFα/β-TFIIAγ complexes, but also the full combination of hybrid complexes between the TFIIA/ALF α and/or β subunits (Catena et al. 2005). The TRF2-TFIIA/ALF interaction suggests that TRF2 may function as a transcription factor. However, TRF2 does not bind to TATA sequences and no alternative TRF2 binding motif or response element has been identified on the promoters of germ cell expressed genes. A demonstration that murine TRF2 acts as a bona fide transcription factor will require the identification of such elements and the demonstration that TRF2 is bound in vivo to the target promoters.

TRF2 may also have a more direct role on chromatin as two hybrid experiments have shown an interaction with TIPT (TRF2 interacting protein in testis), a novel uncharacterized testis-specific protein that is specifically expressed in late pachytene and haploid round spermatids (Brancorsini et al. 2008). TIPT also interacts with the heterochromatin proteins HP1α and HP1γ and colocalizes with HP1γ in the euchromatin compartment. Although the function of TIPT in testis remains to be determined, these observations provide a potential direct link between TRF2 and chromatin organization.

It is noteworthy that in *Drosophila*, TRF2 is also required for spermatogenesis. Mutation of TRF2 leads to defects in meiotic chromatin condensation and chromosome segregation (Kopytova et al. 2006). TRF2 forms a high molecular weight protein complex with the DNA binding factor DREF and proteins such as ISWI involved in chromatin remodeling providing further evidence that TRF2 may directly modulate chromatin organization (Hochheimer et al. 2002).

The specific role of TRF2 in male germ cells is mirrored by that of TRF3 in female germ cells. Oogenesis in mammals, comprises a complex series of events within the ovarian follicles. Oocytes are generated from oogonia and then undergo the early stages of meiotic prophase I. During reproductive life, oocytes are continuously selected from a pool of primordial follicles to develop via a growth phase during which they are arrested at diplotene of prophase I. Studies of TBP and TRF3 expression show a remarkable developmental switch. TBP is expressed in the primordial follicles, but is rapidly down-regulated prior to arrest in metaphase I (Gazdag et al. 2007). In contrast, TRF3 is strongly expressed throughout folliculogenesis, declines at the oocyte stage and disappears following fertilization. These results suggest that TRF3 replaces TBP to direct the specific gene expression program required for oocyte development. This is indeed the case as inactivation of the TRF3 gene results in female infertility, deregulation of oocyte-specific genes and altered chromatin organization (Gazdag et al. 2009). In contrast, inactivation of TBP in the female germ line does not affect oocyte formation demonstrating that it is TRF3 and not TBP that is essential for oocyte development.

There is thus, a striking parallel between the roles of TRF2 and TRF3 in the male and female germ lines. Both proteins show a specific and dynamically regulated expression in male or female germ cells, they are required to direct specific gene expression programs and they play a role in chromatin organization. Interestingly, however, these factors play more general roles in embryonic development in other vertebrates such as the zebrafish *Danio rerio* or *Xenopus* (Veenstra et al. 2000; Muller et al. 2001; Bartfai et al. 2004; Hart et al. 2007; Jacobi et al. 2007). Evolution has therefore selected a more specialized role for these proteins in germ cell function in mammals, where embryonic development takes place within the mother's body.

Evidence for "Promiscuous" Transcription in Male Haploid Germ Cells

Although there is a remarkable parallel between the roles of TRF2 and TRF3 in male and female germ cells, there is also a major difference. In female germ cells, TRF3 replaces TBP, but in male germ cells, TBP is strongly up-regulated and coexpressed with TRF2 (Schmidt and Schibler 1995; Martianov et al. 2002). TBP mRNA levels are almost 100-fold up-regulated in haploid cells compared to somatic tissues, principally because of use of alternate promoters (Schmidt et al. 1997; Schmidt and Schibler 1997). TBP protein levels are 8- to 10-fold higher in testis than in somatic cells. Similarly, TFIIB, RNA polymerase II, TFIIA, and ALF are also strongly expressed in haploid cells (Schmidt and Schibler 1995). It has been proposed that overexpression of the general transcription machinery at this stage is required to produce the large amounts of some mRNAs that have to be synthesized and stored for translation during the elongation and remodeling phase. There is also evidence that overall mRNA levels are generally higher in haploid

Cite this article as *Cold Spring Harb Perspect Biol* doi: 10.1101/cshperspect.a002626

cells than in somatic cells because of "promiscuous" transcription where in addition to transcription of highly expressed testis-specific genes and of normal "housekeeping" genes, many other genes are also transcribed albeit at lower levels (Schmidt and Schibler 1995). It has been suggested that this may facilitate DNA repair of the transcribed genes before transmission to the next generation, because transcription and DNA repair are coupled (Citterio et al. 2000; Laine and Egly 2006).

In agreement with this idea, ChIP-seq experiments show widespread occupancy by CREM of target genes that have no specific role in spermatogenesis, but whose promoters are active in haploid cells (Martianov et al. 2010). The euchromatin of round spermatids therefore seems rather permissive for transcription factor binding which, together with high levels of the basal transcription machinery, may lead to the expression of a large set of genes and generally higher mRNA levels. This suggests a model in which haploid cells express germ cell–specific paralogs of the general factors for two distinct purposes, to orchestrate the expression of a specific gene expression program and at the same time to ensure sufficient levels of basal factors to promote elevated levels of transcription.

Additional evidence for the above hypothesis comes from the study of TAF7L, a germ cell–specific paralog of the TFIID subunit TAF7 that plays an important role in transcription initiation (Gegonne et al. 2006). TAF7L was identified in a screen for spermatogonial expressed genes located on the X chromosome (Wang et al. 2001). The TAF7L protein is localized in the cytoplasm in spermatogonia and early primary spermatocytes, but from mid-pachytene stage onward, TAF7L is imported into the nucleus and accumulates strongly in postmeiotic cells (Pointud et al. 2003). The import of TAF7L into the nucleus is coordinated with both a loss of TAF7 expression, and a potent up-regulation of TBP. In contrast, the TAF4 subunit of TFIID that is essential for TFIID stability is strongly down-regulated at the same stage. Haploid cells therefore appear to contain an abundance of the TBP-TAF1-TAF7 module, but low amounts of the core domain complex and hence low levels of TFIID. Furthermore, the genomes of old world monkeys, apes, and humans include a retrotransposed copy of TAF1 encoding TAF1L that displays the same functional properties as TAF1 (Wang and Page 2002). This may represent a strategy for human germ cells to evade meiotic sex chromosome inactivation of TAF1 on the X chromosome, and maintain high levels of TAF1 activity in haploid cells. These observations suggest that high levels of specific and/or promiscuous gene expression are mediated via the TBP-TAF1-TAF7L module of TFIID.

This model has been tested by TAF7L inactivation in mouse (Cheng et al. 2007) in which its loss does not lead to arrested spermatogenesis, but to a strongly reduced sperm count and reduced fertility. Sperm from mutant animals are characterized by abnormal morphology, reduced motility, and folded tails. Transcriptome analysis, however, identified only six transcripts with mildly reduced expression. These relatively minor changes in gene expression are unlikely to fully account for the observed phenotype. Although further transcriptome studies on purified germ cell populations may reveal more profound effects on gene expression indicative of a specific function of TAF7L for a subset of target genes, these observations are open to an alternative interpretation. Rather than regulating expression of specific genes in haploid cells, TAF7L is required to ensure generally elevated transcription. An overall reduction in transcript levels is more difficult to observe in transcriptome array experiments compared to the selective loss or induction of a set of genes. This interpretation would explain the paucity of changes on the Affymetrix array, and account for a phentotype with a generalized reduction in spermatogenesis rather than arrest at a specific stage.

The above results lend weight to the idea that paralogs of the general transcription machinery are required in male germ cells for two different, but related functions. Overexpressed components of the basal transcription machinery, including ALF, act together with the TBP-TAF1-TAF7l sub-module of TFIID to direct high levels of mRNA transcription in haploid cells required for normal spermatogenesis. TRF2 on the other hand has evolved to direct a specific aspect of the gene expression program and/or modulate chromatin structure in haploid cells (summarized in Fig. 1). These ideas can be tested by, for example, knockout of ALF, which one would predict to have a general effect on transcription and a phenotype similar to that of TAF7L rather than leading to arrest at a specific stage. Further experiments of this type will confirm or infirm the above model and lead to a better understanding of the mechansims required for correct spermatogenesis and male fertility.

CONCLUDING REMARKS

The studies described above provide insights into transcriptional regulatory mechanisms that are unique to male germ cells. TAF paralogs play important, but mechanistically distinct roles in both *Drosophila* and mouse spermatogenesis. In *Drosophila*, these proteins act to sequester the PRC1 repressor complex to facilitate expression of genes required for spermatogenesis in spermatocytes. In contrast, in mice, they may be required to ensure general high levels of mRNA synthesis in haploid cells. A recurrent theme, that has now also found an echo in female germ cells, is the importance of germ cell–specific paralogs of somatically expressed factors and of components of the general transcription machinery illustrated by the tTAFs, TRF2, TRF3, and ALF. Consider also TCEA2, a murine paralog of TFIIS that is selectively expressed at the late pachytene and round spermatid stages (Ito et al. 1996; Umehara et al. 1997). TFIIS was long considered as a transcription elongation factor, however, it has now been shown to be also generally required at the preinitiation stage (Guglielmi et al. 2007; Kim et al. 2007; Sikorski and Buratowski 2009). Here is yet another example of a germ cell–specific paralog of the general transcription machinery, whose function remains to be determined. An unresolved question is why male and female germ cells have adopted this strategy to direct their specific gene expression programs. The diversity of mechanisms discussed above suggests that there is probably no single answer. Future studies should provide further insights into this question.

ACKNOWLEDGMENTS

H.W.-C. is a Royal Society University Research Fellow. Funding for the analysis of tMAC has been from the Wellcome Trust. Funding for in situ hydridizations and FlyTED has been from the BBSRC. The laboratory of I.D. is supported by grants from the CNRS, the INSERM, the ANR, the Association pour la Recherche contre le Cancer, the Ligue Nationale et Départementale Région Alsace contre le Cancer. I.D. is an équipe labéllisée of the Ligue Nationale contre le Cancer.

REFERENCES

Albert M, Peters AH. 2009. Genetic and epigenetic control of early mouse development. *Curr Opin Genet Dev* **19:** 113–121.

Alphey L, Jimenez J, White-Cooper H, Dawson I, Nurse P, Glover DM. 1992. *twine*, a cdc25 homolog that functions in the male and female germline of *Drosophila*. *Cell* **69:** 977–988.

Ayyar S, Jiang J, Collu A, White-Cooper H, White RA. 2003. *Drosophila* TGIF is essential for developmentally regulated transcription in spermatogenesis. *Development* **130:** 2841–2852.

Barreau C, Benson E, Gudmannsdottir E, Newton F, White-Cooper H. 2008. Post-meiotic transcription in *Drosophila* testes. *Development* **135:** 1897–1902.

Bartfai R, Balduf C, Hilton T, Rathmann Y, Hadzhiev Y, Tora L, Orban L, Muller F. 2004. TBP2, a vertebrate-specific member of the TBP family, is required in embryonic development of zebrafish. *Curr Biol* **14:** 593–598.

Beall EL, Bell M, Georlette D, Botchan MR. 2004. Dm-*myb* mutant lethality in *Drosophila* is dependent upon *mip130*: Positive and negative regulation of DNA replication. *Genes Dev* **18:** 1667–1680.

Beall EL, Lewis PW, Bell M, Rocha M, Jones DL, Botchan MR. 2007. Discovery of tMAC: A *Drosophila* testis-specific meiotic arrest complex paralogous to Myb-Muv B. *Genes Dev* **21:** 904–919.

Beissbarth T, Borisevich I, Horlein A, Kenzelmann M, Hergenhahn M, Klewe-Nebenius A, Klaren R, Korn B, Schmid W, Vingron M, et al. 2003. Analysis of CREM-dependent gene expression during mouse spermatogenesis. *Mol Cell Endocrinol* **212:** 29–39.

Blendy JA, Kaestner KH, Weinbauer GF, Nieschlag E, Schutz G. 1996. Severe impairment of spermatogenesis in mice lacking the CREM gene. *Nature* **380:** 162–165.

Brancorsini S, Davidson I, Sassone-Corsi P. 2008. TIPT, a male germ cell–specific partner of TRF2, is chromatin-associated and interacts with HP1. *Cell Cycle* **7:** 1415–1422.

Burley SK. 1996. The TATA box binding protein. *Curr Opin Struct Biol* **6:** 69–75.

Catena R, Argentini M, Martianov I, Parello C, Brancorsini S, Parvinen M, Sassone-Corsi P, Davidson I. 2005. Proteolytic cleavage of ALF into α- and β-subunits that form homologous and heterologous complexes with somatic TFIIA and TRF2 in male germ cells. *FEBS Lett* **579:** 3401–3410.

Chen X, Hiller M, Sancak Y, Fuller MT. 2005. Tissue-specific TAFs counteract Polycomb to turn on terminal differentiation. *Science* **310:** 869–872.

Cheng Y, Buffone MG, Kouadio M, Goodheart M, Page DC, Gerton GL, Davidson I, Wang PJ. 2007. Abnormal sperm in mice lacking the *taf7l* gene. *Mol Cell Biol* **27:** 2582–2589.

Chennathukuzhi V, Morales CR, El-Alfy M, Hecht NB. 2003. The kinesin KIF17b and RNA-binding protein TB-RBP transport specific cAMP-responsive element modulator-regulated mRNAs in male germ cells. *Proc Natl Acad Sci* **100:** 15566–15571.

Chintapalli VR, Wang J, Dow JA. 2007. Using FlyAtlas to identify better *Drosophila melanogaster* models of human disease. *Nat Genet* **39:** 715–720.

Chrivia JC, Kwok RP, Lamb N, Hagiwara M, Montminy MR, Goodman RH. 1993. Phosphorylated CREB binds specifically to the nuclear protein CBP. *Nature* **365:** 855–859.

Citterio E, Vermeulen W, Hoeijmakers JH. 2000. Transcriptional healing. *Cell* **101:** 447–450.

Cler E, Papai G, Schultz P, Davidson I. 2009. Recent advances in understanding the structure and function of general transcription factor TFIID. *Cell Mol Life Sci* **66:** 2123–2134.

Conkright MD, Canettieri G, Screaton R, Guzman E, Miraglia L, Hogenesch JB, Montminy M. 2003. TORCs: Transducers of regulated CREB activity. *Mol Cell* **12:** 413–423.

Cui M, Chen J, Myers TR, Hwang BJ, Sternberg PW, Greenwald I, Han M. 2006. SynMuv genes redundantly inhibit *lin-3/EGF* expression to prevent inappropriate vulval induction in *C. elegans*. *Dev Cell* **10:** 667–672.

Dantonel JC, Wurtz JM, Poch O, Moras D, Tora L. 1999. The TBP-like factor: An alternative transcription factor in metazoa? *Trends Biochem Sci* **24:** 335–339.

de Rooij DG. 2001. Proliferation and differentiation of spermatogonial stem cells. *Reproduction* **121:** 347–354.

Dorus S, Busby SA, Gerike U, Shabanowitz J, Hunt DF, Karr TL. 2006. Genomic and functional evolution of the *Drosophila melanogaster* sperm proteome. *Nat Genet* **38:** 1440–1445.

Falender AE, Freiman RN, Geles KG, Lo KC, Hwang K, Lamb DJ, Morris PL, Tjian R, Richards JS. 2005. Maintenance of spermatogenesis requires TAF4b, a gonad-specific subunit of TFIID. *Genes Dev* **19:** 794–803.

Fay DS, Yochem J. 2007. The SynMuv genes of *Caenorhabditis elegans* in vulval development and beyond. *Dev Biol* **306:** 1–9.

Fimia GM, De Cesare D, Sassone-Corsi P. 1999. CBP-independent activation of CREM and CREB by the LIM-only protein ACT. *Nature* **398:** 165–169.

Foulkes NS, Sassone-Corsi P. 1992. More is better: Activators and repressors from the same gene. *Cell* **68:** 411–414.

Foulkes NS, Mellstrom B, Benusiglio E, Sassone-Corsi P. 1992. Developmental switch of CREM function during spermatogenesis: From antagonist to activator. *Nature* **355:** 80–84.

Foulkes NS, Schlotter F, Pevet P, Sassone-Corsi P. 1993. Pituitary hormone FSH directs the CREM functional switch during spermatogenesis. *Nature* **362:** 264–267.

Fuller MT. 1998. Genetic control of cell proliferation and differentiation in *Drosophila* spermatogenesis. *Semin Cell Dev Biol* **9:** 433–444.

Gangloff Y, Romier C, Thuault S, Werten S, Davidson I. 2001. The histone fold is a key structural motif of transcription factor TFIID. *Trends Biochem Sci* **26:** 250–257.

Gatewood JM, Cook GR, Balhorn R, Schmid CW, Bradbury EM. 1990. Isolation of four core histones from human sperm chromatin representing a minor subset of somatic histones. *J Biol Chem* **265:** 20662–20666.

Gazdag E, Rajkovic A, Torres-Padilla ME, Tora L. 2007. Analysis of TATA-binding protein 2 (TBP2) and TBP expression suggests different roles for the two proteins in regulation of gene expression during oogenesis and early mouse development. *Reproduction* **134:** 51–62.

Gazdag E, Santenard A, Ziegler-Birling C, Altobelli G, Poch O, Tora L, Torres-Padilla ME. 2009. TBP2 is essential for germ cell development by regulating transcription and chromatin condensation in the oocyte. *Genes Dev* **23:** 2210–2223.

Gegonne A, Weissman JD, Zhou M, Brady JN, Singer DS. 2006. TAF7: A possible transcription initiation checkpoint regulator. *Proc Natl Acad Sci* **103:** 602–607.

Georlette D, Ahn S, MacAlpine DM, Cheung E, Lewis PW, Beall EL, Bell SP, Speed T, Manak JR, Botchan MR. 2007. Genomic profiling and expression studies reveal both positive and negative activities for the *Drosophila* Myb MuvB/dREAM complex in proliferating cells. *Genes Dev* **21:** 2880–2896.

Gonzalez C, Casal J, Ripoll P. 1988. Functional monopolar spindles caused by mutation in *mgr*, a cell division gene of *Drosophila melanogaster*. *J Cell Sci* **89:** 39–47.

Guglielmi B, Soutourina J, Esnault C, Werner M. 2007. TFIIS elongation factor and Mediator act in conjunction during transcription initiation in vivo. *Proc Natl Acad Sci* **104:** 16062–16067.

Harrison MM, Ceol CJ, Lu X, Horvitz HR. 2006. Some *C. elegans* class B synthetic multivulva proteins encode a conserved LIN-35 Rb-containing complex distinct from a NuRD-like complex. *Proc Natl Acad Sci* **103:** 16782–16787.

Hart DO, Raha T, Lawson ND, Green MR. 2007. Initiation of zebrafish haematopoiesis by the TATA-box-binding protein-related factor Trf3. *Nature* **450:** 1082–1085.

Hiller M, Chen X, Pringle MJ, Suchorolski M, Sancak Y, Viswanathan S, Bolival B, Lin TY, Marino S, Fuller MT. 2004. Testis-specific TAF homologs collaborate to control a tissue-specific transcription program. *Development* **131:** 5297–5308.

Hiller M, Lin T-Y, Wood C, Fuller MT. 2001. Developmental regulation of transcription by a tissue-specific TAF homolog. *Genes Dev* **15:** 1021–1030.

Hochheimer A, Zhou S, Zheng S, Holmes MC, Tjian R. 2002. TRF2 associates with DREF and directs promoter-selective gene expression in *Drosophila. Nature* **420:** 439–445.

Hummler E, Cole TJ, Blendy JA, Ganss R, Aguzzi A, Schmid W, Beermann F, Schutz G. 1994. Targeted mutation of the CREB gene: Compensation within the CREB/ATF family of transcription factors. *Proc Natl Acad Sci* **91:** 5647–5651.

Ito T, Xu Q, Takeuchi H, Kubo T, Natori S. 1996. Spermatocyte-specific expression of the gene for mouse testis-specific transcription elongation factor S-II. *FEBS Lett* **385:** 21–24.

Jacobi UG, Akkers RC, Pierson ES, Weeks DL, Dagle JM, Veenstra GJ. 2007. TBP paralogs accommodate metazoan- and vertebrate-specific developmental gene regulation. *EMBO J* **26:** 3900–3909.

Jallow Z, Jacobi UG, Weeks DL, Dawid IB, Veenstra GJ. 2004. Specialized and redundant roles of TBP and a vertebrate-specific TBP paralog in embryonic gene regulation in Xenopus. *Proc Natl Acad Sci* **101:** 13525–13530.

Jayaramaiah Raja S, Renkawitz-Pohl R. 2005. Replacement by *Drosophila melanogaster* protamines and Mst77F of histones during chromatin condensation in late spermatids and role of sesame in the removal of these proteins from the male pronucleus. *Mol Cell Biol* **25:** 6165–6177.

Jiang J, White-Cooper H. 2003. Transcriptional activation in *Drosophila* spermatogenesis involves the mutually dependent function of aly and a novel meiotic arrest gene cookie monster. *Development* **130:** 563–573.

Jiang J, Benson E, Bausek N, Doggett K, White-Cooper H. 2007. Tombola, a tesmin/TSO1-family protein, regulates transcriptional activation in the *Drosophila* male germline and physically interacts with always early. *Development* **134:** 1549–1559.

Johannessen M, Delghandi MP, Moens U. 2004. What turns CREB on? *Cell Signal* **16:** 1211–1227.

Kim B, Nesvizhskii AI, Rani PG, Hahn S, Aebersold R, Ranish JA. 2007. The transcription elongation factor TFIIS is a component of RNA polymerase II preinitiation complexes. *Proc Natl Acad Sci* **104:** 16068–16073.

Kolthur-Seetharam U, Martianov I, Davidson I. 2008. Specialization of the general transcriptional machinery in male germ cells. *Cell Cycle* **7:** 3493–3498.

Kopytova DV, Krasnov AN, Kopantceva MR, Nabirochkina EN, Nikolenko JV, Maksimenko O, Kurshakova MM, Lebedeva LA, Yerokhin MM, Simonova OB, et al. 2006. Two isoforms of *Drosophila* TRF2 are involved in embryonic development, premeiotic chromatin condensation, and proper differentiation of germ cells of both sexes. *Mol Cell Biol* **26:** 7492–7505.

Korenjak M, Taylor-Harding B, Binne UK, Satterlee JS, Stevaux O, Aasland R, White-Cooper H, Dyson N, Brehm A. 2004. Native E2F/RBF complexes contain Myb-interacting proteins and repress transcription of developmentally controlled E2F target genes. *Cell* **119:** 181–193.

Kotaja N, Sassone-Corsi P. 2007. The chromatoid body: A germ-cell-specific RNA-processing centre. *Nat Rev Mol Cell Biol* **8:** 85–90.

Kotaja N, Macho B, Sassone-Corsi P. 2005. Microtubule-independent and protein kinase A–mediated function of kinesin KIF17b controls the intracellular transport of activator of CREM in testis (ACT). *J Biol Chem* **280:** 31739–31745.

Kotaja N, De Cesare D, Macho B, Monaco L, Brancorsini S, Goossens E, Tournaye H, Gansmuller A, Sassone-Corsi P. 2004. Abnormal sperm in mice with targeted deletion of the *act* (activator of cAMP-responsive element modulator in testis) gene. *Proc Natl Acad Sci* **101:** 10620–10625.

Kotaja N, Lin H, Parvinen M, Sassone-Corsi P. 2006. Interplay of PIWI/Argonaute protein MIWI and kinesin KIF17b in chromatoid bodies of male germ cells. *J Cell Sci* **119:** 2819–2825.

Laine JP, Egly JM. 2006. When transcription and repair meet: A complex system. *Trends Genet* **22:** 430–436.

Lewis PW, Beall EL, Fleischer TC, Georlette D, Link AJ, Botchan MR. 2004. Identification of a *Drosophila* Myb-E2F2/RBF transcriptional repressor complex. *Genes Dev* **18:** 2929–2940.

Lin TY, Viswanathan S, Wood C, Wilson PG, Wolf N, Fuller MT. 1996. Coordinate developmental control of the meiotic cell cycle and spermatid differentiation in *Drosophila* males. *Development* **122:** 1331–1341.

Litovchick L, Sadasivam S, Florens L, Zhu X, Swanson SK, Velmurugan S, Chen R, Washburn MP, Liu XS, DeCaprio JA. 2007. Evolutionarily conserved multisubunit RBL2/p130 and E2F4 protein complex represses human cell cycle-dependent genes in quiescence. *Mol Cell* **26:** 539–551.

Macho B, Brancorsini S, Fimia GM, Setou M, Hirokawa N, Sassone-Corsi P. 2002. CREM-dependent transcription in male germ cells controlled by a kinesin. *Science* **298:** 2388–2390.

Martianov I, Brancorsini S, Gansmuller A, Parvinen M, Davidson I, Sassone-Corsi P. 2002. Distinct functions of TBP and TLF/TRF2 during spermatogenesis: Requirement of TLF for heterochromatic chromocenter formation in haploid round spermatids. *Development* **129:** 945–955.

Martianov I, Choukrallah M-A, Legras S, Rijkers E, Van Ijcken W, Jost B, Sassone-Corsi P, Davidson I. 2010. Developmentally regulated occupancy of an extended repertoire of CREB and CREM binding loci in male germ cells. *BMC Genomics* **11:** 530.

Martianov I, Fimia GM, Dierich A, Parvinen M, Sassone-Corsi P, Davidson I. 2001. Late arrest of spermiogenesis and germ cell apoptosis in mice lacking the TBP-like TLF/TRF2 gene. *Mol Cell* **7:** 509–515.

Matangkasombut O, Auty R, Buratowski S. 2004. Structure and function of the TFIID complex. *Adv Protein Chem* **67:** 67–92.

Metcalf CE, Wassarman DA. 2007. Nucleolar colocalization of TAF1 and testis-specific TAFs during *Drosophila* spermatogenesis. *Dev Dyn* **236:** 2836–2843.

Muller F, Lakatos L, Dantonel J, Strahle U, Tora L. 2001. TBP is not universally required for zygotic RNA polymerase II transcription in zebrafish. *Curr Biol* **11:** 282–287.

Nakajima T, Uchida C, Anderson SF, Parvin JD, Montminy M. 1997. Analysis of a cAMP-responsive activator reveals a two-component mechanism for transcriptional induction via signal-dependent factors. *Genes Dev* **11:** 738–747.

Nantel F, Sassone-Corsi P. 1996. CREM: A transcriptional master switch during the spermatogenesis differentiation program. *Front Biosci* **1:** pd266–269.

Nantel F, Monaco L, Foulkes NS, Masquilier D, LeMeur M, Henriksen K, Dierich A, Parvinen M, Sassone-Corsi P. 1996. Spermiogenesis deficiency and germ-cell apoptosis in CREM-mutant mice. *Nature* **380:** 159–162.

Osterloh L, von Eyss B, Schmit F, Rein L, Hubner D, Samans B, Hauser S, Gaubatz S. 2007. The human synMuv-like protein LIN-9 is required for transcription of G2/M genes and for entry into mitosis. *EMBO J* **26:** 144–157.

Ozer J, Moore PA, Lieberman PM. 2000. A testis-specific transcription factor IIA (TFIIA τ) stimulates TATA- binding protein-DNA binding and transcription activation. *J Biol Chem* **275:** 122–128.

Papai G, Tripathi MK, Ruhlmann C, Werten S, Crucifix C, Weil PA, Schultz P. 2009. Mapping the initiator binding Taf2 subunit in the structure of hydrated yeast TFIID. *Structure* **17:** 363–373.

Parisi M, Nuttall R, Edwards P, Minor J, Naiman D, Lu J, Doctolero M, Vainer M, Chan C, Malley J, et al. 2004. A survey of ovary-, testis-, and soma-biased gene expression in *Drosophila melanogaster* adults. *Genome Biol* **5:** R40.

Perezgasga L, Jiang J, Bolival B Jr, Hiller M, Benson E, Fuller MT, White-Cooper H. 2004. Regulation of transcription of meiotic cell cycle and terminal differentiation genes by the testis-specific Zn-finger protein matotopetli. *Development* **131:** 1691–1702.

Persengiev SP, Zhu X, Dixit BL, Maston GA, Kittler EL, Green MR. 2003. TRF3, a TATA-box-binding protein-related factor, is vertebrate-specific and widely expressed. *Proc Natl Acad Sci* **100:** 14887–14891.

Pilkinton M, Sandoval R, Song J, Ness SA, Colamonici OR. 2007. Mip/LIN-9 regulates the expression of B-Myb and the induction of cyclin A, cyclin B, and CDK1. *J Biol Chem* **282:** 168–175.

Pointud JC, Mengus G, Brancorsini S, Monaco L, Parvinen M, Sassone-Corsi P, Davidson I. 2003. The intracellular localisation of TAF7L, a paralogue of transcription factor TFIID subunit TAF7, is developmentally regulated during male germ-cell differentiation. *J Cell Sci* **116:** 1847–1858.

Potrzebowski L, Vinckenbosch N, Marques AC, Chalmel F, Jegou B, Kaessmann H. 2008. Chromosomal gene movements reflect the recent origin and biology of therian sex chromosomes. *PLoS Biol* **6:** e80.

Rabenstein MD, Zhou S, Lis JT, Tjian R. 1999. TATA box-binding protein (TBP)-related factor 2 (TRF2), a third member of the TBP family. *Proc Natl Acad Sci* **96:** 4791–4796.

Sassone-Corsi P. 1995. Transcription factors responsive to cAMP. *Annu Rev Cell Dev Biol* **11:** 355–377.

Sassone-Corsi P. 2002. Unique chromatin remodeling and transcriptional regulation in spermatogenesis. *Science* **296:** 2176–2178.

Saurin AJ, Shao Z, Erdjument-Bromage H, Tempst P, Kingston RE. 2001. A *Drosophila* Polycomb group complex includes Zeste and dTAFII proteins. *Nature* **412:** 655–660.

Schmidt EE, Schibler U. 1995. High accumulation of components of the RNA polymerase II transcription machinery in rodent spermatids. *Development* **121:** 2373–2383.

Schmidt EE, Schibler U. 1997. Developmental testis-specific regulation of mRNA levels and mRNA translational efficiencies for TATA-binding protein mRNA isoforms. *DevBiol* **184:** 138–149.

Schmit F, Korenjak M, Mannefeld M, Schmitt K, Franke C, von Eyss B, Gagrica S, Hanel F, Brehm A, Gaubatz S. 2007. LINC, a human complex that is related to pRB-containing complexes in invertebrates regulates the expression of G2/M genes. *Cell Cycle* **6:** 1903–1913.

Schmidt EE, Ohbayashi T, Makino Y, Tamura T, Schibler U. 1997. Spermatid-specific overexpression of the TATA-binding protein gene involves recruitment of two potent testis-specific promoters. *J Biol Chem* **272:** 5326–5334.

Schultz N, Hamra FK, Garbers DL. 2003. A multitude of genes expressed solely in meiotic or postmeiotic spermatogenic cells offers a myriad of contraceptive targets. *Proc Natl Acad Sci* **100:** 12201–12206.

Sikorski TW, Buratowski S. 2009. The basal initiation machinery: beyond the general transcription factors. *Curr Opin Cell Biol* **21:** 344–351.

Teichmann M, Wang Z, Martinez E, Tjernberg A, Zhang D, Vollmer F, Chait BT, Roeder RG. 1999. Human TATA-binding protein-related factor-2 (hTRF2) stably associates with hTFIIA in HeLa cells. *Proc Natl Acad Sci* **96:** 13720–13725.

Timmers HT, Tora L. 2005. SAGA unveiled. *Trends Biochem Sci* **30:** 7–10.

Tora L. 2002. A unified nomenclature for TATA box binding protein (TBP)-associated factors (TAFs) involved in RNA polymerase II transcription. *Genes Dev* **16:** 673–675.

Umehara T, Kida S, Hasegawa S, Fujimoto H, Horikoshi M. 1997. Restricted expression of a member of the transcription elongation factor S-II family in testicular germ cells during and after meiosis. *J Biochem* **121:** 598–603.

Upadhyaya AB, Lee SH, DeJong J. 1999. Identification of a general transcription factor TFIIAα/β homolog selectively expressed in testis. *J Biol Chem* **274:** 18040–18048.

van der Heijden GW, Ramos L, Baart EB, van den Berg IM, Derijck AA, van der Vlag J, Martini E, de Boer P. 2008. Sperm-derived histones contribute to zygotic chromatin in humans. *BMC Dev Biol* **8:** 34.

Veenstra GJ, Weeks DL, Wolffe AP. 2000. Distinct roles for TBP and TBP-like factor in early embryonic gene transcription in xenopus. *Science* **290:** 2312–2315.

Wang Z, Mann RS. 2003. Requirement for two nearly identical TGIF-related homeobox genes in *Drosophila* spermatogenesis. *Development* **130:** 2853–2865.

Wang PJ, Page DC. 2002. Functional substitution for TAF(II)250 by a retroposed homolog that is expressed in human spermatogenesis. *Hum Mol Genet* **11:** 2341–2346.

Wang PJ, McCarrey JR, Yang F, Page DC. 2001. An abundance of X-linked genes expressed in spermatogonia. *Nat Genet* **27:** 422–426.

Wang Y, Wang L, Wang Z. 2008. Transgenic analyses of TGIF family proteins in *Drosophila* imply their role in cell growth. *J Genet Genomics* **35:** 457–465.

White-Cooper H. 2009. Molecular mechanisms of gene regulation during *Drosophila* spermatogenesis. *Reproduction* **139:** 11–21.

White-Cooper H, Leroy D, MacQueen A, Fuller MT. 2000. Transcription of meiotic cell cycle and terminal differentiation genes depends on a conserved chromatin associated protein, whose nuclear localisation is regulated. *Development* **127:** 5463–5473.

White-Cooper H, Schafer MA, Alphey LS, Fuller MT. 1998. Transcriptional and post-transcriptional control mechanisms coordinate the onset of spermatid differentiation with meiosis I in *Drosophila. Development* **125:** 125–134.

Xiao L, Kim M, DeJong J. 2006. Developmental and cell type-specific regulation of core promoter transcription factors in germ cells of frogs and mice. *Gene Expr Patterns* **6:** 409–419.

Xu W, Kasper LH, Lerach S, Jeevan T, Brindle PK. 2007. Individual CREB-target genes dictate usage of distinct cAMP-responsive coactivation mechanisms. *EMBO J* **26:** 2890–2903.

Zalensky AO, Allen MJ, Kobayashi A, Zalenskaya IA, Balhorn R, Bradbury EM. 1995. Well-defined genome architecture in the human sperm nucleus. *Chromosoma* **103:** 577–590.

Zhang D, Penttila TL, Morris PL, Teichmann M, Roeder RG. 2001. Spermiogenesis deficiency in mice lacking the *Trf2* gene. *Science* **292:** 1153–1155.

Translational Control in Oocyte Development

Joel D. Richter[1] and Paul Lasko[2]

[1]Program in Molecular Medicine, University of Massachusetts Medical School, Worcester, Massachusetts 01606

[2]Department of Biology, McGill University, Montreal, Quebec H3A 1B1, Canada

Correspondence: joel.richter@umassmed.edu

Translational control of specific mRNAs is a widespread mechanism of gene regulation, and it is especially important in pattern formation in the oocytes of organisms in which the embryonic axes are established maternally. *Drosophila* and *Xenopus* have been especially valuable in elucidating the relevant molecular mechanisms. Here, we comprehensively review what is known about translational control in these two systems, focusing on examples that illustrate key concepts that have emerged. We focus on protein-mediated translational control, rather than regulation mediated by small RNAs, as the former appears to be predominant in controlling these developmental events. Mechanisms that modulate the ability of the specific mRNAs to be recruited to the ribosome, that regulate polyadenylation of specific mRNAs, or that control the association of particular mRNAs into translationally inert ribonucleoprotein complexes will all be discussed.

Since it was first described in sea urchin and *Drosophila* oocytes more than five decades ago, masked or maternal mRNA has served as a preeminent model for the study of translational control. The key questions that have been addressed over the years in a variety of organisms, perhaps most vividly exemplified by *Drosophila* and *Xenopus*, are how do mRNAs come to be translationally repressed, how are they activated, and how is translational regulation regionally restricted? *Drosophila* genetics has identified a variety of mRNAs whose translational control is essential for embryonic development (e.g., *oskar*, *nanos*, and *gurken*). Moreover, genetics, as well as biochemistry, has revealed some of the factors that control the expression of these mRNAs, such as the RNA binding proteins Cup and Pumilio. *Xenopus*, on the other hand, is not an organism particularly well suited to genetic analysis. However, *Xenopus* oocytes are an excellent biochemical system owing to their abundance, large size for the injection of molecules, and their synchronous entry into the meiotic divisions when cultured ex vivo in medium containing progesterone. Indeed, *Xenopus* oocytes have been instrumental in revealing how the translation of maternal mRNAs during oocyte maturation is regulated by changes in poly(A) tail length. Lessons learned in these systems have been important for revealing not only general principals of translational control, but also for serving as paradigms for translational control in other cells and tissues.

Cite this article as *Cold Spring Harb Perspect Biol* doi: 10.1101/cshperspect.a002758

Here, we focus our attention on select examples of translational control in *Drosophila* and *Xenopus* oocytes. As the literature on this subject is now vast, we believe that the reader is best served by focusing on translational control in oocytes of these two organisms, where many of the key concepts in the field were first developed and have been most completely elaborated.

TRANSLATIONAL REGULATION UNDERLIES ESTABLISHMENT OF EMBRYONIC PATTERN IN *DROSOPHILA*

Molecular asymmetries underlying embryonic axis patterning and germ cell specification are established in *Drosophila* largely by position-dependent translational regulation of maternally transcribed mRNAs that are deposited into the oocyte. This translational regulation is often coupled with localization mechanisms that concentrate the mRNA in the spatial domain of the oocyte where the corresponding protein will be expressed and that reduce its concentration in areas where its expression would be deleterious to development. A detailed discussion of mRNA localization mechanisms is beyond the scope of this article, and the relevant work has been recently reviewed (Becalska and Gavis 2009). Translational repression in the developing oocyte is especially complex, and as will be discussed below, different mechanisms silence translation of specific mRNAs at different developmental stages, thus restricting their potential translation to particular spatial regions. Comparatively less is known about repression is alleviated and translation is activated for these regulated mRNAs, although increasing evidence implicates the DEAD-box helicase Vasa (Vas) as a positive regulator of specific target mRNAs.

TRANSLATIONAL REGULATION OF *oskar*

Localization of *oskar* (*osk*) mRNA and its targeted translation initiates assembly of polar granules and specification of the posterior soma and germ cells (Ephrussi and Lehmann 1992). Two different isoforms, termed Long Osk and Short Osk, are expressed from different initiation codons (Markussen et al. 1995). Short Osk is sufficient to induce pole plasm assembly and rescue the developmental functions of *osk*, whereas Long Osk induces F-actin projections that are required for anchoring its mRNA in the posterior pole plasm (Markussen et al. 1995; Vanzo et al. 2007).

Drosophila oogenesis is conventionally divided into 14 developmental stages (Spradling 1993). Until stage 9, translation of *osk* is silenced throughout the nurse cells and oocyte, whereas afterward, *osk* translation is activated in the posterior pole plasm, while remaining repressed elsewhere. Repression of *osk* translation in early oogenesis involves RNA interference (RNAi) (Findley et al. 2003; Cook et al. 2004; Tomari et al. 2004; Chen et al. 2007; Lim and Kai 2007; Pane et al. 2007). There is, however, little specific information as to the mechanisms involved. miRNA-mediated translational repression is believed to act after initiation (Bushati and Cohen 2007).

As *osk* mRNA moves to the pole plasm, it is translationally repressed by Cup (Fig. 1). Cup, like Maskin (see below), competes with eIF4G for binding to eIF4E (Filardo and Ephrussi 2003; Wilhelm et al. 2003; Nakamura et al. 2004; Nelson et al. 2004; Zappavigna et al. 2004). Cup interacts with Bruno (Bru), an RNA binding protein that negatively regulates *osk* during its localization. Bru contains three RNA recognition motifs (RRMs) that interact directly with specific sequences (Bru-response elements, or BREs) in the *osk* 3′ UTR, and repress its translation (Kim-Ha et al. 1995; Webster et al. 1997; Lie and Macdonald 1999; Snee et al. 2008), most likely by recruiting Cup. The third RRM of Bru has recently been shown to have a noncanonical structure, and 40 amino acids that extend it amino-terminally significantly enhance RNA binding (Lyon et al. 2009).

Bru also packages *osk* RNA into large particles that render it inaccessible to the translational machinery (Chekulaeva et al. 2006). These silencing particles contain Cup and Maternal expression at 31B (Me31B) (Nakamura et al. 2001). Although Cup and Me31B

 Cite this article as *Cold Spring Harb Perspect Biol* doi: 10.1101/cshperspect.a002758

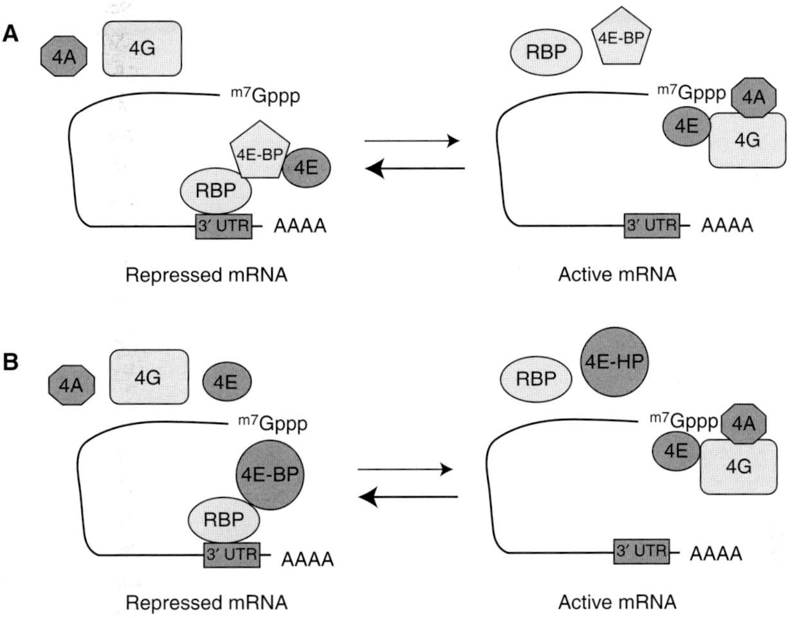

Figure 1. Translational repression by eIF4E binding proteins. (*A*) An eIF4E binding protein (4E-BP) competes with eIF4G (4G) for interaction with the cap-binding protein eIF4E (4E). In general eIF4G has a higher affinity for eIF4E than does the regulatory eIF4E binding protein and an active cap-binding complex including eIF4A (4A) can assemble, but that equilibrium is reversed if the regulatory protein is recruited to a target mRNA by an RNA binding protein (RBP) that binds to the 3′ UTR. Then eIF4E is sequestered away from the cap-binding complex and translation is repressed. 4E-BPs that operate in this way include *Drosophila* Cup and *Xenopus* Maskin. RBPs that recruit these 4E-BPs to specific mRNAs include *Drosophila* Bru, *Drosophila* Smg, and *Xenopus* CPEB. (*B*) Translational repression by 4E homology proteins. 4E homology protein (4E-HP) can bind the 5′ cap structure of the mRNA but not eIF4G. It competes with eIF4E for cap binding and represses translation of mRNAs with which it is associated. The affinity of 4E-HP for the 5′ cap structure is less than that of eIF4E, but that equilibrium is reversed if 4E-HP is recruited to target mRNAs by RBPs that bind the 3′ UTR. In *Drosophila*, 4E-HP can be recruited to target mRNAs by Bicoid or by the Pumilio/Nanos/Brain Tumor complex, whereas in mammalian cells 4E-HP can be recruited by Prep1.

can be copurified, *Me31B* mutations affect *osk* translation at an earlier stage of oogenesis than do *cup* mutations, suggesting that their functions are not mutually dependent. Me31B associates with Bru, implying it may be required for Cup-dependent *osk* silencing at stage 9, but as *Me31B*-null egg chambers degenerate around stage 6, it has not been possible to directly test this (Nakamura et al. 2001). Polypyrimidine tract binding protein (PTB) also promotes the formation of *osk*-containing silencing particles (Besse et al. 2009). To accomplish this, PTB binds directly to multiple sites in the *osk* 3′ UTR, cross-linking many *osk* and PTB molecules into aggregates. It is unknown whether PTB recognizes mRNA targets other than *osk*, but it seems likely that silencing particles containing PTB would include more than one species of mRNA. A detailed ultrastructural analysis of *osk* mRNA localization and its assembly into ribonucleoprotein complexes (RNPs) has recently been completed (Trucco et al. 2009). This study showed that the average size of *osk*-containing RNPs increases drastically from the anterior to the posterior of stage 8–10 oocytes, and that these particles form large aggregates in the pole plasm. This study also documented when and where proteins involved in *osk* localization become associated with these RNPs. An extension of this approach could reveal a great deal about the dynamics of the association of PTB, Me31B, and Bru with *osk* and potentially other mRNAs.

Hrp48, an abundant RNA-binding protein that interacts with elements in both the 5′ and 3′ UTRs of *osk*, also contributes to its translational repression (Huynh et al. 2004; Yano et al. 2004; Norvell et al. 2005). Hrp48 is recruited to *osk*-containing RNPs in the nurse cell nuclei and remains associated with these particles in the cytoplasm (Trucco et al. 2009). Hrp48 interacts with Squid (Sqd) and Ovarian Tumor (Otu) (Norvell et al. 1999; Goodrich et al. 2004), and all three have been functionally implicated in *osk* mRNA localization and translational repression. Cup also copurifies with Hrp48 and Sqd (Clouse et al. 2008), but this association requires RNA, making it unlikely that these proteins operate in an analogous manner to Bru. Glorund, an hnRNP F/H family member, may be another component of this complex (Kalifa et al. 2009). Another RNA binding protein, Bicaudal-C (Bic-C), has been implicated genetically as a negative regulator of *osk* translation (Saffman et al. 1998). Bic-C directly recruits the CCR4 deadenylase complex to target mRNAs through an association with its NOT3/5 subunit (Chicoine et al. 2007). These targets could potentially include *osk*.

Derepression and activation of *osk* translation in the pole plasm may be distinct processes (Gunkel et al. 1998). A key activator of *osk* translation is Orb, the *Drosophila* homolog of cytoplasmic polyadenylation element binding protein (CPEB) (see below). Orb interacts with the *osk* 3′ UTR, and in *orb* mutants *osk* translation is reduced, suggesting that cytoplasmic polyadenylation might underlie activation of *osk* translation (Chang et al. 1999). Orb interacts physically with Bru, yet the addition of a long poly(A) tail to a chimeric *osk-lacZ* mRNA does not overcome Bru-mediated repression in vitro (Castagnetti and Ephrussi 2003). Orb directly associates with PAP and Wispy, two poly(A) polymerases that act at different developmental times: PAP is required during mid-oogenesis to promote Osk expression, whereas Wispy functions only during late oogenesis and in the early embryo (Juge et al. 2002; Benoit et al. 2008). Bic-C interacts with Orb, PAP, and Wispy, and possibly inhibits their association with target mRNAs (Castagnetti and Ephrussi 2003; Chicoine et al. 2007; Benoit et al. 2008; Cui et al. 2008). Consistent with this, *Bic-C* and *orb* display antagonistic genetic interactions.

osk translation in the germ plasm also requires Staufen (Stau), a dsRBD type RNA binding protein essential as well for its localization (Micklem et al. 2000). Unlike Hrp48, Stau first becomes associated with *osk*-containing mRNPs in the nurse cell cytoplasm (Trucco et al. 2009). The mechanism through which Stau activates *osk* translation is unknown, but it involves more than simply relieving Bru-mediated repression, as its activity is BRE-independent (Kim-Ha et al. 1995). Also, Aub activates *osk* translation in the pole plasm, and, as for Stau, its function does not require the BRE (Wilson et al. 1996; Harris and Macdonald 2001). The basis for the contrasting roles for Aub with respect to unlocalized *osk* in early oogenesis and localized *osk* in the pole plasm remains unexplained.

TRANSLATIONAL REGULATION OF *nos*

Posterior targeting of Nos protein also involves RNA localization and translational regulation. A 90 nt region of the *nos* 3′ UTR, termed the translational control element (TCE), forms a complex secondary structure, and mutations that disrupt any portion of it prevent the binding of repressors and render the entire element inactive (Crucs et al. 2000; Forrest et al. 2004). Different parts of the TCE interact with different *trans*-acting factors at different developmental stages to ensure translational repression of unlocalized *nos* mRNA. During late oogenesis, repression of *nos* is mediated by Glorund (Glo), which binds to the stem of stem-loop III of the TCE (Kalifa et al. 2006).

The loop of stem-loop II of the TCE contains a Smaug Recognition Element (SRE), the binding site for another translational repressor called Smaug (Smg) (Smibert et al. 1996). Smg interacts with Cup, and Cup binding is required for Smg-mediated repression of

SRE-containing mRNAs in embryo extracts (Nelson et al. 2004). Smg is therefore believed to function in a manner analogous to Bru. However, Smg has a second more widespread function in mRNA metabolism. It interacts directly with the POP2 subunit of the CCR4 deadenylase complex, and recruits it to a large set of maternal mRNAs in the early embryo, including *nos*, targeting them for decay (Semotok et al. 2005; Zaessinger et al. 2006; Tadros et al. 2007; Benoit et al. 2009). Thus, *nos* mRNA is repressed in two distinct ways by Smg: by cap-dependent translational repression and by deadenylation of the silenced transcript. Osk relieves Smg/CCR4-dependent deadenylation of *nos*, thus enabling its translation in the pole plasm (Zaessinger et al. 2006).

A recent study (Rangan et al. 2009) compares the translational regulation of 10 other mRNAs with similar localization kinetics to *nos* with the regulation of *nos* itself. In all cases, translational regulation elements resided in the 3′ UTRs. Often, translational activation correlated with an increase in poly(A) tail length, but surprisingly for *pgc* and *gcl* mRNAs, reduction of *orb* activity had little effect on translation. As several translational regulators target the CCR4 deadenylase, it appears it may be more effective to regulate deadenylation rather than polyadenylation in this system.

TRANSLATIONAL REGULATION OF *grk*

As for *osk* and *nos*, *grk* expression is regulated through mRNA localization coupled with translational control. Bru, K10, and Sqd have all been implicated as translational repressors of *grk*, and Bru binds directly to the *grk* 3′ UTR (Saunders and Cohen 1999; Filardo and Ephrussi 2003; Cáceres and Nilson 2009), although a link to Cup has not yet been established. *grk* translation is positively regulated by the DEAD-box helicase Vas. Severe *vas* mutations largely block Grk accumulation in early oocytes (Styhler et al. 1998; Tomancak et al. 1998). Vas interacts physically and functionally with eIF5B, a general translation factor required for recruitment of the 60S ribosomal subunit to the initiation complex (Carrera et al. 2000). Oocytes that contain only a form of Vas (VasΔ617) that is specifically compromised for eIF5B binding largely fail to accumulate Grk, implying that Vas positively regulates *grk* by recruiting eIF5B (Fig. 2) (Johnstone and Lasko 2004). Vas has recently been shown to bind specifically to a U-rich motif present in the 3′ UTR of another mRNA, *mei-P26*, and to positively regulate its translation in stem cells and dividing cystocytes through that interaction (Liu et al. 2009). Activation of *grk* mRNA translation at the antero-dorsal cortex of the developing oocyte is also mediated by poly(A) binding

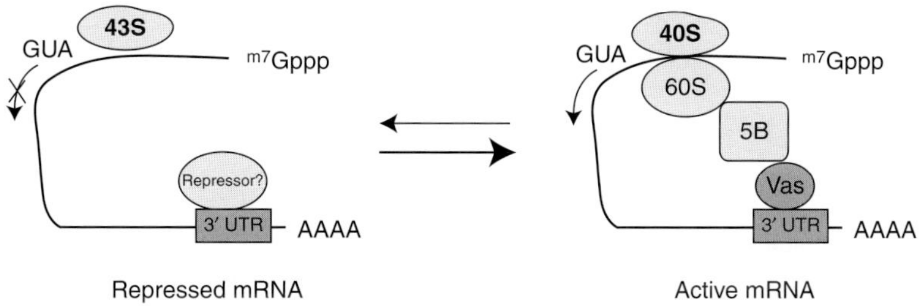

Figure 2. Proposed mechanism for Vas-mediated translational activation. eIF5B (5B) is required for recruitment of the large ribosomal subunit (60S) to the initiation complex, so that the elongation phase of translation can proceed. When this process is blocked, translation is stalled with the small ribosomal subunit and initiator tRNA-Met (43S) stalled at the start codon (AUG). Vas binds a sequence element in the 3′ UTR and eIF5B (5B), thus promoting the recruitment of the 60S subunit and activating translation. Regulation of an mRNA expressed in erythroid cells, r15-LOX, at the level of subunit joining involves a 3′ UTR binding repressor that inhibits eIF5B recruitment (Ostareck et al. 2003). It is unknown whether Vas-mediated activation involves displacement of such a repressor.

protein 55D (PABP55D) in association with Encore (Clouse et al. 2008).

4EHP AND TRANSLATIONAL REPRESSION OF *caudal* AND *hunchback*

The homeodomain-containing protein Bicoid (Bcd) is a transcriptional regulator essential for anterior patterning in *Drosophila* (Berleth et al. 1988; Driever and Nüsslein-Volhard 1989). Bcd also associates with RNA and regulates translation, repressing *caudal* (*cad*) RNA, and producing a posterior-to-anterior gradient of Cad protein. Bcd binds to the 3′ UTR of *cad* RNA, and *cad* repression depends on the presence of a 5′ cap structure (Dubnau and Struhl 1996; Rivera-Pomar et al. 1996; Niessing et al. 1999, 2002).

4E homology protein (4EHP) is an eIF4E-related cap-binding protein that cannot bind eIF4G. Females homozygous for a hypomorphic allele of this gene produce embryos with anterior defects similar to embryos produced by *bcd* mutant mothers (Cho et al. 2005). These embryos also fail to repress *cad* translation in anterior regions. These phenotypes could be rescued by *4EHP* transgenic constructs, but not by constructs producing mutant forms of 4EHP that were abrogated for cap binding or for Bcd binding. Similarly, transgenically produced forms of Bcd that were abrogated for 4EHP binding could not repress *cad* translation. These results showed that simultaneous interactions of 4EHP with the cap structure and of Bcd with the *cad* 3′ UTR renders *cad* mRNA translationally inactive. Recent evidence indicates that, like Bcd, a mammalian homeodomain-containing transcription factor, Prep1, also negatively regulates translation of a target gene, Hoxb4, through a mechanism that requires 4EHP and the Hoxb4 3′ UTR (Villaescusa et al. 2009).

In *Drosophila*, 4EHP was also shown to be involved in *hb* translational repression (Cho et al. 2006). In wild-type embryos, translation of the uniformly distributed maternal *hunchback* (*hb*) mRNA is inhibited at the posterior to form an anterior-to-posterior protein concentration gradient. This inhibition requires assembly of an mRNP complex (the NRE-complex) that consists of Nanos (Nos), Pumilio (Pum) and Brain tumor (Brat) proteins, and the Nos response element (NRE) present in the 3′ UTR of *hb* mRNA (Wharton and Struhl 1991; Chagnovich and Lehmann 2001; Sonoda and Wharton 2001). The Hb gradient extends more posteriorly in embryos produced by *4EHP* mutant females, and 4EHP is recruited to the NRE-complex through binding to the dorsal surface of the NHL domain within Brat. Consistently, mutant forms of Brat abrogated for 4EHP binding do not fully support *hb* repression.

TRANSLATIONAL CONTROL BY CYTOPLASMIC POLYADENYLATION IN VERTEBRATES

Looking back, 1987–1989 were very good years for insights into posttranscriptional control of gene expression in vertebrates. Vassalli, Strickland, and their colleagues (Huarte et al. 1987) found that in maturing mouse oocytes, tissue plasminogen activator (tPA) mRNA underwent polyadenylation and translational activation. Similar observations were made in 1988 by the Dworkin laboratory (Smith et al. 1988a,b), which showed that a set of mRNAs underwent cytoplasmic polyadenylation and translational activation in maturing *Xenopus* oocytes. The year 1988 was also when Vassalli, Strickland and coworkers (Strickland et al. 1988) showed that "information" in the tPA 3′ UTR promoted cytoplasmic polyadenylation and translation. Finally, in 1989 specific mRNAs in *Xenopus* oocytes were shown to contain specific *cis* elements that directed polyadenylation-induced translation (Fox and Wickens 1989; McGrew et al. 1989). Since these early observations got the ball rolling, many biochemical studies using *Xenopus* oocytes have revealed the details of how cytoplasmic polyadenylation works, how this process mediates translation, and how it controls oocyte development.

OOCYTE MATURATION

To place the translational control in a proper biological context, it is first necessary to consider the process of oocyte maturation particularly in *Xenopus laevis*, because most of what we know about maternal mRNA expression in vertebrates comes from studies of that species (see Brook et al. 2009). Primordial germ cells (PGCs) divide mitotically while migrating to the developing gonad; once there, they enter meiosis, proceed through synapsis and recombination, and arrest at the end of prophase I, also known as diplotene, which resembles mitotic G2. The oocytes grow enormously by taking up vitellogenin from the bloodstream; during this time, they also synthesize and store large amounts of mRNA, tRNA, ribosomes, and all the factors necessary for translation. However, as little as 5% by mass of the mRNA is actively engaged in protein synthesis. The nontranslating or stored mRNAs are generally referred to as masked or maternal mRNA. Some of these mRNAs are translated when the oocytes reenter meiosis, a period known as oocyte maturation. The cells proceed through metaphase I (MI) and arrest at MII, where they await fertilization. Several other mRNAs are translated between MI and MII, whereas others are translated during fertilization or particular times of embryogenesis. Although no single mechanism regulates the translation of all mRNAs, one that is particularly important, and one that is understood in some biochemical detail, is cytoplasmic polyadenylation.

POLYADENYLATION-INDUCED TRANSLATION

The story of how poly(A) elongation controls translation during oocyte maturation begins, not surprisingly, in the nucleus. The synthesis of nuclear premRNAs proceeds far downstream from where the mature 3′ end of the mRNA is located. A signal in the premRNA, AAUAAA, is usually found about 20 bases upstream of the mature 3′ end (Richard and Manley 2009). This cleavage and polyadenylation signal, which is found on virtually all noncore histone mRNAs, is recognized by the cleavage and polyadenylation specificity factor CPSF. CPSF is composed of four subunits with sizes (in mammalian cells) of 150, 100, 73, and 30 kDa. Symplekin is another protein often intimately associated with CPSF (Shatkin and Manley 2000). Together with stimulatory factors, the CPSF 73-kDa subunit cleaves the RNA about 20 bases downstream from the AAUAAA (Mandel et al. 2006), whereupon a long poly(A) tail is added.

Some premRNAs have another key sequence that is often 20–100 bases upstream of AAUAAA, and that is the cytoplasmic polyadenylation element (CPE), which has the general structure of UUUUUAU. Following splicing but still in the nucleus, the CPE is bound by the shuttling cytoplasmic polyadenylation element binding protein CPEB (Lin et al. 2010). Together with other proteins such as symplekin and the exon-junction protein eIF4AIII (see below), the ribonucleoprotein (RNP) complex is exported to the cytoplasm where it is joined by two new critical factors: Gld2 and PARN (Barnard et al. 2004; Kim and Richter 2006; Lin et al. 2010). Gld2 (germline development 2) is an unusual cytoplasmic poly(A) polymerase discovered in 2002 by three laboratories working in *Caenorhabditis elegans* and yeast (Read et al. 2002; Saitoh et al. 2002; Wang et al. 2002). PARN (poly(A) ribonuclease) is the second factor; as the name suggests, it is a deadenylating enzyme (Korner et al. 1998; Copeland and Wormington 2001). Both of these enzymes are active, but because PARN is the more active, the poly(A) tail originally acquired in the nucleus is now shortened, to about 20–40 bases. At this stage, Gld2 continues to catalyze polyadenylation but the adenylate residues are quickly removed by PARN. This "running in place" model for polyadenylation can theoretically take place for months on end (Kim and Richter 2006).

Progesterone secretion from the follicle cells induces oocyte entry into the meiotic divisions. This steroid interacts with a surface-associated receptor, the result of which is a cascade of events that ultimately leads to another meiotic arrest at MII (Tunquist and Mallet 2003). More germane to this discussion is the observation that progesterone induces a decrease in GSK-3β levels that in turn induces the activation of the kinase Aurora A (Mendez et al. 2000a; Sarkissian et al. 2004). Aurora A, and/or MAP kinase and a GTP exchange factor XGef

(Reverte et al. 2003; Martinez et al. 2005; Keady et al. 2007), which is also activated by progesterone, phosphorylates CPEB on serine 174, an event that not only induces a close association between CPEB and CPSF (Mendez et al. 2000b), but also expels PARN from the RNP complex (Kim and Richter 2006). Because PARN is no longer present, Gld2 polyadenylates the RNA by default (Fig. 3). In other words, polyadenylation is controlled by the presence or absence of PARN, and not by the recruitment of Gld2 or another poly(A) polymerase as originally suggested (Richter 1999).

Is a short poly(A) tail in and of itself sufficient to keep mRNAs silent until the tails are elongated? Apparently, this is not the case because another factor, Maskin, is a key regulator of translation. Maskin, which probably shuttles to the nucleus with CPEB and is exported with it and mRNA into the cytoplasm (Lin et al. 2010), not only binds CPEB, but the cap-binding factor eIF4E as well. Like *Drosophila* Cup, the binding of Maskin to eIF4E precludes the binding of eIF4G to eIF4E, thus Maskin inhibits translation because eIF4G, indirectly through eIF3, recruits the 40S ribosomal subunit to the 5′ end of the mRNA (Stebbins-Boaz et al. 1999; Cao and Richter 2006).

How is Maskin-mediated repression alleviated by polyadenylation? The answer to this question lies with another translationally dormant mRNA encoding the protein RINGO (Ferby et al. 1999). Oocytes contain no detectable RINGO protein but instead contain inactive RINGO mRNA that is translated on progesterone stimulation. RINGO mRNA translation is not controlled by CPEB and/or Maskin, but instead it is regulated by Pum2 (Pumilio 2), DAZL (deleted in azoospermia-like), and ePAB (embryonic poly(A) binding protein) (Padmanabhan and Richter 2006). Pum2 binds both a Pumilio binding element (PBE) in the RINGO 3′ UTR as well as the 5′ cap, thereby prevent eIF4E from binding the cap and repressing translation (Cao et al. 2010). On progesterone stimulation, Pum2

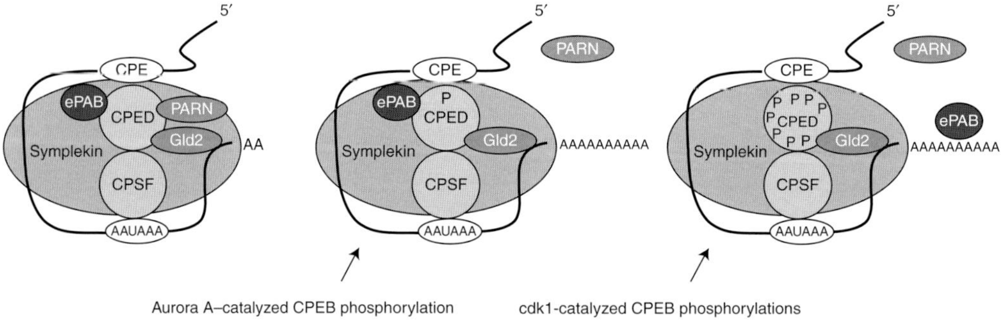

Figure 3. Cytoplasmic polyadenylation. In immature oocytes, certain maternal mRNAs contain a CPE (general structure of UUUUUAU) that is bound by CPEB, and the polyadenylation hexanucleotide AAUAAA that is bound by CPSF (for simplicity, only the 3′ UTR is depicted in the figure). CPEB is also bound by Gld2, a poly(A) polymerase; PARN, a deadenylating enzyme; and ePAB, a poly(A) binding protein. The entire complex is assembled on symplekin, a scaffold protein. When exported from the nucleus, the CPE-containing RNA has a long poly(A) tail. Once assembled with CPEB and the other factors noted above, PARN shortens the poly(A) tail to ∼20–40 nucleotides. Gld2 is also activated at this time and catalyzes poly(A) addition, but because PARN is more active, the poly(A) tail is maintained in a shortened state. Progesterone secretion from follicle cells sets off a signaling cascade in the oocyte that results in activation of the kinase Aurora A, which phosphorylates CPEB on a single site (serine 174). This event causes the expulsion of PARN from the ribonucleoprotein (RNP) complex, resulting in default Gld2-catalyzed polyadenylation. Next, the kinase cdk1 is activated, which phosphorylates CPEB on six additional sites; these events cause ePAB to dissociate from CPEB and bind the newly elongated poly(A) tail. ePAB not only protects the poly(A) tail from hydrolysis by exonucleases, but it binds the initiation factor eIF4G and helps stimulate translation.

dissociates from the mRNA and RINGO mRNA is translated (Padmanabhan and Richter 2006). RINGO protein, which is functionally similar to cyclin B1 (Nebreda and Ferby 2000; Karaiskou et al. 2001; Gutierrez et al. 2006), binds to and activates cdk1. The now active kinase cdk1 phosphorylates CPEB on six sites (excluding the Aurora A/MAP kinase site) (Mendez and Richter 2002; Kim and Richter 2007). These phosphorylation events cause the dissociation of another CPEB-bound factor, ePAB, which in turn now binds the newly elongated poly(A) tail (Fig. 3). ePAB bound to the poly(A) tail performs two tasks: It prevents deadenylation of PARN as well as other deadenylating enzymes such as CCR4/Not1 that are present in the oocyte cytoplasm, and it serves to dock eIF4G. At this point, PABP may potentiate the interaction of eIF4G with eIF4E, thus displacing Maskin (Cao et al. 2002; Kim and Richter 2007). Maskin is also phosphorylated, which also induces its dissociation from eIF4E (Barnard et al. 2005; Cao et al. 2006). Thus, the loop is now closed: eIF4E-bound eIF4G positions the 40S subunit on the 5′ end of the mRNA and translation is initiated.

Although these events are probably necessary to enter MI, additional events are important for progression to MII. For example, it has been known for some time that polyadenylation does not occur en masse, but instead there is sequence specificity as to when it occurs (Sheets et al. 1994; Stebbins-Boaz and Richter 1994, 1996; Mendez et al. 2002). Pique et al. (2008) have found that the number of CPEs, and their proximity to PBEs, determines, at least in part, the timing of when the polyadenylation occurs (see also Nakahata et al. 2001, 2003). Moreover, this polyadenylation may also be controlled by the amount of CPEB that is present because the majority of this protein undergoes ubiquitin-mediated destruction early during maturation (Mendez et al. 2002; Reverte et al. 2001).

The oocyte's arrest at MII also requires changes in poly(A) tail length, particularly of the mRNA encoding Emi2, an inhibitor of the anaphase promoting complex/cyclosome (Tunquist and Maller 2003). Emi2 is required specifically at MII to arrest meiotic progression, and this is controlled by CPEB-mediated cytoplasmic polyadenylation of Emi2 mRNA (Tung et al. 2007). This process of meiotic arrest is more complex, for it involves negative feedback loops that mediate not only the expression of CPE-containing RNAs, but ARE (AU-rich element)-containing mRNAs as well (Belloc and Mendez 2008).

Several other proteins have been found to be associated with the CPEB complex, in particular p54/DDX6/RCK1 (Minshall et al. 2001). This protein, a DEAD box helicase that is the ortholog of *Drosophila* Me31B, is a component of p-bodies. It resides in a complex with CPEB, and in RNA tethering experiments, represses translation (MInshall et al. 2009). 4E-T, a protein with Maskin-like functions, also binds CPEB as well as eIF4E (Minshall et al. 2007). Because 4E-T appears to bind CPEB specifically in small, early-stage oocytes, there may be a developmental switch between 4E-T and Maskin as the oocytes grow.

Additional factors are also involved in a developmental switch during oogenesis. Musashi is an RNA binding protein that was initially shown to control neural stem-cell fate, be involved in certain cancers, and influence oocyte meiotic progression (MacNicol et al. 2008). Musashi, together with CPEB and Pumilio, may help control the timing of when particular mRNAs are polyadenylated and translated during maturation (Charlesworth et al. 2006; Arumugam et al. 2009). The biochemical mechanism by which Musashi accomplishes these tasks is unclear.

CONCLUDING REMARKS

Drosophila and *Xenopus* oocytes have served as model systems not only for gamete development, but also for general principles of translational control. Moreover, when these evolutionarily distant organisms have conserved mechanisms of translational control, one might imagine that such mechanisms occur not only in gametes, but in other species, other tissues, or in other developmental events. Consider, for example, cytoplasmic polyadenylation-induced translation. This process controls translation in *Xenopus* and *Drosophila* oocytes (reviewed above; see also Dworkin and Hershey 1981; Salles et al. 1994; for early studies), in sea urchin eggs (Wilt 1973; Duncan and Humphreys 1984), in mouse oocytes (Tay and Richter 2001), and probably in every other metazoan as well. Cytoplasmic polyadenylation-induced translation is not restricted to oocytes; it occurs in the brain where it mediates neuronal synaptic plasticity, and learning and memory (Wu et al. 1998; Alarcon et al. 2004; Berger-Sweeney et al. 2006), and in mitotic mammalian somatic cells where it controls cellular senescence (Groisman and Richter 2006; Burns and Richter 2008). Thus, cytoplasmic polyadenylation as a regulatory process for translation is conserved both phylogenetically as well as ontogenically.

When discussing conservation of regulatory proteins among animal groups, one usually thinks of primary sequence conservation. However, in the case of *Drosophila* and *Xenopus* oocytes, there is also functional conservation of a key factor that controls translation. Maskin was identified in *Xenopus* oocytes as the protein that bridges the CPEB-controlled cytoplasmic polyadenylation machinery with the translational regulatory apparatus (i.e., the cap-binding complex of eIF4E and eIF4G). However, a clear ortholog of Maskin is not present in *Drosophila*. Its closest fly counterpart is TACC (transforming acidic coiled coil protein), but as TACC does not contain the eIF4E binding domain it is unlikely to influence translation through a mechanism similar to that used by Maskin. However, *Drosophila* contains Cup, a protein that is functionally equivalent to Maskin in that on the one hand, it interacts with a protein that binds a specific 3′ UTR element while on the other hand, it binds eIF4E. Another *Drosophila* protein related to Cup, CG32016, has not yet been well studied but is likely to have a similar function. Recently, two other proteins functionally equivalent to Maskin/Cup have been identified in the nervous system of vertebrates, neuroguidin (Jung et al. 2006) and CYFIP1 (Napoli et al. 2008). It seems unlikely that the activities of neuroguidin or CYFIP1 would not have been revealed in a timely manner without the precedential observations of Maskin and Cup. Once again, it is easy to see why oocyte molecular biology is not just for developmental biologists anymore.

REFERENCES

Alarcon JM, Hodgman R, Theis M, Huang YS, Kandel ER, Richter JD. 2004. Selective modulation of some forms of schaffer collateral-CA1 synaptic plasticity in mice with a disruption of the *CPEB-1* gene. *Learn Mem* **11:** 318–327.

Arumugam K, Wang Y, Hardy LL, Macnicol MC, Macnicol AM. 2009. Enforcing temporal control of maternal mRNA translation during oocyte cell-cycle progression. *EMBO* **29:** 387–397.

Barnard DC, Ryan K, Manley JL, Richter JD. 2004. Symplekin and xGLD-2 are required for CPEB-mediated cytoplasmic polyadenylation. *Cell* **119:** 641–651.

Barnard DC, Cao Q, Richter JD. 2005. Differential phosphorylation controls Maskin association with eukaryotic translation initiation factor 4E and localization on the mitotic apparatus. *Mol Cell Biol* **25:** 7605–7615.

Becalska AN, Gavis ER. 2009. Lighting up mRNA localization in *Drosophila* oogenesis. *Development* **136:** 2493–2503.

Belloc E, Méndez R. 2008. A deadenylation negative feedback mechanism governs meiotic metaphase arrest. *Nature* **452:** 1017–1021.

Benoit P, Papin C, Kwak JE, Wickens M, Simonelig M. 2008. PAP- and GLD-2-type poly(A) polymerases are required sequentially in cytoplasmic polyadenylation and oogenesis in *Drosophila*. *Development* **135:** 1969–1979.

Benoit B, He CH, Zhang F, Votruba SM, Tadros W, Westwood JT, Smibert CA, Lipshitz HD, Theurkauf WE. 2009. An essential role for the RNA-binding protein Smaug during the *Drosophila* maternal-to-zygotic transition. *Development* **136:** 923–932.

Berger-Sweeney J, Zearfoss NR, Richter JD. 2006. Reduced extinction of hippocampal-dependent memories in CPEB knockout mice. *Learn Mem* **13:** 4–7.

Berleth T, Burri M, Thoma G, Bopp D, Richstein S, Frigerio G, Noll M, Nüsslein-Volhard C. 1988. The role of localization of *bicoid* RNA in organizing the anterior pattern of the *Drosophila* embryo. *EMBO J* **7:** 1749–1756.

Besse F, López de Quinto S, Marchand V, Trucco A, Ephrussi A. 2009. *Drosophila* PTB promotes formation of high-order RNP particles and represses *oskar* translation. *Genes Dev* **23:** 195–207.

Brook M, Smith JW, Gray NK. 2009. The DAZL and PABP families: RNA-binding proteins with interrelated roles in translational control in oocytes. *Reproduction* **137:** 595–617.

Burns DM, Richter JD. 2008. CPEB regulation of human cellular senescence, energy metabolism, and p53 mRNA translation. *Genes Dev* **22:** 3449–3460.

Bushati N, Cohen SM. 2007. microRNA functions. *Annu Rev Cell Dev Biol* **23:** 175–205.

Cáceres L, Nilson LA. 2009. Translational repression of *gurken* mRNA in the *Drosophila* oocyte requires the hnRNP Squid in the nurse cells. *Dev Biol* **326:** 327–334.

Cao Q, Richter JD. 2002. Dissolution of the maskin-eIF4E complex by cytoplasmic polyadenylation and poly(A)-binding protein controls cyclin B1 mRNA translation and oocyte maturation. *EMBO J* **21:** 3852–3862.

Cao Q, Kim JH, Richter JD. 2006. CDK1 and calcineurin regulate Maskin association with eIF4E and translational control of cell cycle progression. *Nat Struct Mol Biol* **13:** 1128–1134.

Cao Q, Padmanabhan K, Richter JD. 2010. Pumilio 2 controls translation by competing with eIF4E for 7-methyl guanosine cap recognition. *RNA* **16:** 221–227.

Carrera P, Johnstone O, Nakamura A, Casanova J, Jäckle H, Lasko P. 2000. VASA mediates translation through interaction with a *Drosophila* yIF2 homolog. *Mol Cell* **5:** 181–187.

Castagnetti S, Ephrussi A. 2003. Orb and a long poly(A) tail are required for efficient *oskar* translation at the posterior pole of the *Drosophila* oocyte. *Development* **130:** 835–843.

Chagnovich D, Lehmann R. 2001. Poly(A)-independent regulation of maternal *hunchback* translation in the *Drosophila* embryo. *Proc Natl Acad Sci* **98:** 11359–11364.

Chang JS, Tan L, Schedl P. 1999. The *Drosophila* CPEB homolog, Orb, is required for Oskar protein expression in oocytes. *Dev Biol* **215:** 91–106.

Charlesworth A, Wilczynska A, Thampi P, Cox LL, MacNicol AM. 2006. Musashi regulates the temporal order of mRNA translation during *Xenopus* oocyte maturation. *EMBO J* **25:** 2792–2801.

Chekulaeva M, Hentze MW, Ephrussi A. 2006. Bruno acts as a dual repressor of *oskar* translation, promoting mRNA oligomerization and formation of silencing particles. *Cell* **124:** 521–533.

Chen Y, Pane A, Schüpbach T. 2007. *cutoff* and *aubergine* mutations result in retrotransposon upregulation and checkpoint activation in *Drosophila. Curr Biol* **17:** 637–642.

Chicoine J, Benoit P, Gamberi C, Paliouras M, Simonelig M, Lasko P. 2007. Bicaudal-C recruits CCR4-NOT deadenylase to target mRNAs and regulates oogenesis, cytoskeletal organization, and its own expression. *Dev Cell* **13:** 691–704.

Cho P, Poulin F, Cho-Park YA, Cho-Park IB, Chicoine JD, Lasko P, Sonenberg N. 2005. A new paradigm for translational control: Inhibition of *caudal* via 5′-3′ tethering by Bicoid and the eIF4E cognate 4EHP. *Cell* **121:** 411–423.

Cho P, Gamberi C, Cho-Park YA, Cho-Park IB, Lasko P, Sonenberg N. 2006. Cap-dependent translational inhibition establishes two opposing morphogen gradients in the *Drosophila* embryo. *Curr Biol* **16:** 2035–2041.

Clouse KN, Ferguson SB, Schüpbach T. 2008. Squid, Cup, and PABP55B function together to regulate *gurken* translation in *Drosophila. Dev Biol* **313:** 713–724.

Cook HA, Koppetsch BS, Wu J, Theurkauf WE. 2004. The *Drosophila* SDE3 homolog Armitage is required for *oskar* mRNA silencing and embryonic axis specification. *Cell* **116:** 817–829.

Copeland PR, Wormington M. 2001. The mechanism and regulation of deadenylation: Identification and characterization of *Xenopus* PARN. *RNA* **7:** 875–886.

Crucs S, Chatterjee S, Gavis ER. 2000. Overlapping but distinct RNA elements control repression and activation of *nanos* translation. *Mol Cell* **5**, 457–467.

Cui J, Sackton KL, Horner VL, Kumar KE, Wolfner MF. 2008. Wispy, the *Drosophila* homolog of GLD-2, is required during oogenesis and egg activation. *Genetics* **178:** 2017–2029.

Driever W, Nüsslein-Volhard C. 1989. The Bicoid protein is a positive regulator of *hunchback* transcription in the early *Drosophila* embryo. *Nature* **337:** 138–143.

Dubnau J, Struhl G. 1996. RNA recognition and translational regulation by a homeodomain protein. *Nature* **379:** 694–699.

Duncan R, Humphreys T. 1984. The poly(A)$^+$RNA sequence complexity is also represented in poly(A)$^-$RNA in sea-urchin embryos. *Differentiation* **28:** 24–29.

Dworkin MB, Hershey JW. 1981. Cellular titers and subcellular distributions of abundant polyadenylate-containing ribonucleic acid species during early development in the frog *Xenopus laevis. Mol Cell Biol* **1:** 983–993.

Ephrussi A, Lehmann R. 1992. Induction of germ cell formation by *oskar. Nature* **358:** 387–392.

Ferby I, Blazquez M, Palmer A, Eritja R, Nebreda AR. 1999. A novel p34^{cdc2}-binding and activating protein that is necessary and sufficient to trigger G_2/M progression in *Xenopus* oocytes. *Genes Dev* **13:** 2177–2189.

Filardo P, Ephrussi A. 2003. Bruno regulates *gurken* during *Drosophila* oogenesis. *Mech Dev* **120:** 289–297.

Findley SD, Tamanaha M, Clegg NJ, Ruohola-Baker H. 2003. *maelstrom*, a *Drosophila spindle*-class gene, encodes a protein that colocalizes with Vasa and RDE1/AGO1 homolog, Aubergine, in nuage. *Development* **130:** 859–871.

Forrest KM, Clark IE, Jain RA, Gavis ER. 2004. Temporal complexity within a translational control element in the *nanos* mRNA. *Development* **131:** 5849–5857.

Fox CA, Sheets MD, Wickens MP. 1989. Poly(A) addition during maturation of frog oocytes: distinct nuclear and cytoplasmic activities and regulation by the sequence UUUUUAU. *Genes Dev* **3:** 2151–2162.

Goodrich JS, Clouse KN, Schüpbach T. 2004. Hrb27C, Sqd, and Otu cooperatively regulate *gurken* RNA localization and mediate nurse cell chromosome dispersion in *Drosophila* oogenesis. *Development* **131:** 1949–1958.

Groisman I, Ivshina M, Marin V, Kennedy NJ, Davis RJ, Richter JD. 2006. Control of cellular senescence by CPEB. *Genes Dev* **20:** 2701–2712

Gunkel N, Yano T, Markussen F-H., Olsen LC, Ephrussi A. 1998. Localization-dependent translation requires a functional interaction between the 5′ and 3′ ends of *oskar* mRNA. *Genes Dev* **12:** 1652–1664

Gutierrez GJ, Vögtlin A, Castro A, Ferby I, Salvagiotto G, Ronai Z, Lorca T, Nebreda AR. 2006. Meiotic regulation of the CDK activator RINGO/Speedy by ubiquitin-proteasome-mediated processing and degradation. *Nat Cell Biol* **8:** 1084–1094.

Harris AN, Macdonald PM. 2001. *aubergine* encodes a *Drosophila* polar granule component required for pole cell formation and related to eIF2C. *Development* **128:** 2823–2832.

Huarte J, Belin D, Vassalli A, Strickland S, Vassalli JD. 1987. Meiotic maturation of mouse oocytes triggers the translation and polyadenylation of dormant tissue-type plasminogen activator mRNA. *Genes Dev* **1:** 1201–1211.

Huynh J-R., Munro TP, Smith-Litière K, Lepesant J-A. St Johnston D. 2004. The *Drosophila* hnRNPA/B homolog, Hrp48, is specifically required for a distinct step in *osk* mRNA localization. *Dev Cell* **6:** 625–635.

Johnstone O, Lasko P. 2004. Interaction with eIF5B is essential for Vasa function during development. *Development* **131:** 4167–4178.

Juge F, Zaessinger S, Temme C, Wahle E, Simonelig M. 2002. Control of poly(A) polymerase level is essential to cytoplasmic polyadenylation and early development in *Drosophila*. *EMBO J* **21:** 6603–6613.

Jung MY, Lorenz L, Richter JD. 2006. Translational control by neuroguidin, a eukaryotic initiation factor 4E and CPEB binding protein. *Mol Cell Biol* **26:** 4277–4287.

Kalifa Y, Huang T, Rosen LN, Chatterjee S, Gavis ER. 2006. Glorund, a *Drosophila* hnRNP F/H homolog, is an ovarian repressor of *nanos* translation. *Dev Cell* **10:** 291–301.

Kalifa Y, Armenti ST, Gavis ER. 2009. Glorund interactions in the regulation of *gurken* and *oskar* mRNAs. *Dev Biol* **326:** 68–74.

Karaiskou A, Perez LH, Ferby I, Ozon R, Jessus C, Nebreda AR. 2001. Differential regulation of Cdc2 and Cdk2 by RINGO and cyclins. *J Biol Chem* **276:** 36028–36034.

Keady BT, Kuo P, Martínez SE, Yuan L, Hake LE. 2007. MAPK interacts with XGef and is required for CPEB activation during meiosis in *Xenopus* oocytes. *J Cell Sci* **120:** 1093–1103.

Kim JH, Richter JD. 2006. Opposing polymerase-deadenylase activities regulate cytoplasmic polyadenylation. *Mol Cell* **24:** 173–183.

Kim JH, Richter JD. 2007. RINGO/cdk1 and CPEB mediate poly(A) tail stabilization and translational regulation by ePAB. *Genes Dev* **21:** 2571–2579.

Kim-Ha J, Kerr K, Macdonald PM. 1995. Translational regulation of *oskar* mRNA by Bruno, an ovarian RNA-binding protein, is essential. *Cell* **81:** 403–412.

Körner CG, Wormington M, Muckenthaler M, Schneider S, Dehlin E, Wahle E. 1998. The deadenylating nuclease (DAN) is involved in poly(A) tail removal during the meiotic maturation of *Xenopus* oocytes. *EMBO J* **17:** 5427–5437.

Lie YS, Macdonald PM. 1999. Translational regulation of *oskar* mRNA occurs independent of the cap and poly(A) tail in *Drosophila* ovarian extracts. *Development* **126:** 4989–4996.

Lim AK, Kai T. 2007. Unique germ-line organelle, nuage, functions to repress selfish genetic elements in *Drosophila melanogaster*. *Proc Natl Acad Sci* **104:** 6714–6719.

Lin CL, Evans V, Shen S, Xing Y, Richter JD. 2010. The nuclear experience of CPEB: implications for RNA processing and translational control. *RNA* in press.

Liu N, Han H, Lasko P. 2009. Vasa promotes *Drosophila* germ line stem cell differentiation by activating *mei-P26* translation through a direct interaction with a U-rich motif in its 3' UTR. *Genes Dev* **23:** 2742–2752.

Lyon AM, Reveal BS, Macdonald PM, Hoffman DW. 2009. Bruno protein contains an expanded RNA recognition motif. *Biochemistry* **48:** 12202–12212.

MacNicol AM, Wilczynska A, MacNicol MC. 2008. Function and regulation of the mammalian Musashi mRNA translational regulator. *Biochem Soc Trans* **36:** 528–530.

Mandel CR, Kaneko S, Zhang H, Gebauer D, Vethantham V, Manley JL, Tong L. 2006. Polyadenylation factor CPSF-73 is the pre-mRNA 3′-end-processing endonuclease. *Nature* **444:** 953–956.

Markussen F-H, Michon AM, Breitwieser W, Ephrussi A. 1995. Translational control of *oskar* generates Short OSK, the isoform that induces pole plasma assembly. *Development* **121:** 3723–3732.

Martínez SE, Yuan L, Lacza C, Ransom H, Mahon GM, Whitehead IP, Hake LE. 2005. XGef mediates early CPEB phosphorylation during *Xenopus* oocyte meiotic maturation. *Mol Biol Cell* **16:** 1152–1164.

McGrew LL, Dworkin-Rastl E, Dworkin MB, Richter JD. 1989. Poly(A) elongation during *Xenopus* oocyte maturation is required for translational recruitment and is mediated by a short sequence element. *Genes Dev* **3:** 803–815.

Mendez R, Hake LE, Andresson T, Littlepage LE, Ruderman JV, Richter JD. 2000a. Phosphorylation of CPE binding factor by Eg2 regulates translation of c-mos mRNA. *Nature* **404:** 302–307.

Mendez R, Murthy KG, Ryan K, Manley JL, Richter JD. 2000b. Phosphorylation of CPEB by Eg2 mediates the recruitment of CPSF into an active cytoplasmic polyadenylation complex. *Mol Cell* **6:** 1253–1259.

Mendez R, Barnard D, Richter JD. 2002. Differential mRNA translation and meiotic progression require Cdc2-mediated CPEB destruction. *EMBO J* **21:** 1833–1844.

Micklem DR, Adams J, Grünert S, St Johnston D. 2000. Distinct roles of two conserved Staufen domains in *oskar* mRNA localization and translation. *EMBO J* **19:** 1366–1377.

Minshall N, Thom G, Standart N. 2001. A conserved role of a DEAD box helicase in mRNA masking. *RNA* **7:** 1728–1742

Minshall N, Reiter MH, Weil D, Standart N. 2007. CPEB interacts with an ovary-specific eIF4E and 4E-T in early *Xenopus* oocytes. *J Biol Chem* **282:** 37389–37401

Minshall N, Kress M, Weil D, Standart N. 2009. Role of p54 RNA helicase activity and its C-terminal domain in translational repression, P-body localization and assembly. *Mol Biol Cell* **20:** 2464–2472

Nakahata S, Katsu Y, Mita K, Inoue K, Nagahama Y, Yamashita M. 2001. Biochemical identification of *Xenopus* Pumilio as a sequence-specific cyclin B1 mRNA-binding protein that physically interacts with a Nanos homolog, Xcat-2, and a cytoplasmic polyadenylation element-binding protein. *J Biol Chem* **276:** 20945–20953.

Nakahata S, Kotani T, Mita K, Kawasaki T, Katsu Y, Nagahama Y, Yamashita M. 2003. Involvement of *Xenopus* Pumilio in the translational regulation that is specific to cyclin B1 mRNA during oocyte maturation. *Mech Dev* **120:** 865–880.

Nakamura A, Amikura R, Hanyu K, Kobayashi S. 2001. Me31B silences translation of oocyte-localizing RNAs through the formation of cytoplasmic RNP complex during *Drosophila* oogenesis. *Development* **128:** 3233–3242.

Nakamura A, Sato K, Hanyu-Nakamura K. 2004. *Drosophila* Cup is an eIF4E binding protein that associates with Bruno and regulates *oskar* mRNA translation in oogenesis. *Dev Cell* **6:** 69–78.

Napoli I, Mercaldo V, Boyl P, Eleuteri B, Zalfa F, De Rubeis S, Di Marino D, Mohr E, Massimi M, Falconi M, et al. 2008. The fragile X syndrome protein represses activity-dependent translation through CYFIP1, a new 4E-BP. *Cell* **134:** 1042–1054.

Nebreda AR, Ferby I. 2000. Regulation of the meiotic cell cycle in oocytes. *Curr Opin Cell Biol* **12:** 666–675.

Nelson MR, Leidal AM, Smibert CA. 2004. *Drosophila* Cup is an eIF4E-binding protein that functions in Smaug-mediated translational repression. *EMBO J* **23:** 150–159.

Niessing D, Dostatni N, Jäckle H, Rivera-Pomar R. 1999. Sequence interval within the PEST motif of Bicoid is important for translational repression of *caudal* mRNA in the anterior region of the *Drosophil*a embryo. *EMBO J* **18:** 1966–1973.

Niessing D, Blanke S, Jäckle H. 2002. Bicoid associates with the 5′ cap-bound complex of *caudal* mRNA and represses translation. *Genes Dev* **16:** 2576–2582.

Norvell AKR, Wehr K, Schüpbach T. 1999. Specific isoforms of Squid, a *Drosophila* hnRNP, perform distinct roles in Gurken localization during oogenesis. *Genes Dev* **13:** 864–876.

Norvell A, Debec A, Finch D, Gibson L, Thoma B. 2005. Squid is required for efficient posterior localization of *oskar* mRNA during *Drosophila* oogenesis. *Dev Genes Evol* **215:** 340–349.

Ostareck DH, Ostareck-Lederer A, Shatsky IN, Hentze MW. 2003. Lipoxygenase mRNA silencing in erythroid differentiation: The 3′ UTR regulatory complex controls 60S ribosomal subunit joining. *Cell* **104:** 281–290.

Padmanabhan K, Richter JD. 2006. Regulated Pumilio-2 binding controls RINGO/Spy mRNA translation and CPEB activation. *Genes Dev* **20:** 199–209.

Pane A, Wehr K, Schüpbach T. 2007. *zucchini* and *squash* encode two putative nucleases required for rasiRNA production in the *Drosophila* germline. *Dev Cell* **12:** 851–862.

Piqué M, López JM, Foissac S, Guigó R, Méndez R. 2008. A combinatorial code for CPE-mediated translational control. *Cell* **132:** 434–448.

Rangan P, DeGennaro M, Jamie-Bustamante K, Coux RX, Martinho RG, Lehmann R. 2009. Temporal and spatial control of germ-plasm RNAs. *Curr Biol* **19:** 72–77.

Read RL, Martinho RG, Wang SW, Carr AM, Norbury CJ. 2002. Cytoplasmic poly(A) polymerases mediate cellular responses to S phase arrest. *Proc Natl Acad Sci* **99:** 12079–12084.

Reverte CG, Ahearn MD, Hake LE. 2001. CPEB degradation during *Xenopus* oocyte maturation requires a PEST domain and the 26S proteasome. *Dev Biol* **231:** 447–458.

Reverte CG, Yuan L, Keady BT, Lacza C, Attfield KR, Mahon GM, Freeman B, Whitehead IP, Hake LE. 2003. XGef is a CPEB-interacting protein involved in *Xenopus* oocyte maturation. *Dev Biol* **255:** 383–398.

Richard P, Manley JL. 2009. Transcription termination by nuclear RNA polymerases. *Genes Dev* **23:** 1247–1269.

Richter JD. 1999. Cytoplasmic polyadenylation in development and beyond. *Microbiol Mol Biol Rev* **63:** 446–456

Rivera-Pomar R, Niessing D, Schmidt-Ott U, Gehring WJ, Jäckle H. 1996. RNA binding and translational suppression by Bicoid. *Nature* **379:** 746–749.

Saffman EE, Styhler S, Rother K, Li W, Richard S, Lasko P. 1998. Premature translation of *oskar* in oocytes lacking the RNA-binding protein Bicaudal-C. *Mol Cell Biol* **18:** 4855–4862.

Saitoh S, Chabes A, McDonald WH, Thelander L, Yates JR, Russell P. 2002. Cid13 is a cytoplasmic poly(A) polymerase that regulates ribonucleotide reductase mRNA. *Cell* **109:** 563–573.

Sallés FJ, Lieberfarb ME, Wreden C, Gergen JP, Strickland S. 1994. Coordinate initiation of *Drosophila* development by regulated polyadenylation of maternal messenger RNAs. *Science* **266:** 1996–1999.

Sarkissian M, Mendez R, Richter JD. 2004. Progesterone and insulin stimulation of CPEB-dependent polyadenylation is regulated by Aurora A and glycogen synthase kinase-3. *Genes Dev* **18:** 48–61.

Saunders C, Cohen RS. 1999. The role of oocyte transcription, the 5′ UTR, and translation repression and derepression in *Drosophila gurken* mRNA and protein localization. *Mol Cell* **3:** 43–54.

Semotok JL, Cooperstock RL, Pinder BD, Vari HK, Lipshitz HD, Smibert CA. 2005. Smaug recruits the CCR4/POP2/NOT deadenylase complex to trigger maternal transcript localization in the early *Drosophila* embryo. *Curr Biol* **15:** 284–294.

Shatkin AJ, Manley JL. 2000. The ends of the affair: Capping and polyadenylation. *Nature Struc Mol Biol* **7:** 838–842.

Sheets MD, Fox CA, Hunt T, Vande Woude G, Wickens M. 1994. The 3′-untranslated regions of c-mos and cyclin mRNAs stimulate translation by regulating cytoplasmic polyadenylation. *Genes Dev* **8:** 926–938.

Smibert CA, Wilson JE, Kerr K, Macdonald PM. 1996. Smaug protein represses translation of unlocalized *nanos* mRNA in the *Drosophila* embryo. *Genes Dev* **10:** 2600–2609.

Smith RC, Dworkin-Rastl E, Dworkin MB. 1988a. Expression of a histone H1-like protein is restricted to early *Xenopus* development. *Genes Dev* **2:** 1284–1295.

Smith RC, Dworkin MB, Dworkin-Rastl E. 1988b. Destruction of a translationally controlled mRNA in *Xenopus* oocytes delays progesterone-induced maturation. *Genes Dev* **2:** 1296–1306.

Snee M, Benz D, Jen J, Macdonald PM. 2008. Two distinct domains of Bruno bind specifically to the *oskar* mRNA. *RNA Biol* **5:** 1–9.

Sonoda J, Wharton RP. 2001. *Drosophila* Brain Tumor is a translational repressor. *Genes Dev* **15:** 762–773.

Spradling AC. 1993. Developmental genetics of oogenesis. In *The development of* Drosophila melanogaster (ed. Bate M, Martinez-Arias A), pp. 1–70. Cold Spring Harbor Laboratory Press, Cold Spring Harbor, NY.

Stebbins-Boaz B, Richter JD. 1994. Multiple sequence elements and a maternal mRNA product control cdk2 RNA polyadenylation and translation during early *Xenopus* development. *Mol Cell Biol* **14:** 5870–5880.

Stebbins-Boaz B, Hake LE, Richter JD. 1996. CPEB controls the cytoplasmic polyadenylation of cyclin, Cdk2 and c-mos mRNAs and is necessary for oocyte maturation in *Xenopus*. *EMBO J* **15:** 2582–2592.

Stebbins-Boaz B, Cao Q, de Moor CH, Mendez R, Richter JD. 1999. Maskin is a CPEB-associated factor that transiently interacts with elF-4E. *Mol Cell* **4:** 1017–1027.

Strickland S, Huarte J, Belin D, Vassalli A, Rickles RJ, Vassalli JD. 1988. Antisense RNA directed against the 3′ noncoding region prevents dormant mRNA activation in mouse oocytes. *Science* **241:** 680–684.

Styhler S, Nakamura A, Swan A, Suter B, Lasko P. 1998. *vasa* is required for GURKEN accumulation in the oocyte, and is involved in oocyte differentiation and germline cyst development. *Development* **125:** 1569–1578.

Tadros W, Goldman AL, Babak T, Menzies F, Vardy L, Orr-Weaver T, Hughes TR, Westwood JT, Smibert CA, Lipshitz HD. 2007. SMAUG is a major regulator of maternal mRNA destabilization in *Drosophila* and its translation is activated by the PAN GU kinase. *Dev Cell* **12:** 143–155.

Tay J, Richter JD. 2001. Germ cell differentiation and synaptonemal complex formation are disrupted in CPEB knockout mice. *Dev Cell* **1:** 201–213.

Tomancak P, Guichet A, Zavorszky P, Ephrussi A. 1998. Oocyte polarity depends on regulation of *gurken* by Vasa. *Development* **125:** 1723–1732.

Tomari Y, Du T, Haley B, Schwarz DS, Bennett R, Cook HA, Koppetsch BS, Theurkauf WE, Zamore PD. 2004. RISC assembly defects in the *Drosophila* RNAi mutant *armitage*. *Cell* **119:** 831–841.

Trucco A, Gaspar I, Ephrussi A. 2009. Assembly of endogenous *oskar* mRNA particles for motor-dependent transport in the *Drosophila* oocyte. *Cell* **139:** 983–998.

Tung JJ, Padmanabhan K, Hansen DV, Richter JD, Jackson PK. 2007. Translational unmasking of Emi2 directs cytostatic factor arrest in meiosis II. *Cell Cycle* **6:** 725–731.

Tunquist BJ, Maller JL. 2003. Under arrest: Cytostatic factor (CSF)-mediated metaphase arrest in vertebrate eggs. *Genes Dev* **17:** 683–710.

Vanzo N, Oprins A, Xanthakis D, Ephrussi A, Rabouille C. 2007. Stimulation of endocytosis and actin dynamics by Oskar polarizes the *Drosophila* oocyte. *Dev Cell* **12:** 543–555.

Villaescusa JC, Buratti C, Penkov D, Mathiasen L, Planagumà J, Ferretti E, Blasi F. 2009. Cytoplasmic Prep1 interacts with 4EHP inhibiting *Hoxb4* translation. *PLoS ONE* **4:** e5213.

Wang L, Eckmann CR, Kadyk LC, Wickens M, Kimble J. 2002. A regulatory cytoplasmic poly(A) polymerase in *Caenorhabditis elegans*. *Nature* **419:** 312–316.

Webster PJ, Liang L, Berg CA, Lasko P, Macdonald PM. 1997. Translational repressor Bruno plays multiple roles in development and is widely conserved. *Genes Dev* **11:** 2510–2521.

Wharton RP, Struhl G. 1991. RNA regulatory elements mediate control of *Drosophila* body pattern by the posterior morphogen nanos. *Cell* **67:** 955–967.

Wilhelm JE, Hilton M, Amos Q, Henzel WJ. 2003. Cup is an eIF4E binding protein required for both the translational repression of *oskar* and the recruitment of Barentsz. *J Cell Biol* **163:** 1197–1204.

Wilson JE, Connell JE, Macdonald PM. 1996. *Aubergine* enhances *oskar* translation in the *Drosophila* ovary. *Development* **122:** 1631–1639.

Wilt FH. 1973. Polyadenylation of maternal RNA of sea urchin eggs after fertilization. *Proc Natl Acad Sci* **70:** 2345–2349.

Wu L, Wells D, Tay J, Mendis D, Abbott MA, Barnitt A, Quinlan E, Heynen A, Fallon JR, Richter JD. 1998. CPEB-mediated cytoplasmic polyadenylation and the regulation of experience-dependent translation of α-CaMKII mRNA at synapses. *Neuron* **21:** 1129–1139.

Yano T, Lopez de Quinto S, Matsui Y, Shevchenko A, Shevchenko A, Ephrussi A. 2004. Hrp48, a *Drosophila* hnRNPA/B homolog, binds and regulates translation of *oskar* mRNA. *Dev Cell* **6:** 637–648.

Zaessinger S, Busseau I, Simonelig M. 2006. Oskar allows *nanos* mRNA translation in *Drosophila* embryos by preventing its deadenylation by Smaug/CCR4. *Development* **133:** 4573–4583.

Zappavigna V, Piccioni F, Villaescusa JC, Verrotti AC. 2004. Cup is a nucleocytoplasmic shuttling protein that interacts with the eukaryotic translation initiation factor 4E to modulate *Drosophila* ovary development. *Proc Natl Acad Sci* **101:** 14800–14805.

RNA Granules in Germ Cells

Ekaterina Voronina[1], Geraldine Seydoux[1], Paolo Sassone-Corsi[2], and Ippei Nagamori[2]

[1]Department of Molecular Biology and Genetics and Howard Hughes Medical Institute, Center for Cell Dynamics, Johns Hopkins University School of Medicine, Baltimore, Maryland

[2]Department of Pharmacology, School of Medicine, University of California, Irvine, Irvine, California

Correspondence: evoronina@jhmi.edu (E.V.); gseydoux@jhmi.edu (G.S.); psc@uci.edu (P.S.-C.); inagamor@uci.edu (I.N.)

"Germ granules" are cytoplasmic, nonmembrane-bound organelles unique to germline. Germ granules share components with the P bodies and stress granules of somatic cells, but also contain proteins and RNAs uniquely required for germ cell development. In this review, we focus on recent advances in our understanding of germ granule assembly, dynamics, and function. One hypothesis is that germ granules operate as hubs for the post-transcriptional control of gene expression, a function at the core of the germ cell differentiation program.

"Germ granules" is a general term used to refer to the RNA-rich cytoplasmic bodies of germ cells. By electron microscopy, germ granules appear as compact granulo-fibrillar aggregates with no surrounding membrane (Fig. 1A,D) (Eddy 1975; Guraya 1979). Germ granules contain proteins and RNAs required for germ cell development, and may carry out posttranscriptional regulation specific to germ cells. In some organisms (i.e., *Drosophila*, *Xenopus*, *Caenorhabditis elegans*, and zebrafish), germ granules are present continuously throughout development, with the exception of mature sperm. In those organisms, germ granules are transmitted from oocyte to embryo as part of the germ plasm, a specialized cytoplasm that segregates with the germ lineage and is sufficient to specify germ cell fate. In other organisms, notably mammals, germ granules are not detected in oocytes or early embryos, but are formed de novo in primordial germ cells shortly after their specification. The current view is that, while not all organisms require germ granules to specify germ cell fate, all organisms depend on germ granules for germ cell function. Molecular studies have revealed intriguing similarities between germ granules and RNA granules present in somatic cells. In this review, we focus on recent advances in our understanding of germ granule composition, variety, dynamics, and function.

TYPES OF GERM GRANULES

Cytoplasmic aggregates that stained with RNA dyes were first detected in the germ cells of insects in 1890 (Ritter 1890), and have since been observed by cytochemistry or electron microscopy in over 80 species across the animal kingdom from rotifers to mammals (Eddy 1975). Today, antibodies against the Vasa family of RNA helicases are commonly used to identify

Cite this article as *Cold Spring Harb Perspect Biol* doi: 10.1101/cshperspect.a002774

germ granules. Many other conserved components of germ granules have also been identified (Tables 1 and 2). Some, like Vasa family members, are present in most germ granules, others are found only in specific stages. The morphology and localization of germ granules also changes during development. Here we describe the major classes of germ granules found in immature and differentiating germ cells (nuage), gametes (sponge, balbiani, and chromatoid bodies), and embryos (germ plasm) (Fig. 1) (Eddy 1975; Guraya 1979; also see de Sousa Lopes and Roelen 2010).

Perinuclear Granules

During most of germline development, germ granules appear as rounded fibrillar aggregates (or nuage, from the French for "cloud") that cluster around nuclei. This arrangement has been documented in the primordial germ cells (PGCs) of *Drosophila* (Mahowald 1968, 1971), *C. elegans* (Strome and Wood 1982, 1983), *Xenopus* (Ikenishi et al. 1996), and zebrafish (Knaut et al. 2000). In mice, small granules surrounded by fibrillar matrix and mitochondria become apparent around the nuclei of PGCs two days after their formation at 9–9.5 days of gestation (Spiegelman and Bennett 1973; Clark and Eddy 1975). Perinuclear granules persist after the PGCs become incorporated into the somatic gonad. Electron micrographic studies of *C. elegans* germ cells in meiotic prophase have shown that germ granules overlie dense clusters of nuclear pores, corresponding to the main sites of mRNA export from the nucleus (Fig. 1A) (Strome and Wood 1982; Pitt et al. 2000; Sheth et al. 2010). Close association of the nuage with nuclear pores has also been reported in ultrastructural studies of *Xenopus* (Czolowska 1969), zebrafish (Knaut et al. 2000; reviewed in Kloc et al. 2004), and mouse (reviewed in Chuma et al. 2009).

Balbiani Body

In differentiating gametes, germ granules often adopt unique morphologies. The Balbiani body, or mitochondrial cloud, is a transient structure, containing mitochondria, Golgi, endoplasmic reticulum (ER) and RNA, that forms in the young (previtellogenic) oocytes of insects and vertebrates (Cox and Spradling 2003; Kloc et al. 2004; Wilk et al. 2005; Pepling et al. 2007). In *Drosophila*, ER membranes associated with the Balbiani body are enriched for the ER resident protein reticulon, suggesting that these membranes represent a specialized subset of smooth ER (Roper 2007). The Balbiani body initially forms near the nucleus and spreads to the cytoplasm during oocyte growth. In *Drosophila* and *Xenopus*, the aggregates eventually localize to a subregion of the oocyte cortex, in which they incorporate with other germ granule material to form the germ plasm that will be transmitted to the germline of the next generation (see below). It has been proposed that the Balbiani body selects the best mitochondria for incorporation in the germline (Cox and Spradling 2003). In *Xenopus*, mitochondria in the germ plasm have lower respiratory activity than mitochondria destined for the soma (Kogo et al. 2011). Lower respiration is hypothesized to decrease oxidative stress and mutational damage. A *Drosophila* mutant that disrupts mitochondrial transport and association with the Balbiani body, however, had no apparent effect on fertility (Cox and Spradling 2006).

Balbiani bodies, consisting of ER and mitochondria interspersed with nuage surrounding Golgi stacks, have also been reported in mouse oocytes, which do not form germ plasm (Pepling et al. 2007). The mouse Balbiani bodies are present in the cyst and primordial follicle stages, and disappear during follicular growth (Fig. 1B) (Pepling et al. 2007). Their function is not yet known.

Sponge Bodies

During *Drosophila* oogenesis, each growing oocyte is connected to 15 nurse cells through intercellular bridges or "ring canals" (Fig. 2A). In addition to perinuclear nuage, nurse cells and the oocyte contain sponge bodies, which are stacks of endoplasmic reticulum (ER) cisternae embedded in an electron-dense matrix scattered through the cytoplasm (Fig. 1C)

Table 1. Conserved components of germ granules—listed are components localizing to germ granules in two or more species

	Function	Organism: Protein name and germ granule type	References
Proteins			
Vasa and related DEAD box RNA helicases	RNP remodeling	*Drosophila*: Vasa in nuage and polar granules; *C. elegans*: GLH-1-4 in P granules; *Xenopus*: XVLG1, nuage in oogonia, perinuclear in PGCs; Zebrafish: Vasa, perinuclear in PGCs; Mouse: MVH in chromatoid body; Humans: Vasa, perinuclear in fetal oocytes	Hay et al. 1988; Liang et al. 1994 Gruidl et al. 1996a Ikenishi et al. 1996; Bilinski et al. 2004 Knaut et al. 2000 Toyooka et al. 2000 Castrillon et al. 2000
Argonaute family	Small RNA (miRNA/ piRNA) regulation	*Drosophila*: Aubergine in nuage and polar granules, PIWI in polar granules, Ago3 in nuage; Mouse: Miwi, Mili, Ago2, Ago3 in chromatoid body; Zebrafish: Ziwi, Zili in germ granules, perinuclear in PGC and germ cells; *C. elegans*: PRG-1, CSR-1 in P granules	Harris and Macdonald 2001; Megosh et al. 2006; Nishida et al. 2007; Li et al. 2009 Kotaja et al. 2006a,b Houwing et al. 2007, 2008; Batista et al. 2008; Claycomb et al. 2009
Dicer	miRNA production	Mouse: Dicer in chromatoid body; *C. elegans*: DRH-3, DCR-1 in P granules	Kotaja et al. 2006a Claycomb et al. 2009 Beshore et al. 2011
Maelstrom (HMG box)	Chromatin and piRNA regulation	*Drosophila*: Maelstrom in nuage; Mouse: Maelstrom in chromatoid body	Findley et al. 2003 Costa et al. 2006
Tudor domain proteins (methyl-binding modules)	Recruitment of methylated proteins to germ granules; recruitment of mitochondrial ribosomal RNA to polar granules in *Drosophila*	*Drosophila*: Tudor in nuage and polar granules, Krimp, Tejas, and PAPI in nuage; Mouse: Mtr (TDRD1-7;9) in intermitochondrial cement and chromatoid body, Tdrd4/ Rnf17 in Rnf17-granule	Bardsley et al. 1993 Arkov et al. 2006; Lim and Kai 2007; Patil and Kai 2010; Liu et al. 2011 Chuma et al. 2003; Pan et al. 2005; Hosokawa et al. 2007; Vasileva et al. 2009 Aravin et al. 2009; Shoji et al. 2009
WD repeat proteins	Cofactors of protein arginine methyltransferase; targeting of methylation?	*Drosophila*: Valois (homolog of MEP50) in polar granules and nuage; Mouse: WDR77 in chromatoid bodies	Anne and Mechler 2005 Vagin et al. 2009b
Sm proteins	Splicing in the spliceosome, germ granule localization and germ cell fate in *C. elegans*	*C. elegans*: Sm in P granules; *Xenopus*: Sm in nuage in oocytes; Mouse: Sm in chromatoid body; *Drosophila*: SmB and SmD3 in polar granules	Barbee et al. 2002 Bilinski et al. 2004 Moussa et al. 1994; Chuma et al. 2003 Anne 2010; Gonsalvez et al. 2010

Continued

Table 1. *Continued*

	Function	Organism: Protein name and germ granule type	References
Dcp1	Decapping of mRNAs in P bodies, localization of Oskar RNA to the germ plasm in *Drosophila*	*Drosophila*: dDcp1 in sponge bodies; Mouse: Dcp1a in chromatoid body, and perinuclear spermatocyte nuage	Lin et al. 2006 Kotaja et al. 2006a Aravin et al. 2009
Dcp2	Decapping of mRNAs in P bodies	*Drosophila*: dDcp2 in sponge bodies; *C. elegans*: DCP-2 in P granules	Lin et al. 2006 Lall et al. 2005
Dhh1p/Rck (DEAD box RNA helicase)	Repression of translation (also in P bodies)	*Drosophila*: Me31b in sponge bodies, and polar granules; *C. elegans*: CGH-1 in P granules	Nakamura et al. 2001; Thomson et al. 2008 Navarro et al. 2001
Scd6p/Rap55 (Sm-like domain)	Repression of translation (also in P bodies)	*Drosophila*: Trailer Hitch in Balbiani body (early oocyte), nuage and sponge bodies (oogenesis); *C. elegans*: CAR-1 in P granules; mouse: Trailer hitch in Balbiani body (early oocytes), perinuclear granule (spermatocytes)	Wilhelm et al. 2005 Liu et al. 2011 Audhya et al. 2005; Boag et al. 2005; Squirrell et al. 2006 Pepling et al. 2007
Xrn1	Exonuclease degrading mRNA	*Drosophila*: PCM in nuage; Mouse: Xrn1 in perinuclear spermatocyte nuage	Lim et al. 2009 Aravin et al. 2009
Nanos	Repression of translation	Zebrafish: in perinucelar germ granules of PGCs; *C. elegans*: in P granules of embryos	Koprunner et al. 2001 Subramaniam and Seydoux 1999
eIF4F complex	Translational regulation by components or binding partners of eIF4F	*Drosophila*: eIF4A polar granules; eIF4E-binding protein Cup in Babliani body and sponge bodies (oogenesis); *C. elegans*: IFE-1(eIF4E) in P granules	Cox and Spradling 2003; Wilhelm et al. 2003; Nakamura et al. 2004; Thomson et al. 2008 Amiri et al. 2001
RNAs			
Nanos (mRNA)	Repression of translation	*Drosophila*: *Nanos* RNA enriched in germ plasm; *C. elegans*: *nos-2* RNA enriched in P granules; *Xenopus*: *Xcat2* RNA in balbiani body; Zebrafish: *nos1* RNA enriched in germ plasm; Pea Aphid: nanos RNA in nuage-like structure in oocytes	Forrest and Gavis 2003 Subramaniam and Seydoux 1999 Mosquera et al. 1993; Bilinski et al. 2004 Koprunner et al. 2001 Chang et al. 2006
Mitochondrial large and small ribosomal RNAs (noncoding RNAs)	Translation of *germ cell less* (*gcl*) in *Drosophila*	*Drosophila*: on surface of polar granules of syncytial embryo; *Xenopus*: in germinal granules; planaria: on surface of chromatoid bodies	Kashikawa et al. 1999 Kashikawa et al. 2001 Sato et al. 2001
piRNA	Repression of transposons	Zebrafish: oocyte germ granules; mouse: chromatoid body	Houwing et al. 2007 Melkar et al. 2010

Table 2. Components of the mouse chromatoid body

Component	Function	References
MVH	Dead box RNA helicase that is required for spermatogenesis	Chuma et al. 2003; Kotaja et al. 2006a,b
MIWI	Argonaute/PIWI family RNA binding protein that is required for spermatogenesis. MIWI and MILI repress transposone element mediated by piRNA	Deng and Lin 2002; Grivna et al. 2006; Kotaja et al. 2006a,b
RanBPM	RanGTP binding protein that is involved in microtubule nucleation. It associates with MVH	Shibata et al. 2004
Kif17b	Kinesin motor protein that is involved in the transport of the coactivator protein ACT and mRNAs	Kotaja et al. 2006a
Tdrd1	Tudor domain containing protein that is required for spermatogenesis. It interacts with MILI	Hosokawa et al. 2007; Kojima et al. 2009; Wang et al. 2009
Tdrd6	Tudor domain containing protein that is required for spermatogenesis and CB formation	Vasileva et al. 2009
Tdrd7	Tudor domain containing protein	Hosokawa et al. 2007
p48 and p52	Germ cell specific RNA binding proteins	Oko et al. 1996
Ago2 and Ago3	Main catalytic engine of RISC	Kotaja et al 2006b
Dicer	Cytoplasmic RNase that produces siRNA and miRNA from their precursors	Kotaja et al. 2006b
Dcp1a	Decapping enzyme that remove 5'cap of mRNA	van Dijk et al. 2002
GW182	P-body localized RNA binding protein that is required for RNAi	Kotaja et al. 2006b
mRNA	Translation template	Kotaja et al. 2006b
MiRNA	Small RNA that represses complement target mRNA expression	Kotaja et al. 2006b
SnRNP	Essential component of the spliceosomal complex that function in pre-mRNA processing	Paniagua et al. 1985; Moussa et al. 1994; Biggiogera et al. 1990
GRTH	Dead box RNA helicase that is required for spermatogenesis. Deficient of GRTH shows impaired CB formation	Tsai-Morris et al. 2004
GW182	Essential component for si/miRNA induced gene silencing	Kotaja et al. 2006b
H2B	Chromatin formation	Haraguchi et al. 2005
H4	Chromatin formation	Werner and Werner 1995
Acetylated H3	Chromatin regulation	Haraguchi et al. 2005
Acetylated H4	Chromatin regulation	Haraguchi et al. 2005
Actin	Cytoskeletal protein	Walt and Armbruster 1984; Haraguchi et al. 2005
Vimentin	Intermediate filament	Haraguchi et al. 2005
Ubiquitin	Targeting signal for several pathways	Haraguchi et al. 2005
E2	Ubiquitin conjugating enzyme	Haraguchi et al. 2005

Continued

Table 2. *Continued*

Component	Function	References
p52	26S proteasome subunit	Haraguchi et al. 2005
pA700	Proteasome activator	Haraguchi et al. 2005
COX1	Cytochrome C oxidase subunit 1	Haraguchi et al. 2005
HSP70	protein chaperon	Haraguchi et al. 2005
F1α	ATP synthase subunit α	Haraguchi et al. 2005
F1β	ATP synthase subunit β	Haraguchi et al. 2005
LDH	Dehydrogenation of 2-hydroxybutylate	Haraguchi et al. 2005
Enolase	Conversion from 2-phosphogylcerate to phosphoenoylpyruvate	Haraguchi et al. 2005
PHGPx	Reduces lipid hydroperoxides in membranes	Haraguchi et al. 2005
Cytochrome C	Electron transfer component	Hess et al. 1993
Ca^{2+}, Mg^{2+}		Andonov and Chaldakov 1989; Rouelle-Rossier et al. 1993
Histocompatibility antigen	Immune system	Head and Kresge 1985
LAMP1 and LAM2	Lysosome localized glycoproteins	Haraguchi et al. 2005
Acid phosphatase	Lysosome localized phosphatase	Anton 1983
Cathepsins B,D, H, and L	Protease	Haraguchi et al. 2005
LAP	Leucin aminopeptidase	Haraguchi et al. 2005
DNase	hydrolysis of DNA	Haraguchi et al. 2005
RNase	hydrolysis of RNA	Haraguchi et al. 2005
NADPase	hydrolysis of NADP	Tang et al. 1982; Thorne-Tjomsland et al. 1988
CMPase	hydrolysis of monophosphate polysaccharides	Tang et al. 1982; Thorne-Tjomsland et al. 1988 Krimer and Esponda 1980

Many more components have been identified for the individual model organisms, and may later prove conserved. For component lists reflecting component diversity in popular model organisms see: mouse chromatoid body Table 2 in this manuscript; *Drosophila* sponge bodies (Snee and Macdonald 2009); *C. elegans* P granules (Updike and Strome 2009b).

(Wilsch-Brauninger et al. 1997). Sponge bodies appear to derive from fragments of perinuclear nuage that associate with ER membranes, detach from the nucleus, and intermingle in the cytoplasm with other P-body-like granules (see below) (Wilsch-Brauninger et al. 1997; Jaglarz et al. 2011). Sponge bodies accumulate in the oocyte, apparently by movement along ER membranes that extend from the nurse cells through the ring canals into the oocyte (Jaglarz et al. 2011). Sponge bodies movement has been hypothesized to transport perinuclear nuage material from the nurse cells to the germ (pole) plasm of the oocyte (Nakamura et al. 2001). The composition of sponge bodies, however, is dynamic and differs between nurse cells and oocytes, suggesting that sponge bodies are not simply a mobile version of nuage, but represent a range of heterogeneous RNP assemblies (Snee and Macdonald 2004, 2009).

Membrane-rich structures reminiscent of sponge bodies have also been reported in *C. elegans* oocytes (Fig. 2B). During the oocyte growth phase, the perinuclear germ granules detach from the nuclear envelope, apparently taking with them membrane fragments containing clusters of nuclear pores (Pitt et al. 2000). Stressed and aging oocytes assemble even larger cytoplasmic RNP complexes, whose formation has been correlated with dramatic nuclear membrane blebbing (Jud et al. 2008; Patterson et al. 2011). *Drosophila* sponge bodies also change morphology under different environmental conditions (Snee and McDonald 2009), suggesting a role for these membranous structures in protecting oocytes from stress.

 Cite this article as *Cold Spring Harb Perspect Biol* doi: 10.1101/cshperspect.a002774

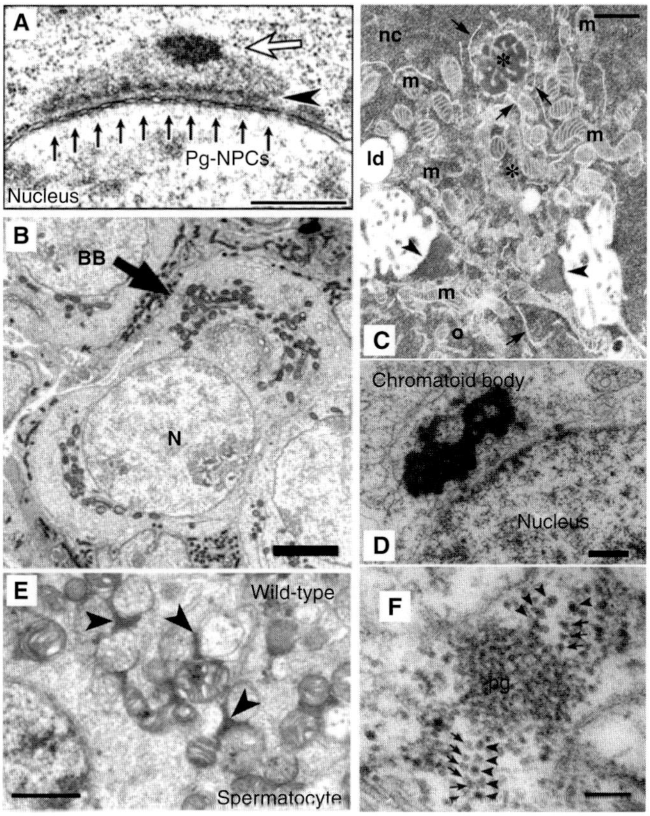

Figure 1. Electron micrographs of the major germ granule classes. (*A*) Perinuclear P granule in a meiotic germ cell of *C. elegans*. Distinct subdomains are visible: the crest (white arrow) and base (arrowhead) overlying a cluster of nuclear pores (black arrows). Scale bar: 500 nm. (Panel *A* is adapted from Sheth et al. [2010] and reproduced, with permission, from The Company of Biologists © 2010.) (*B*) Balbiani body in a mouse oocyte. Black arrow points to mitochondria clustered around Golgi membranes (BB). *N*, nucleus. Scale bar: 5 um. (Panel *B* is adapted from Pepling et al. [2007] and reproduced, with permission, from National Academy of Sciences, USA © 2007.) (*C*) Sponge bodies in the *Drosophila* nurse cell. Sponge bodies (asterisks) intermingled with mitochondria (*m*) and ER cisternae (arrows) near an intercellular bridge (black arrowheads) connecting the nurse cell (*nc*) to the oocyte (*o*). *ld*, lipid droplet. Scale bar: 500 nm. (Panel *C* adapted from Jaglarz et al. [2011] and reproduced, with permission, from Springer © 2011.) (*D*) Perinuclear chromatoid body in a mouse stage I round spermatid. Scale bar: 500 nm. (Panel *D* is adapted from Vasileva et al. [2009] and reproduced here, with permission, from Elsevier © 2009.) (*E*) Intermitochondrial cement in a mouse spermatocyte. IMC (arrowheads) forms in the spaces between clustered mitochondria. Scale bar: 1 um. (Panel *E* is adapted from Chuma et al. [2006] and reproduced, with permission, from National Academy of Sciences, USA © 2006.) (*F*) Polar granules (*pg*) in the cortical cytoplasm of a *Drosophila* embryo. Polysomes (arrows) extend from the surface of the granule. Scale bar: 100 nm. (Panel *F* is adapted from Amikura et al. [2001] and reproduced, with permission, from the Company of Biologists 2001.)

Germ Granules in Germ Plasm

In mature oocytes of animals with "germ plasm," materials transported in the Balbiani and sponge bodies coalesce in a subregion of the oocyte cortex at a position that predicts the site of PGC formation, as seen in *Xenopus* (Czolowska 1969), zebrafish (Knaut et al. 2000), and *Drosophila* (Illmensee and Mahowald 1974). Immediately after fertilization, in *Drosophila*, germ granules appear in small clusters surrounded by ribosomes (Fig. 1F)

(Amikura et al. 2001; Mahowald 2001) near membranes of the endoplasmic reticulum (Thomson et al. 2008). During germ (pole) cell formation, the germ granules are transported on microtubules toward the nuclei of the future pole cells (Lerit and Gavis 2011). In zebrafish embryos, the germ plasm forms on the distal ends of the first and second cleavage furrows, in a complex process involving microtubules, microfilaments, and two distinct RNA segregation pathways (Knaut et al. 2000; Theusch et al. 2006). By the 32-cell stage, the germ plasm aggregates are found in four cells that will give rise to the germline (Knaut et al. 2000).

In *C. elegans*, germ granules detach from nuclei during oogenesis and remain uniformly distributed in the cytoplasm through fertilization. In the zygote, after embryonic polarity is specified, the germ granules become enriched in the posterior cytoplasm and, as a result, are inherited primarily by the posterior daughter, the germline blastomere P_1. This process is repeated for another three asymmetric divisions separating soma from germline. By the time of the birth of the germline founder cells, the germ granules have returned to their perinuclear position (Strome and Wood 1982).

Chromatoid Body

During mouse spermatogenesis (Fig. 2C), germ granules undergo a number of morphological changes (Eddy 1975). Perinuclear nuage and cytoplasmic granules associated with mitochondria ("intermitochondrial cement") are found in early male germ cells (Fig. 1E) (Chuma et al. 2006; reviewed in Chuma et al. 2009). During meiosis, spermatocytes assemble a unique type of germ granule called the chromatoid body (Fig. 1D) (Kotaja and Sassone-Corsi 2007). The chromatoid body arises from intermitochondrial fibers that disperse during the meiotic divisions and consolidate postmeiosis into a single sphere apposed against the nucleus. The site of chromatoid body/nucleus contact is rich in nuclear pores (Fawcett et al. 1970) and labeling of the nucleus with tritiated uridine also labels the chromatoid body, suggesting that RNA synthesized in the nucleus is transferred to the chromatoid body (Soderstrom and Parvinen 1976). Remarkably, the chromatoid body is motile, moving both around the nucleus and into neighboring cells through cytosolic bridges. This movement has been proposed to mediate sharing of RNAs between the haploid spermatids (Ventela et al. 2003). The chromatoid body relocalizes to the base of the flagellum in elongating spermatids (Fig. 2C), and disassembles later during spermiogenesis (Parvinen 2005). No germ granules have been reported in the mature sperm of any organism.

Chromatoid-like bodies have also been described in the neoblasts of planaria; the totipotent stem cells that give rise to both somatic and germ cells (Coward 1974; Hori 1997). Chromatoid bodies are lost when neoblast progeny differentiate into somatic lineages.

P Bodies

Two main classes of cytoplasmic RNA granules have been described in somatic cells: processing bodies and stress granules (Balagopal and Parker 2009). Processing bodies (P bodies) and stress granules share components implicated in RNA storage, turnover, and translational regulation. Each granule type also possesses unique features: P bodies contain components of the mRNA decay machinery, and stress granules contain TIA proteins, poly-A binding protein, and stalled translation initiation complexes (Anderson and Kedersha 2009). Several P body and stress granule components have been detected in germ granules (Tables 1 and 2). This overlap has generated some speculation as to whether germ granules might correspond to P bodies and/or stress granules. Although functional parallels are likely, a direct correspondence remains unclear. So far, stalled initiation complexes have not been reported in germ granules, suggesting that germ granules may not be directly equivalent to stress granules. In contrast, most components of the RNA decay machinery (DCP1, DCP2, CGH-1/Me31B/Rck, CAR-1/Tral/RAP55, and PCM/Xrn1) have been detected in the perinuclear nuage, sponge bodies, germ plasm, and/or chromatoid bodies, suggesting a strong connection to P

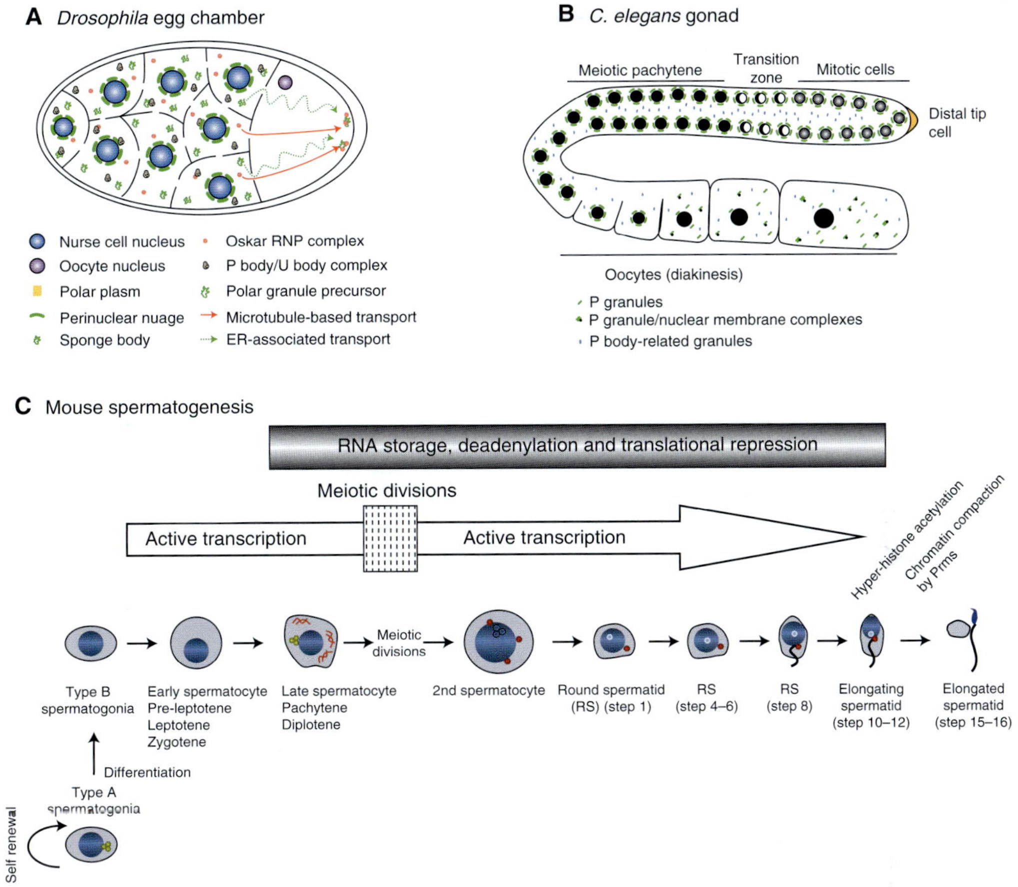

Figure 2. Germ granules during gametogenesis. (*A*) Transport of germ granule material in a *Drosophila* egg chamber. Various germ granule RNPs intermingle in the nurse cells' cytoplasm. Green dashed lines represent migration of sponge bodies, red lines indicate the *oskar* RNP transport pathway. (Panel *A* adapted from Jaglarz et al. [2011] and reproduced, with permission, from Springer © 2011.) (*B*) Germ granules in the germline of an adult *C. elegans* hermaphrodite. Germ cells progress along an "assembly line" of development from mitotic stem cells at the distal end to the oocytes at the proximal end. Germ cells share a common cytoplasm, and are individualized late in oogenesis. P granules (green) are docked at the germ cell nuclei until diplotene, and transition to the cytoplasm along with nuclear pore complexes (green/black). P body-related granules (blue) accumulate in the shared cytoplasm (rachis) and in oocytes. (*C*) Mammalian spermatogenesis. Intermitochondrial cement (yellow dots) is present in type A spermatogonia, and disappears in type B spermatogonia until meiosis. The chromatoid body (CB; red filaments then becoming dots) appears in late pachytene cells as thick cytoplasmic fibers. After the first meiotic division, the CB starts condensing as one or two round foci. Except during meiotic divisions, transcription is active until the elongating spermatid stage, and many transcripts are translationally silenced and located in the CB. In elongated spermatids, transcripts are released from translational arrest.

bodies (Nakamura et al. 2001; Navarro et al. 2001; Audhya et al. 2005; Boag et al. 2005; Lall et al. 2005; Wilhelm et al. 2005; Kotaja et al. 2006a; Lin et al. 2006; Squirrell et al. 2006; Kotaja and Sassone-Corsi 2007; Pepling et al. 2007; Thomson et al. 2008; Aravin et al. 2009; Lim et al. 2009; Yabuta et al. 2011). Careful examination of the distribution of P body proteins in germ cells has revealed, however, that these factors also accumulate in foci that do not contain core germ granule proteins. For example, mature mouse oocytes, which lack

VASA-positive granules, contain foci positive for the decapping enzyme Dcp1a (Swetloff et al. 2009). Mouse spermatocytes also contain canonical P bodies distinct from the germ granules (Aravin et al. 2009; Shoji et al. 2009). In *C. elegans*, several types of P-body related granules have been observed in pachytene germ cells and in maturing oocytes (Navarro et al. 2001; Audhya et al. 2005; Boag et al. 2005; Noble et al. 2008). In embryos, granules containing the decapping activator PATR-1 and other P body proteins, but lacking the germ granule component PGL-1, dock around the germ granules (Audhya et al. 2005; Lall et al. 2005; Gallo et al. 2008). Unlike germ granules, the PATR-1 granules are partitioned symmetrically during mitosis and recruit additional P body proteins after segregation into somatic blastomeres (Gallo et al. 2008). Similarly, in the sponge bodies of *Drosophila* nurse cells, structures resembling the P body/U body complexes of somatic cells (Liu and Gall 2007) can be distinguished from nuage (Lim et al. 2009; Jaglarz et al. 2011). Together, these observations suggest that germ cells contain bona fide P-bodies, or P body-related granules, which interact and may even exchange components with germ granules, but are maintained as distinct RNP assemblies. Whether germ granules represent specialized P-bodies, P-body/stress granule hybrids, or a distinct granule type unique to the germline, remains to be determined.

DYNAMICS

In electron micrographs, germ granules appear as dense, fibrillar structures that in some cases appear to have distinct subcompartments (Fig. 1A,D) (Fawcett et al. 1970; Sheth et al. 2010). In live recordings, perinuclear germ granules in *Drosophila* nurse cells (nuage) and in *C. elegans* meiotic cells (P granules) have been observed to maintain their shape and relative position for up to 40 minutes, the maximum recording time (Snee and Macdonald 2004; Sheth et al. 2010). Despite this apparent stability, experiments using fluorescence-recovery after photobleaching (FRAP) have revealed that individual components are in fact highly dynamic. GFP::Vasa is replaced in the *Drosophila* nuage with a half-time of 60 sec (Snee and Macdonald 2004). Similarly, GFP::PGL-1 diffuses within seconds inside the perinuclear granules of *C. elegans* (Brangwynne et al. 2009), and whole perinuclear clusters recover from photobleaching with a half-time of 20 sec (Sheth et al. 2010). Whether germ granules also contain stable components that form static scaffolds is not known.

Under some circumstances, whole granules can rapidly change size or position. During the asymmetric cleavages of the *C. elegans* embryo, germ granules in cytoplasm destined for the germline grow and fuse into larger granules, whereas germ granules in cytoplasm destined for the soma shrink in size until no longer visible (Brangwynne et al. 2009). These dynamics have been compared to those of liquid droplets undergoing condensation and dissolution (Brangwynne et al. 2009). In zygotes, a subset of germ granules have also been observed to be move directionally over several microns (Hird et al. 1996). This movement is microtubule-independent (Hill and Strome 1990) and may be mediated by cytoplasmic flows that develop in response to actomyosin contractions in the zygote's cortex (Marston and Goldstein 2006).

Microtubules have been implicated in germ granules dynamics in *Drosophila* and zebrafish. During *Drosophila* oogenesis, transport of *oskar* RNP particles from nurse cell to the posterior pole occurs along microtubules in kinesin-dependent manner (reviewed in Becalska and Gavis 2009). In *Drosophila* embryos, translocation of VASA-containing granules from the cortex to nuclei occurs along astral microtubules in a process that requires dynein (Lerit and Gavis 2011). In zebrafish, microtubules promote germ granule aggregation in early embryos (Theusch et al. 2006), and have been implicated in germ granule partitioning during mitosis in primordial germ cells (Strasser et al. 2008). Disruption of microtubules or dynein causes the germ granules to coalesce into larger granules, suggesting a role in regulating both granule dynamics and size (Strasser et al. 2008).

In mouse spermatids, the chromatoid body is highly mobile, tracking around the nucleus and shuttling from cell to cell via cytoplasmic bridges (Fawcett et al. 1970; Ventela et al. 2003). Microtubule-depolymerizing drugs disrupt these movements and cause the chromatoid body to disperse (Ventela et al. 2003). The testis-specific kinesin KIF17b is a possible candidate for mediating chromatoid body movement. KIF17b was first identified as a nucleus-cytoplasm shuttling protein that transports ACT (activator of CREM in testis), a transcriptional coregulator of postmeiotic transcription (Fimia et al. 1999). KIF17b also contributes to the transport of TB-RBP, Testis-brain RNA-binding protein, which associates with CREM regulated mRNAs (Chennathukuzhi et al. 2003a,b). Importantly, KIF17b is also found in the chromatoid body (Kotaja et al. 2006b), where it may participate in RNA transport and/or mediate chromatoid body movement.

ASSEMBLY OF GERM GRANULES

Unlike other cellular organelles, germ granules are not bound by membranes, yet maintain distinct boundaries and shapes. What molecular interactions underlie germ granule form and integrity? Although no single assembly pathway accounts for the wide range of germ granule morphologies, several have been documented.

Nucleation by Seed Proteins

Genetic analyses in *Drosophila* have led to a hierarchical model for the organization of germ plasm: a single protein, Oskar, is necessary and sufficient to initiate a cascade of interactions that assemble germ plasm at the site of Oskar localization (Mahowald 2001). Oskar exists in two forms: a short form required for germ granule assembly (Markussen et al. 1995), and a long form required to anchor the granules to the cell cortex (Vanzo and Ephrussi 2002). Short Oskar initiates granule formation likely through direct interactions with Vasa, which in turn recruits Tudor (Breitwieser et al. 1996; Anne 2010), Valois, and Piwi-family Argonautes (Kirino et al. 2009; Nishida et al. 2009). Germ granule mRNAs (i.e., *nanos*, *germ-cell-less*, and *polar granule component*) are synthesized in the nurse cells and transferred to the oocyte during nurse cell cytoplasm "dumping" (reviewed in Becalska and Gavis 2009). Localization to the germ plasm depends on sequences in the 3′UTR and an actin-dependent, cortex-anchoring mechanism initiated by Oskar (Becalska and Gavis 2009; Rangan et al. 2009). How the long form of Oskar anchors the germ plasm to the cortex is not fully understood, but it involves recruitment of endosomal proteins and stimulation of endocytosis, which eventually reorganizes F-actin in the oocyte cortex (Tanaka and Nakamura 2008).

Oskar homologs have not been found outside of insects, but proteins with similar germ plasm "seed" functions have been identified in zebrafish and *C. elegans*. In zebrafish, the Balbiani body component Bucky Ball is necessary and sufficient to induce germ plasm aggregation (Bontems et al. 2009). Bucky ball is conserved among vertebrates but does not possess recognizable functional domains; its mode of action remains unknown (Bontems et al. 2009). In *C. elegans*, the RGG-domain proteins PGL-1 and PGL-3 are required for germ granule integrity in embryos, and can form cytoplasmic granules when ectopically expressed in tissue culture cells or in intestinal cells (Hanazawa et al. 2011; Updike et al. 2011). Ectopic PGL granules recruit RNA and other germ granule components. Granule formation depends on a domain in the PGL proteins that mediates self-association, whereas recruitment of RNA and RNA-binding proteins requires the RGG domain (Hanazawa et al. 2011). PGL-1 and PGL-3 also form transient granules in the somatic blastomeres of early embryos, but in those cells, granule formation depends on the autophagy machinery, which ultimately targets the PGL granules for degradation (Zhang et al. 2009).

Nucleation by Organelles

Germ granules often form near mitochondria, raising the possibility that nucleating factors also exist on the membranes of these organelles. Two recent reports implicate such mechanism in generation of the intermitochondrial cement

(IMC) in mouse spermatocytes. Mutants lacking the mitochondrial phospholipase D MitoPLD do not form IMC and mislocalize IMC components (Huang et al. 2011; Watanabe et al. 2011). MitoPLD generates the signaling lipid phosphatidic acid. Remarkably, a deficiency in the phosphatidic acid-metabolizing enzyme, Lipin 1 causes hyper-aggregation of electron-dense nuage around mitochondria (Huang et al. 2011). Phosphatidic acid may therefore function as the organizing signal for IMC.

Binding of Tudor Domains to Methylated Arginines

The Tudor domain is a small (~55 amino acid) protein–protein interaction module that recognizes methylated arginines or lysines. Tudor domains have been found in several proteins localized to perinuclear nuage, germ plasm, and/or chromatoid bodies (see Table 1 for references). The founding member of the family, *Drosophila* Tudor, has 11 Tudor domains, but Tudor domains can also be found in single copy or combined with other conserved motifs, such as in SpnE/TDRD9 (reviewed in Arkov and Ramos 2010).

The Tudor domain forms an aromatic cage that interacts directly with the methylated amino acid (reviewed in Arkov and Ramos 2010; Vourekas et al. 2010). Crystallographic studies of Tudor proteins complexed with methyl-arginine-containing peptides show that the Tudor module is embedded in a larger 170 amino-acid domain that contacts the peptide at other sites (Liu et al. 2010; Vourekas et al. 2010). These additional interactions may help Tudor-containing proteins discriminate among potential partners.

Piwi-family Argonautes, Vasa helicases and Sm proteins in *Drosophila*, *Xenopus*, and mouse all contain methylated arginines and thus are potential targets for Tudor domain-containing proteins (Kuramochi-Miyagawa et al. 2004; Gonsalvez et al. 2006; Thomson et al. 2008; Chen et al. 2009; Kirino et al. 2009; Kojima et al. 2009; Nishida et al. 2009; Reuter et al. 2009; Shoji et al. 2009; Vagin et al. 2009a,b; Wang et al. 2009; Kirino et al. 2010a,b; Patil and Kai 2010; Liu et al. 2011). Symmetrical dimethylation of arginines (sDMA) is performed in flies and mice by the methyltransferase Capsuleen/PRMT and its cofactor Valois/MEP50/WDR77 (Gonsalvez et al. 2006; Anne et al. 2007). In *Drosophila*, sDMA modifications of the Argonaute Aubergine are recognized by Tudor, and are necessary for Aubergine-Tudor interaction and germ granule assembly (Anne et al. 2007; Kirino et al. 2009; Nishida et al. 2009). sDMA modifications of the spliceosomal Sm proteins SmB and SmD3 also depend on Capsuleen/Valois, and are required for their localization to germ granules; however, the relevant Tudor-domain partners of these proteins have not been identified (Anne 2010; Gonsalvez et al. 2010). In the mouse, several Tudor domain proteins interact with the sDMA-modified Piwi family Argonautes and are required for spermatogenesis (reviewed in Arkov and Ramos 2010; Vourekas et al. 2010). Lack of interactions causes both the Tudor domain proteins and Piwi-family Argonautes to become mislocalized, suggesting a mutual dependency for localization.

Low Specificity Interactions

Tudor-sDMA interactions have the potential to create high affinity/specificity complexes that are stable and long-lived. A different type of interaction, however, may be needed to bring these complexes into higher-order dynamic structures. Studies in *C. elegans* has led to the proposal that germ granules show liquid-like behaviors (Brangwynne et al. 2009). In response to pressure, germ granules in the nuage of meiotic germ cells appeared to "drip" off nuclei and fuse into large drops; germ granules also "condense" and "dissolve" during the asymmetric divisions of the embryo (Brangwynne et al. 2009). Liquid behaviors are thought to be mediated by weak, fast-rearranging molecular interactions, as could arise from low-affinity protein–RNA interactions. Consistent with a role for RNA in maintaining germ granule integrity, depletion of the VegT and Xlsirts RNAs disrupts the organization of germ granules in the vegetal cortex of *Xenopus* oocytes

 Cite this article as *Cold Spring Harb Perspect Biol* doi: 10.1101/cshperspect.a002774

(Kloc and Etkin 1994; Kloc et al. 2005, 2007). Inhibition of transcription causes loss of PGL-1 and GLH-2 from the *C. elegans* nuage (Sheth et al. 2010) and loss of the electron-dense core of the mouse chromatoid body (Soderstrom 1977). Mutations in genes implicated in the biogenesis of small RNAs (*ego-1*, *csr-1*, *drh-3*) cause a dramatic increase in the size and number of germ granules in meiotic germ cells (Updike and Strome 2009a).

Other studies in *C. elegans* have implicated weak hydrophobic interactions involving phenylalanine-glycine (FG)-repeats. FG-repeats are naturally unfolded and flexible, and thus well suited for dynamic binding to multiple partners, as in the liquid droplet model above. FG-repeats are found in nucleoporins that mediate active transport across the nuclear pore and prevent passive diffusion of molecules greater than >30 kD (Elad et al. 2009; Terry and Wente 2009). Perinuclear germ granules exclude dextrans larger than 40 kDa and are sensitive to treatments predicted to disrupt FG-repeat interactions (Updike et al. 2011). Consistent with a role for FG-repeats, three *C. elegans* VASA family members, GLH-1, 2, and 4, contain FG repeats (Kuznicki et al. 2000), and the FG-containing nucleoporin Nup98 and mRNA export factor DDX-19 colocalize with germ granules (Sheth et al. 2010; Voronina and Seydoux 2010). RNAi screens have also identified several nucleoporins required for germ granule integrity in embryos (Updike and Strome 2009a; Voronina and Seydoux 2010) and for the assembly of sponge body-like aggregates in stressed oocytes (Patterson et al. 2011).

Recently, FG-repeat proteins have been recognized as a subtype of prions (Ader et al. 2010). Remarkably, overexpression in mouse neuroblastoma N2a cells of the cytosolic form of the prion PrP causes the formation of a large RNP aggregate near the cell nucleus, resembling a chromatoid body (Beaudoin et al. 2009). The "PrP body" associates with mitochondria, and is able to recruit mRNAs, FG-repeat nucleoporins, Dicer, Dcp1a, DDX6, and SmB proteins reminiscent of germ granules (Beaudoin et al. 2009; Roucou 2009). Nucleation of stress granules in somatic cells also depends on a prion-domain containing protein, TIA-1 (Gilks et al. 2004). An intriguing possibility is that prion-like templating properties give germ granules not only their unique structural characteristics, but also the ability to propagate from germ cell to germ cell, and from generation to generation in animals with germ plasm.

HYPOTHESES FOR FUNCTION OF GERM GRANULES

Germ Cell Fate Specification and Differentiation

The first proposed function for germ granules was to support the specification of the germline in embryos where germ granules are inherited maternally as a part of the germ plasm. In *Xenopus* and *Drosophila* embryos, irradiation of a specific region of the cytoplasm (germ plasm) results in embryos that lack germ cells (Geigy 1931; Bounoure 1937; Smith 1966). In *Drosophila*, transplantation of germ plasm to an ectopic site is sufficient to reprogram cells to become germ cells (Illmensee and Mahowald 1974). In fact, ectopic localization of the germ plasm component Oskar is sufficient to induce ectopic germ granules (see above) and form ectopic germ cells (Ephrussi and Lehmann 1992). Similarly, overexpression of the Balbiani body protein Bucky ball in zebrafish embryos induces extra germ cells (Bontems et al. 2009). These experiments suggest that germ plasm, and perhaps the granule granules therein, contain the factors that specify germ cell fate.

In *Drosophila*, mothers with mutations in *vasa* generate embryos that lack germ cells (Schupbach and Wieschaus 1986; Lasko and Ashburner 1990; Styhler et al. 1998). In *Xenopus* embryos, morpholino treatments that reduce Vasa protein levels decrease germ cell numbers (Ikenishi and Tanaka 1997; Ikenishi et al. 2006). Genetic analyses have identified several other germ granule proteins like Vasa that are required for germ cell specification, but none (besides Oskar) is sufficient to induce germ cell fate on its own (Rongo and Lehmann 1996; Harris and Macdonald 2001; Megosh in Ollinger et al. 2010). Transposition could cause meiotic defects by inducing DNA damage, interfering with homolog recognition, and/or cytotoxicity of retrotransposon-encoded proteins. Another possibility, however, is that the piRNA pathway regulates transposons and meiosis in parallel.

Mitosis and Chromatin Organization

The 22G-RNAs of *C. elegans* target not only repetitive elements, but also unique sequences, including many annotated genes (Gu et al. 2009). 22G-RNAs targeting unique genes associate preferentially with the Argonaute CSR-1 (Claycomb et al. 2009). CSR-1 is a component of the nuage in which it functions with the Dicer-related helicase DRH-3, the RNA-dependent RNA polymerase EGO-1, and the Tudor domain protein EKL-1 to generate 22G-RNAs. Surprisingly, mutations in *csr-1* do not affect the expression of most genes targeted by 22G-RNAs, but cause dramatic defects in chromosome organization during meiosis and mitosis (Claycomb et al. 2009). In embryos, CSR-1 and its partners associate with mitotic chromosomes and are required for the proper distribution of kinetochore proteins (Claycomb et al. 2009). These observation suggest that CSR-1 could function in the delineation of centromeres, as has been shown for Ago1 in *Schizosaccharomyces pombe* (Verdel et al. 2004).

A role for nuage components in chromosome dynamics has also been reported in *Drosophila* (Pek and Kai 2011a). *vasa* mutant germ cells display lagging chromosomes at anaphase that have reduced levels of type I condensins. Vasa coimmunoprecipitates with type I condensins and localizes to perichromosomal foci during mitosis. *aubergine* and *spnE* mutants lack Vasa-positive perichromosomal foci and show mitotic defects similar to those seen in the *vasa* mutants (Pek and Kai 2011b). These findings suggest that small RNAs produced in the nuage also regulate chromatin compaction during mitosis (Pek and Kai 2011b). Interestingly, in the sea urchin, Vasa is also required for chromosome segregation but this function does not appear to be specific to the germline: Vasa associates with mitotic spindles in all early embryonic blastomeres (Yajima and Wessel 2011). These observations suggest that Vasa may have an ancestral role in mitosis, distinct from its role in germ cell development.

Regulation of mRNAs

As germ granules are enriched in RNA-binding proteins and developmentally regulated mRNAs, they have been hypothesized to function as hubs of posttranscriptional regulation; specifically to regulate mRNA localization, levels, and translational activity (Kotaja and Sassone-Corsi 2007; Rangan et al. 2009).

mRNA Localization

Germ granules have been implicated in RNA localization during oogenesis in several organisms (for details see reviews by Kloc et al. 2004; Zhou and King 2004; Becalska and Gavis 2009). Localized mRNAs associate with Balbiani and sponge bodies in the oocytes of *Drosophila*, *Xenopus*, and zebrafish (Ephrussi et al. 1991; Kim-Ha et al. 1991; Wang and Lehmann 1991; Kloc et al. 2002; Cox and Spradling 2003; Wilk et al. 2005; Kosaka et al. 2007). The Balbiani body protein Bucky Ball is required for the Balbiani body aggregation and localization of several mRNAs to the vegetal cortex of zebrafish eggs (Dosch et al. 2004; Marlow and Mullins 2008; Bontems et al. 2009). Several RNA-binding proteins associated with sponge bodies are required for the directed transport of *oskar* to the posterior pole, including Exuperantia, Yps, Rap55/Trailer Hitch, and the hnRNP Hrp48 (Wilhelm et al. 2000, 2005; Yano et al. 2004). *nanos* localization to the germ plasm, in contrast, appears to depend on the trapping of a minority of mRNA molecules in the germ (polar) granules (Wang et al. 1994; Rongo et al. 1997). The Argonaute Aubergine is a candidate tether for *nanos* in the polar granules (Becalska et al. 2011). Aub is in a complex with *nanos* RNA, and *aub*/+ embryos show defects in *nanos* RNA localization but no defect in Oskar localization. Remarkably Aubergine's function in *nanos* localization appears to be

 Cite this article as *Cold Spring Harb Perspect Biol* doi: 10.1101/cshperspect.a002774

independent of its role in the piRNA pathway (Becalska et al. 2011).

mRNA Translation

Germ granules have been implicated in both the repression and activation of translation. Granules implicated in RNA transport such as in Balbiani and sponge bodies typically contain mRNAs that are translationally repressed and become activated after release from the granules or after transition to another granule type.

Several mechanisms of translation repression have been described for germ granule associated mRNAs. In *Drosophila* nurse cells, repressed mRNAs are often found in sponge bodies (Nakamura et al. 2001). Repression of *oskar* mRNA in sponge bodies depends on the sequence-specific translational repressor Bruno (Kim-Ha et al. 1995). Bruno binds to Bruno-response elements in the *oskar* 3′UTR and promotes the formation of large (up to 80S) *oskar* RNA-containing complexes that exclude ribosomes (Chekulaeva et al. 2006). Bruno additionally represses *oskar* translation through interactions with Cup. Cup is a component of sponge bodies that forms a complex with the translation initiation factor eIF4E and the DEAD-box helicase Dhh1/Me31B (Nakamura et al. 2001, 2004; Wilhelm et al. 2003). eIF4E binding proteins (4E-BPs) such as Cup are thought to modulate translational initiation by blocking the interaction between eIF4E and eIF4G in a reversible manner (Gingras et al. 1998). Mutations in Cup that disrupt binding to eIF4E cause premature translation of *oskar* mRNA (Nakamura et al. 2004). A similar mechanism has been proposed for the *C. elegans* germ granule component SPN-2 (Li et al. 2009b). SPN-2 is a 4E-BP protein that is recruited to the *mei-1* mRNA by interaction with the sequence-specific RNA-binding proteins OMA-1 and OMA-2 (Li et al. 2009b).

During *C. elegans* oogenesis, translationally-repressed mRNAs are targeted to P-body like granules (grP bodies), that also contain the sequence-specific RNA-binding protein and translational repressor PUF-5 (Noble et al. 2008). These granules become larger in arrested oocytes and may correspond to the membrane-associated, sponge body-like structures characterized by electron microscopy (Jud et al. 2008; Patterson et al. 2011). Loss of the P-body protein CAR1/Tral/RAP55 disrupts grP bodies, and causes ectopic translation of *glp-1* mRNA in oocytes, as is observed upon loss of PUF-5 and related PUFs (Noble et al. 2008).

Unlike the repressive sponge bodies, germ granules in the *Drosophila* germ (pole) plasm have been implicated in translational activation. For example, *nanos* mRNA in the cytoplasm is maintained translationally repressed by mechanisms that block translation both before and after initiation (Andrews et al. 2011; Jeske et al. 2011). Only mRNA localized with the germ (polar) granules is translated. Electron microscopy studies have shown that polyribosome chains extend from the surface of germ granules (Mahowald 1968, 1971; Amikura et al. 2001), suggesting that the granules could deliver mRNAs directly to the translational machinery. Translational activation of mRNAs in the germ plasm occurs in discrete temporal patterns dictated by specific sequences in the 3'UTR (Rangan et al. 2009), and may involve disassembly of silencing complexes. Translation of *nanos* in the germ plasm requires Oskar, which binds directly to the repressor Smaug and is sufficient in vitro to prevent assembly of a Smaug-dependent silencing complex on the *nanos* 3′UTR (Ephrussi and Lehmann 1992; Zaessinger et al. 2006; Jeske et al. 2011). Another key player is likely to be Vasa. Vasa binds to the initiation factor eIF5B and this interaction is required for germ cell formation in embryos (Carrera et al. 2000; Johnstone and Lasko 2004). Several germ plasm mRNAs bind to Vasa (Liu et al. 2009). Translation of at least one of them, *mei-p26*, is dependent on Vasa, and requires the Vasa-eIF5B interaction (Liu et al. 2009).

In *Xenopus* and mouse, translational activation has been correlated with exit from germ granules. In *Xenopus*, *Xcat2* mRNA is maintained in germ granules in the Balbiani body and in the germ plasm through oogenesis and early embryogenesis (Kloc et al. 2002). Release of *Xcat2* from germ granules correlates with its translational derepression (Kloc et al. 2002). Similarly, in mouse spermatids, translation of *Brd2* and *GCNF* correlates with their exit from the chromatoid body (Nguyen Chi et al. 2009). When in the chromatoid body, the mRNAs cofractionate with Vasa (MVH) in ribonucleoprotein complexes away from polysomes (Grivna et al. 2006; Nguyen Chi et al. 2009). In stage V and older spermatids, mRNAs exit the chromatoid body, associate with polysomes, and become translated (Nguyen Chi et al. 2009). Remarkably, at that time, MVH also becomes associated with polysomes, consistent with a possible role in translational activation (Nguyen Chi et al. 2009).

mRNA Stability

Germ granules have also been implicated in the regulation of mRNA levels, by promoting either mRNA storage or mRNA turnover at specific developmental stages. In the mouse, genes required for spermiogenesis are transcribed several days before the corresponding proteins are needed (Kleene 1993). The untranslated transcripts are stored stably without degradation for up to a week. Some of these transcripts localize to chromatoid body, suggesting that this organelle provides a stable storage site for mRNAs (Soderstrom and Parvinen 1976; Kotaja et al. 2006a; Nguyen Chi et al. 2009). A role for germ granules in mRNA storage has also been proposed in *C. elegans*, based on the function of the DEAD-box RNA helicase Dhh1/CGH-1 (Boag et al. 2008; Noble et al. 2008). In somatic cells, Dhh1/CGH-1 functions with PATR-1 to promote mRNA decapping and degradation in P bodies. In contrast, in germ cells, Dhh1/CGH-1 functions independently from PATR-1, and associates with ~300 mRNAs that are stably maintained during oogenesis and passed on to the embryo (Boag et al. 2008). Loss of Dhh1/CGH-1 causes destabilization of many of these mRNAs and sterility (Boag et al. 2008; Noble et al. 2008).

Other germ granule proteins in contrast have been implicated in mRNA turnover. The *C. elegans* Argonaute family members ALG-3 and ALG-4 generate 26G-RNAs that are required for meiosis and spermiogenesis, especially at elevated temperatures (Conine et al. 2010). Most 26G-RNAs target mRNAs are expressed during spermatogenesis, and 27% of stringently defined targets are up-regulated by twofold or more in *alg-3/4* mutant worms (Conine et al. 2010). Similarly, the RNA-dependent RNA polymerase EGO-1 is required to produce a population of small triphosphorylated antisense RNAs that target hundreds of germline transcripts expressed at different stages of germline development (Maniar and Fire 2011). In *ego-1* mutants, the small RNAs are not made and their target mRNAs are moderately overexpressed (Maniar and Fire 2011). One possibility is that *alg-3/4* and *ego-1* provide temporal fine-tuning of gene expression during germline development by targeting specific mRNAs for degradation.

CONCLUDING REMARKS

Germ granules are remarkable not only for their morphological diversity but also for the many functions they perform during germ cell development. This complexity can be dizzying, but the studies summarized here suggest two important principles and questions for future investigations.

First, many germ granule proteins have been shown to play fundamental roles in RNA regulation. Control of gene expression at the RNA level is common in germ cells, and may be essential to preserve the plasticity of the germline genome (Seydoux and Braun 2006). In fact, Vasa, PIWI, and Nanos homologs have been proposed to form a "conserved multipotency program" used not only by germ cells, but also by multipotent progenitors that retain the ability to generate both somatic and germline tissues (Juliano et al. 2010). An important challenge will be to determine whether specific steps of RNA regulation occur within germ granules, or whether germ granules function primarily as sorting stations where RNAs and regulators meet up before transitioning to the cytoplasm.

Second, unlike other organelles, germ granules are maintained in the cytoplasm without the help of surrounding membranes. The molecular interactions that give germ granules their unique combination of macro-level stability and micro-level dynamics are only beginning to be elucidated. An understanding of the unique microenvironment of germ granules is likely to yield fundamental new insights into the physical principles that organize the flow of genetic information in cells.

ACKNOWLEDGMENTS

We gratefully acknowledge support from NIH grants 5F32GM080923 (E.V.), HD37047 (G.S.), GM081634 (P.S-C.), the Howard Hughes Medical Institute (G.S. and E.V.), and the California Institute for Regenerative Medicine (I.N.).

REFERENCES

Ader C, Frey S, Maas W, Schmidt HB, Gorlich D, Baldus M. 2010. Amyloid-like interactions within nucleoporin FG hydrogels. *Proc Natl Acad Sci* **107:** 6281–6285.

Amikura R, Kashikawa M, Nakamura A, Kobayashi S. 2001. Presence of mitochondria-type ribosomes outside mitochondria in germ plasm of *Drosophila* embryos. *Proc Natl Acad Sci* **98:** 9133–9138.

Amiri A, Keiper BD, Kawasaki I, Fan Y, Kohara Y, Rhoads RE, Strome S. 2001. An isoform of eIF4E is a component of germ granules and is required for spermatogenesis in *C. elegans*. *Development* **128:** 3899–3912.

Anderson P, Kedersha N. 2009. RNA granules: Post-transcriptional and epigenetic modulators of gene expression. *Nat Rev Mol Cell Biol* **10:** 430–436.

Andonov MD, Chaldakov GN. 1989. Morphological evidence for calcium storage in the chromatoid body of rat spermatids. *Experientia* **45:** 377–378.

Andrews S, Snowflack DR, Clark IE, Gavis ER. 2011. Multiple mechanisms collaborate to repress nanos translation in the *Drosophila* ovary and embryo. *RNA* **17:** 967–977.

Anne J. 2010. Targeting and anchoring Tudor in the pole plasm of the *Drosophila* oocyte. *PLoS One* **5:** e14362.

Anne J, Mechler BM. 2005. Valois, a component of the nuage and pole plasm, is involved in assembly of these structures, and binds to Tudor and the methyltransferase Capsuleen. *Development* **132:** 2167–2177.

Anne J, Ollo R, Ephrussi A, Mechler BM. 2007. Arginine methyltransferase Capsuleen is essential for methylation of spliceosomal Sm proteins and germ cell formation in *Drosophila*. *Development* **134:** 137–146.

Anton E. 1983. Association of Golgi vesicles containing acid phosphatase with the chromatoid body of rat spermatids. *Experientia* **39:** 393–394.

Aravin A, Gaidatzis D, Pfeffer S, Lagos-Quintana M, Landgraf P, Iovino N, Morris P, Brownstein MJ, Kuramochi-Miyagawa S, Nakano T, et al. 2006. A novel class of small RNAs bind to MILI protein in mouse testes. *Nature* **442:** 203–207.

Aravin AA, Sachidanandam R, Girard A, Fejes-Toth K, Hannon GJ. 2007. Developmentally regulated piRNA clusters implicate MILI in transposon control. *Science* **316:** 744–747.

Aravin AA, Sachidanandam R, Bourc'his D, Schaefer C, Pezic D, Toth KF, Bestor T, Hannon GJ. 2008. A piRNA pathway primed by individual transposons is linked to de novo DNA methylation in mice. *Mol Cell* **31:** 785–799.

Aravin AA, van der Heijden GW, Castaneda J, Vagin VV, Hannon GJ, Bortvin A. 2009. Cytoplasmic compartmentalization of the fetal piRNA pathway in mice. *PLoS Genet* **5:** e1000764.

Arkov AL, Ramos A. 2010. Building RNA-protein granules: Insight from the germline. *Trends Cell Biol* **20:** 482–490.

Arkov AL, Wand JY, Ramos A, Lehmann R. 2006. The role of Tudor domains in germline development and polar granule architecture. *Development* **133:** 4053–4062.

Audhya A, Hyndman F, McLeod IX, Maddox AS, Yates JR III, Desai A, Oegema K. 2005. A complex containing the Sm protein CAR-1 and the RNA helicase CGH-1 is required for embryonic cytokinesis in *Caenorhabditis elegans*. *J Cell Biol* **171:** 267–279.

Balagopal V, Parker R. 2009. Polysomes, P bodies and stress granules: States and fates of eukaryotic mRNAs. *Curr Opin Cell Biol* **21:** 403–408.

Barbee SA, Lublin AL, Evans TC. 2002. A novel function for the Sm proteins in germ granule localization during *C. elegans* embryogenesis. *Curr Biol* **12:** 1502–1506.

Bardsley A, McDonald K, Boswell RE. 1993. Distribution of tudor protein in the *Drosophila* embryo suggests separation of functions based on site of localization. *Development* **119:** 207–219.

Batista PJ, Ruby JG, Claycomb JM, Chiang R, Fahlgren N, Kasschau KD, Chaves DA, Gu W, Vasale JJ, Duan S, et al. 2008. PRG-1 and 21U-RNAs interact to form the piRNA complex required for fertility in *C. elegans*. *Mol Cell* **31:** 67–78.

Beaudoin S, Vanderperre B, Grenier C, Tremblay I, Leduc F, Roucou X. 2009. A large ribonucleoprotein particle induced by cytoplasmic PrP shares striking similarities with the chromatoid body, an RNA granule predicted to function in posttranscriptional gene regulation. *Biochim Biophys Acta* **1793:** 335–345.

Becalska AN, Gavis ER. 2009. Lighting up mRNA localization in *Drosophila* oogenesis. *Development* **136:** 2493–2503.

Becalska AN, Kim YR, Belletier NG, Lerit DA, Sinsimer KS, Gavis ER. 2011. Aubergine is a component of a nanos mRNA localization complex. *Dev Biol* **349:** 46–52.

Beshore EL, McEwen TJ, Jud MC, Marshall JK, Schisa JA, Bennett KL. 2011. *C. elegans* Dicer interacts with the P-granule component GLH-1 and both regulate germline RNPs. *Dev Biol* **350:** 370–381.

Biggiogera M, Fakan S, Leser G, Martin TE, Gordon J. 1990. Immunoelectron microscopical visualization of ribonucleoproteins in the chromatoid body of mouse spermatids. *Mol Reprod Dev* **26:** 150–158.

Bilinski SM, Jaglarz MK, Szymanska B, Etkin LD, Kloc M. 2004. Sm proteins, the constituents of the spliceosome, are components of nuage and mitochondrial cement in *Xenopus* oocytes. *Exp Cell Res* **299:** 171–178.

Boag PR, Nakamura A, Blackwell TK. 2005. A conserved RNA-protein complex component involved in physiological germline apoptosis regulation in *C. elegans*. *Development* **132:** 4975–4986.

Boag PR, Atalay A, Robida S, Reinke V, Blackwell TK. 2008. Protection of specific maternal messenger RNAs by the P body protein CGH-1 (Dhh1/RCK) during Caenorhabditis elegans oogenesis. *J Cell Biol* **182:** 543–557.

Bontems F, Stein A, Marlow F, Lyautey J, Gupta T, Mullins MC, Dosch R. 2009. Bucky ball organizes germ plasm assembly in zebrafish. *Curr Biol* **19:** 414–422.

Bounoure L. 1937. Le sort de la lignee germinale chez la Grenouille rousse apres l'action des rayons ultraviolets sur le pole inferieur de l'oeuf. *C r hebd Acad Sci* **204:** 1837.

Brangwynne CP, Eckmann CR, Courson DS, Rybarska A, Hoege C, Gharakhani J, Julicher F, Hyman AA. 2009. Germline P granules are liquid droplets that localize by controlled dissolution/condensation. *Science* **324:** 1729–1732.

Breitwieser W, Markussen FH, Horstmann H, Ephrussi A. 1996. Oskar protein interaction with Vasa represents an essential step in polar granule assembly. *Genes Dev* **10:** 2179–2188.

Brennecke J, Aravin AA, Stark A, Dus M, Kellis M, Sachidanandam R, Hannon GJ. 2007. Discrete small RNA-generating loci as master regulators of transposon activity in Drosophila. *Cell* **128:** 1089–1103.

Brennecke J, Malone CD, Aravin AA, Sachidanandam R, Stark A, Hannon GJ. 2008. An epigenetic role for maternally inherited piRNAs in transposon silencing. *Science* **322:** 1387–1392.

Carmell MA, Girard A, van de Kant HJ, Bourc'his D, Bestor TH, de Rooij DG, Hannon GJ. 2007. MIWI2 is essential for spermatogenesis and repression of transposons in the mouse male germline. *Dev Cell* **12:** 503–514.

Carrera P, Johnstone O, Nakamura A, Casanova J, Jackle H, Lasko P. 2000. VASA mediates translation through interaction with a *Drosophila* yIF2 homolog. *Mol Cell* **5:** 181–187.

Castrillon DH, Quade BJ, Wang TY, Quigley C, Crum CP. 2000. The human VASA gene is specifically expressed in the germ cell lineage. *Proc Natl Acad Sci* **97:** 9585–9590.

Chang CC, Lee WC, Cook CE, Lin GW, Chang T. 2006. Germ-plasm specification and germline development in the parthenogenetic pea aphid *Acyrthosiphon pisum*: Vasa and Nanos as markers. *Int J Dev Biol* **50:** 413–421.

Chekulaeva M, Hentze MW, Ephrussi A. 2006. Bruno acts as a dual repressor of *oskar* translation, promoting mRNA oligomerization and formation of silencing particles. *Cell* **124:** 521–533.

Chen C, Jin J, James DA, Adams-Cioaba MA, Park JG, Guo Y, Tenaglia E, Xu C, Gish G, Min J, et al. 2009. Mouse Piwi interactome identifies binding mechanism of Tdrkh Tudor domain to arginine methylated Miwi. *Proc Natl Acad Sci* **106:** 20336–20341.

Chennathukuzhi V, Morales CR, El-Alfy M, Hecht NB. 2003a. The kinesin KIF17b and RNA-binding protein TB-RBP transport specific cAMP-responsive element modulator-regulated mRNAs in male germ cells. *Proc Natl Acad Sci* **100:** 15566–15571.

Chennathukuzhi V, Stein JM, Abel T, Donlon S, Yang S, Miller JP, Allman DM, Simmons RA, Hecht NB. 2003b. Mice deficient for testis-brain RNA-binding protein exhibit a coordinate loss of TRAX, reduced fertility, altered gene expression in the brain, and behavioral changes. *Mol Cell Biol* **23:** 6419–6434.

Chuma S, Hiyoshi M, Yamamoto A, Hosokawa M, Takamune K, Nakatsuji N. 2003. Mouse Tudor Repeat-1 (MTR-1) is a novel component of chromatoid bodies/nuages in male germ cells and forms a complex with snRNPs. *Mech Dev* **120:** 979–990.

Chuma S, Hosokawa M, Kitamura K, Kasai S, Fujioka M, Hiyoshi M, Takamune K, Noce T, Nakatsuji N. 2006. Tdrd1/Mtr-1, a tudor-related gene, is essential for male germ-cell differentiation and nuage/germinal granule formation in mice. *Proc Natl Acad Sci* **103:** 15894–15899.

Chuma S, Hosokawa M, Tanaka T, Nakatsuji N. 2009. Ultrastructural characterization of spermatogenesis and its evolutionary conservation in the germline: Germinal granules in mammals. *Mol Cell Endocrinol* **306:** 17–23.

Clark JM, Eddy EM. 1975. Fine structural observations on the origin and associations of primordial germ cells of the mouse. *Dev Biol* **47:** 136–155.

Claycomb JM, Batista PJ, Pang KM, Gu W, Vasale JJ, van Wolfswinkel JC, Chaves DA, Shirayama M, Mitani S, Ketting RF, et al. 2009. The Argonaute CSR-1 and its 22G-RNA cofactors are required for holocentric chromosome segregation. *Cell* **139:** 123–134.

Conine CC, Batista PJ, Gu W, Claycomb JM, Chaves DA, Shirayama M, Mello CC. 2010. Argonautes ALG-3 and ALG-4 are required for spermatogenesis-specific 26G-RNAs and thermotolerant sperm in *Caenorhabditis elegans*. *Proc Natl Acad Sci* **107:** 3588–3593.

Costa Y, Speed RM, Gautier P, Semple CA, Maratou K, Turner JM, Cooke HJ. 2006. Mouse MAELSTROM: The link between meiotic silencing of unsynapsed chromatin and microRNA pathway? *Hum Mol Genet* **15:** 2324–2334.

Coward SJ. 1974. Chromatoid bodies in somatic cells of the planarian: Observations on their behavior during mitosis. *Anat Rec* **180:** 533–545.

Cox RT, Spradling AC. 2003. A Balbiani body and the fusome mediate mitochondrial inheritance during *Drosophila* oogenesis. *Development* **130:** 1579–1590.

Cox RT, Spradling AC. 2006. Milton controls the early acquisition of mitochondria by *Drosophila* oocytes. *Development* **133:** 3371–3377.

Cox DN, Chao A, Baker J, Chang L, Qiao D, Lin H. 1998. A novel class of evolutionarily conserved genes defined by piwi are essential for stem cell self-renewal. *Genes Dev* **12:** 3715–3727.

Czolowska R. 1969. Observations on the origin of the "germinal cytoplasm" in *Xenopus laevis*. *J Embryol Exp Morphol* **22:** 229–251.

Das PP, Bagijn MP, Goldstein LD, Woolford JR, Lehrbach NJ, Sapetschnig A, Buhecha HR, Gilchrist MJ, Howe KL, Stark R, et al. 2008. Piwi and piRNAs act upstream of an endogenous siRNA pathway to suppress Tc3 transposon mobility in the *Caenorhabditis elegans* germline. *Mol Cell* **31:** 79–90.

Deng W, Lin H. 2002. miwi, a murine homolog of piwi, encodes a cytoplasmic protein essential for spermatogenesis. *Dev Cell* **2:** 819–830.

de Sousa Lopes SM, Roelen BA. 2010. An overview on the diversity of cellular organelles during the germ cell cycle. *Histol Histopathol* **25:** 267–276.

Dosch R, Wagner DS, Mintzer KA, Runke G, Wiemelt AP, Mullins MC. 2004. Maternal control of vertebrate development before the midblastula transition: Mutants from the zebrafish I. *Dev Cell* **6:** 771–780.

Eddy EM. 1975. Germ plasm and the differentiation of the germ cell line. *Int Rev Cytol* **43:** 229–280.

Elad N, Maimon T, Frenkiel-Krispin D, Lim RY, Medalia O. 2009. Structural analysis of the nuclear pore complex by integrated approaches. *Curr Opin Struct Biol* **19:** 226–232.

Ephrussi A, Lehmann R. 1992. Induction of germ cell formation by *oskar. Nature* **358:** 387–392.

Ephrussi A, Dickinson LK, Lehmann R. 1991. Oskar organizes the germ plasm and directs localization of the posterior determinant nanos. *Cell* **66:** 37–50.

Fawcett DW, Eddy EM, Phillips DM. 1970. Observations on the fine structure and relationships of the chromatoid body in mammalian spermatogenesis. *Biol Reprod* **2:** 129–153.

Fimia GM, De Cesare D, Sassone-Corsi P. 1999. CBP-independent activation of CREM and CREB by the LIM-only protein ACT. *Nature* **398:** 165–169.

Findley SD, Tamanaha M, Clegg NJ, Ruohola-Baker H. 2003. *Maelstrom*, a *Drosophila spindle*-class gene, encodes a protein that colocalizes with Vasa and RDE1/AGO1 homolog, Aubergine, in nuage. *Development* **130:** 859–871.

Forrest KM, Gavis ER. 2003. Live imaging of endogenous RNA reveals a diffusion and entrapment mechanism for *nanos* mRNA localization in *Drosophila. Curr Biol* **13:** 1159–1168.

Gallo CM, Munro E, Rasoloson D, Merritt C, Seydoux G. 2008. Processing bodies and germ granules are distinct RNA granules that interact in *C. elegans* embryos. *Dev Biol* **323:** 76–87.

Gallo CM, Wang JT, Motegi F, Seydoux G. 2010. Cytoplasmic partitioning of P granule components is not required to specify the germline in *C. elegans. Science* **330:** 1685–1689.

Geigy R. 1931. Action de l'ultraviolet sur le pole germinal dans l'oeuf de Drosophila melanogaster (Castration et mutabilite). *Rev Suisse Zool* **38:** 187–288.

Gilks N, Kedersha N, Ayodele M, Shen L, Stoecklin G, Dember LM, Anderson P. 2004. Stress granule assembly is mediated by prion-like aggregation of TIA-1. *Mol Biol Cell* **15:** 5383–5398.

Gingras AC, Kennedy SG, O'Leary MA, Sonenberg N, Hay N. 1998. 4E-BP1, a repressor of mRNA translation, is phosphorylated and inactivated by the Akt(PKB) signaling pathway. *Genes Dev* **12:** 502–513.

Gonsalvez GB, Rajendra TK, Tian L, Matera AG. 2006. The Sm-protein methyltransferase, dart5, is essential for germ-cell specification and maintenance. *Curr Biol* **16:** 1077–1089.

Gonsalvez GB, Rajendra TK, Wen Y, Praveen K, Matera AG. 2010. Sm proteins specify germ cell fate by facilitating oskar mRNA localization. *Development* **137:** 2341–2351.

Grivna ST, Pyhtila B, Lin H. 2006. MIWI associates with translational machinery and PIWI-interacting RNAs (piRNAs) in regulating spermatogenesis. *Proc Natl Acad Sci* **103:** 13415–13420.

Gruidl ME, Smith PA, Kuznicki KA, McCrone JS, Kirchner J, Roussell DL, Strome S, Bennett KL. 1996. Multiple potential germ-line helicases are components of the germ-line-specific P granules of *Caenorhabditis elegans. Proc Natl Acad Sci* **93:** 13837–13842.

Gu W, Shirayama M, Conte D Jr, Vasale J, Batista PJ, Claycomb JM, Moresco JJ, Youngman EM, Keys J, Stoltz MJ, et al. 2009. Distinct argonaute-mediated 22G-RNA pathways direct genome surveillance in the *C. elegans* germline. *Mol Cell* **36:** 231–244.

Guraya SS. 1979. Recent advances in the morphology, cytochemistry, and function of Balbiani's vitelline body in animal oocytes. *Int Rev Cytol* **59:** 249–321.

Hanazawa M, Yonetani M, Sugimoto A. 2011. PGL proteins self associate and bind RNPs to mediate germ granule assembly in C. elegans. *J Cell Biol* **192:** 929–937.

Haraguchi CM, Mabuchi T, Hirata S, Shoda T, Hoshi K, Akasaki K, Yokota S. 2005. Chromatoid bodies: Aggresome-like characteristics and degradation sites for organelles of spermiogenic cells. *J Histochem Cytochem* **53:** 455–465.

Harris AN, Macdonald PM. 2001. Aubergine encodes a *Drosophila* polar granule component required for pole cell formation and related to eIF2C. *Development* **128:** 2823–2832.

Hay B, Jan LY, Jan YN. 1988. A protein component of *Drosophila* polar granules is encoded by vasa and has extensive sequence similarity to ATP dependent helicases. *Cell* **55:** 577–587.

Head JR, Kresge CK. 1985. Reaction of the chromatoid body with a monoclonal antibody to a rat histocompatibility antigen. *Biol Reprod* **33:** 1001–1008.

Hess RA, Miller LA, Kirby JD, Margoliash E, Goldberg E. 1993. Immunoelectron microscopic localization of testicular and somatic cytochromes c in the seminiferous epithelium of the rat. *Biol Reprod* **48:** 1299–1308.

Hill DP, Strome S. 1990. Brief cytochalasin-induced disruption of microfilaments during a critical interval in 1-cell *C. elegans* embryos alters the partitioning of developmental instructions to the 2-cell embryo. *Development* **108:** 159–172.

Hird SN, Paulsen JE, Strome S. 1996. Segregation of germ granules in living Caenorhabditis elegans embryos: Cell-type-specific mechanisms for cytoplasmic localisation. *Development* **122:** 1303–1312.

Hori I. 1997. Cytological approach to morphogenesis in the planarian blastema. II. The effect of neuropeptides. *J Submicrosc Cytol Pathol* **29:** 91–97.

Horwich MD, Li C, Matranga C, Vagin V, Farley G, Wang P, Zamore PD. 2007. The *Drosophila* RNA methyltransferase, DmHen1, modifies germline piRNAs and single-stranded siRNAs in RISC. *Curr Biol* **17:** 1265–1272.

Hosokawa M, Shoji M, Kitamura K, Tanaka T, Noce T, Chuma S, Nakatsuji N. 2007. Tudor-related proteins TDRD1/MTR-1, TDRD6 and TDRD7/TRAP: Domain composition, intracellular localization, and function in male germ cells in mice. *Dev Biol* **301:** 38–52.

Houwing S, Kamminga LM, Berezikov E, Cronembold D, Girard A, van den Elst H, Filippov DV, Blaser H, Raz E, Moens CB, et al. 2007. A role for Piwi and piRNAs in germ cell maintenance and transposon silencing in Zebrafish. *Cell* **129:** 69–82.

Houwing S, Berezikov E, Ketting RF. 2008. Zili is required for germ cell differentiation and meiosis in zebrafish. *EMBO J* **27:** 2702–2711.

Huang H, Gao Q, Peng X, Choi SY, Sarma K, Ren H, Morris AJ, Frohman MA. 2011. piRNA-associated germline nuage formation and spermatogenesis require MitoPLD profusogenic mitochondrial-surface lipid signaling. *Dev Cell* **20:** 376–387.

Ikenishi K, Tanaka TS. 1997. Involvement of the protein of *Xenopus vasa* homolog (*Xenopus vasa*-like gene 1, *XVLG1*) in the differentiation of primordial germ cells. *Dev Growth Differ* **39:** 625–633.

Ikenishi K, Tanaka TS, Komiya T. 1996. Spatio-temporal distribution of the protein of *Xenopus vasa* homologue (*Xenopus vasa*-like gene 1, *XVLG1*) in embryos. *Dev Growth Differ* **38:** 527–535.

Ikenishi K, Nishiumi F, Komiya T. 2006. The Xdsg protein in presumptive primordial germ cells (pPGCs) is essential to their differentiation into PGCs in *Xenopus*. *Dev Biol* **297:** 483–492.

Illmensee K, Mahowald AP. 1974. Transplantation of posterior polar plasm in *Drosophila*. Induction of germ cells at the anterior pole of the egg. *Proc Natl Acad Sci* **71:** 1016–1020.

Jaglarz MK, Kloc M, Jankowska W, Szymanska B, Bilinski SM. 2011. Nuage morphogenesis becomes more complex: Two translocation pathways and two forms of nuage coexist in *Drosophila* germline syncytia. *Cell Tissue Res* **344:** 169–181.

Janic A, Mendizabal L, Llamazares S, Rossell D, Gonzalez C. 2010. Ectopic expression of germline genes drives malignant brain tumor growth in *Drosophila*. *Science* **330:** 1824–1827.

Jeske M, Moritz B, Anders A, Wahle E. 2011. Smaug assembles an ATP-dependent stable complex repressing nanos mRNA translation at multiple levels. *EMBO J* **30:** 90–103.

Johnstone O, Lasko P. 2004. Interaction with eIF5B is essential for Vasa function during development. *Development* **131:** 4167–4178.

Jud MC, Czerwinski MJ, Wood MP, Young RA, Gallo CM, Bickel JS, Petty EL, Mason JM, Little BA, Padilla PA, et al. 2008. Large P body-like RNPs form in *C. elegans* oocytes in response to arrested ovulation, heat shock, osmotic stress, and anoxia and are regulated by the major sperm protein pathway. *Dev Biol* **318:** 38–51.

Julaton VT, Reijo Pera RA. 2011. NANOS3 function in human germ cell development. *Hum Mol Genet* **20:** 2238–2250.

Juliano CE, Swartz SZ, Wessel GM. 2010. A conserved germline multipotency program. *Development* **137:** 4113–4126.

Kamminga LM, Luteijn MJ, den Broeder MJ, Redl S, Kaaij LJ, Roovers EF, Ladurner P, Berezikov E, Ketting RF. 2010. Hen1 is required for oocyte development and piRNA stability in zebrafish. *EMBO J* **29:** 3688–3700.

Kashikawa M, Amikura R, Nakamura A, Kobayashi S. 1999. Mitochondrial small ribosomal RNA is present on polar granules in early cleavage embryos of *Drosophila melanogaster*. *Dev Growth Differ* **41:** 495–502.

Kashikawa M, Amikura R, Kobayashi S. 2001. Mitochondrial small ribosomal RNA is a component of germinal granules in Xenopus embryos. *Mech Dev* **101:** 71–77.

Kim-Ha J, Smith JL, Macdonald PM. 1991. oskar mRNA is localized to the posterior pole of the Drosophila oocyte. *Cell* **66:** 23–35.

Kim-Ha J, Kerr K, Macdonald PM. 1995. Translational regulation of *oskar* mRNA by bruno, an ovarian RNA-binding protein, is essential. *Cell* **81:** 403–412.

Kirino Y, Mourelatos Z. 2007a. The mouse homolog of HEN1 is a potential methylase for Piwi-interacting RNAs. *RNA* **13:** 1397–1401.

Kirino Y, Mourelatos Z. 2007b. Mouse Piwi-interacting RNAs are 2′-O-methylated at their 3′ termini. *Nat Struct Mol Biol* **14:** 347–348.

Kirino Y, Kim N, de Planell-Saguer M, Khandros E, Chiorean S, Klein PS, Rigoutsos I, Jongens TA, Mourelatos Z. 2009. Arginine methylation of Piwi proteins catalysed by dPRMT5 is required for Ago3 and Aub stability. *Nat Cell Biol* **11:** 652–658.

Kirino Y, Vourekas A, Kim N, de Lima Alves F, Rappsilber J, Klein PS, Jongens TA, Mourelatos Z. 2010a. Arginine methylation of vasa protein is conserved across phyla. *J Biol Chem* **285:** 8148–8154.

Kirino Y, Vourekas A, Sayed N, de Lima Alves F, Thomson T, Lasko P, Rappsilber J, Jongens TA, Mourelatos Z. 2010b. Arginine methylation of Aubergine mediates Tudor binding and germ plasm localization. *RNA* **16:** 70–78.

Kleene KC. 1993. Multiple controls over the efficiency of translation of the mRNAs encoding transition proteins, protamines, and the mitochondrial capsule selenoprotein in late spermatids in mice. *Dev Biol* **159:** 720–731.

Kloc M, Etkin LD. 1994. Delocalization of Vg1 mRNA from the vegetal cortex in *Xenopus* oocytes after destruction of Xlsirt RNA. *Science* **265:** 1101–1103.

Kloc M, Dougherty MT, Bilinski S, Chan AP, Brey E, King ML, Patrick CW Jr, Etkin LD. 2002. Three-dimensional ultrastructural analysis of RNA distribution within germinal granules of *Xenopus*. *Dev Biol* **241:** 79–93.

Kloc M, Bilinski S, Etkin LD. 2004. The Balbiani body and germ cell determinants: 150 years later. *Curr Top Dev Biol* **59:** 1–36.

Kloc M, Wilk K, Vargas D, Shirato Y, Bilinski S, Etkin LD. 2005. Potential structural role of non-coding and coding RNAs in the organization of the cytoskeleton at the vegetal cortex of *Xenopus* oocytes. *Development* **132:** 3445–3457.

Kloc M, Bilinski S, Dougherty MT. 2007. Organization of cytokeratin cytoskeleton and germ plasm in the vegetal cortex of *Xenopus* laevis oocytes depends on coding and non-coding RNAs: Three-dimensional and ultrastructural analysis. *Exp Cell Res* **313:** 1639–1651.

Knaut H, Pelegri F, Bohmann K, Schwarz H, Nusslein-Volhard C. 2000. Zebrafish vasa RNA but not its protein is a component of the germ plasm and segregates asymmetrically before germline specification. *J Cell Biol* **149:** 875–888.

Kobayashi S, Yamada M, Asaoka M, Kitamura T. 1996. Essential role of the posterior morphogen nanos for germline development in *Drosophila*. *Nature* **380:** 708–711.

Kogo N, Tazaki A, Kashino Y, Morichika K, Orii H, Mochii M, Watanabe K. 2011. Germ-line mitochondria exhibit suppressed respiratory activity to support their accurate transmission to the next generation. *Dev Biol* **349:** 462–469.

Kojima K, Kuramochi-Miyagawa S, Chuma S, Tanaka T, Nakatsuji N, Kimura T, Nakano T. 2009. Associations between PIWI proteins and TDRD1/MTR-1 are critical for integrated subcellular localization in murine male germ cells. *Genes Cells* **14:** 1155–1165.

Koprunner M, Thisse C, Thisse B, Raz E. 2001. A zebrafish *nanos*-related gene is essential for the development of primordial germ cells. *Genes Dev* **15:** 2877–2885.

Kosaka K, Kawakami K, Sakamoto H, Inoue K. 2007. Spatiotemporal localization of germ plasm RNAs during zebrafish oogenesis. *Mech Dev* **124:** 279–289.

Kotaja N, Sassone-Corsi P. 2007. The chromatoid body: A germ-cell-specific RNA-processing centre. *Nat Rev Mol Cell Biol* **8:** 85–90.

Kotaja N, Bhattacharyya SN, Jaskiewicz L, Kimmins S, Parvinen M, Filipowicz W, Sassone-Corsi P. 2006a. The chromatoid body of male germ cells: Similarity with processing bodies and presence of Dicer and microRNA pathway components. *Proc Natl Acad Sci* **103:** 2647–2652.

Kotaja N, Lin H, Parvinen M, Sassone-Corsi P. 2006b. Interplay of PIWI/Argonaute protein MIWI and kinesin KIF17b in chromatoid bodies of male germ cells. *J Cell Sci* **119:** 2819–2825.

Krimer DB, Esponda P. 1980. Presence of polysaccharides and proteins in the chromatoid body of mouse spermatids. *Cell Biol Int Rep* **4:** 265–270.

Kuramochi-Miyagawa S, Kimura T, Ijiri TW, Isobe T, Asada N, Fujita Y, Ikawa M, Iwai N, Okabe M, Deng W, et al. 2004. Mili, a mammalian member of piwi family gene, is essential for spermatogenesis. *Development* **131:** 839–849.

Kuramochi-Miyagawa S, Watanabe T, Gotoh K, Takamatsu K, Chuma S, Kojima-Kita K, Shiromoto Y, Asada N, Toyoda A, Fujiyama A, et al. 2010. MVH in piRNA processing and gene silencing of retrotransposons. *Genes Dev* **24:** 887–892.

Kuznicki KA, Smith PA, Leung-Chiu WM, Estevez AO, Scott HC, Bennett KL. 2000. Combinatorial RNA interference indicates GLH-4 can compensate for GLH-1; these two P granule components are critical for fertility in *C. elegans*. *Development* **127:** 2907–2916.

Lall S, Piano F, Davis RE. 2005. *Caenorhabditis elegans* decapping proteins: Localization and functional analysis of Dcp1, Dcp2, and DcpS during embryogenesis. *Mol Biol Cell* **16:** 5880–5890.

Lasko PF, Ashburner M. 1990. Posterior localization of vasa protein correlates with, but is not sufficient for, pole cell development. *Genes Dev* **4:** 905–921.

Lerit DA, Gavis ER. 2011. Transport of germ plasm on astral microtubules directs germ cell development in Drosophila. *Curr Biol* **21:** 439–448.

Li C, Vagin VV, Lee S, Xu J, Ma S, Xi H, Seitz H, Horwich MD, Syrzycka M, Honda BM, et al. 2009a. Collapse of germline piRNAs in the absence of Argonaute3 reveals somatic piRNAs in flies. *Cell* **137:** 509–521.

Li W, DeBella LR, Guven-Ozkan T, Lin R, Rose LS. 2009b. An eIF4E-binding protein regulates katanin protein levels in *C. elegans* embryos. *J Cell Biol* **187:** 33–42.

Lim AK, Kai T. 2007. Unique germ-line organelle, nuage, functions to repress selfish genetic elements in *Drosophila melanogaster*. *Proc Natl Acad Sci* **104:** 6714–6719.

Liang L, Diehl-Jones W, Lasko P. 1994. Localization of vasa protein to the *Drosophila* pole plasm is independent of its RNA-binding and helicase activities. *Development* **120:** 1201–1211.

Lim AK, Tao L, Kai T. 2009. piRNAs mediate posttranscriptional retroelement silencing and localization to pi-bodies in the Drosophila germline. *J Cell Biol* **186:** 333–342.

Lin H, Spradling AC. 1997. A novel group of pumilio mutations affects the asymmetric division of germline stem cells in the *Drosophila* ovary. *Development* **124:** 2463–2476.

Lin MD, Fan SJ, Hsu WS, Chou TB. 2006. *Drosophila* decapping protein 1, dDcp1, is a component of the *oskar* mRNP complex and directs its posterior localization in the oocyte. *Dev Cell* **10:** 601–613.

Liu JL, Gall JG. 2007. U bodies are cytoplasmic structures that contain uridine-rich small nuclear ribonucleoproteins and associate with P bodies. *Proc Natl Acad Sci* **104:** 11655–11659.

Liu N, Han H, Lasko P. 2009. Vasa promotes *Drosophila* germline stem cell differentiation by activating mei-P26 translation by directly interacting with a (U)-rich motif in its 3′ UTR. *Genes Dev* **23:** 2742–2752.

Liu H, Wang JY, Huang Y, Li Z, Gong W, Lehmann R, Xu RM. 2010. Structural basis for methylarginine-dependent recognition of Aubergine by Tudor. *Genes Dev* **24:** 1876–1881.

Liu L, Qi H, Wang J, Lin H. 2011. PAPI, a novel TUDOR-domain protein, complexes with AGO3, ME31B and TRAL in the nuage to silence transposition. *Development* **138:** 1863–1873.

Ma L, Buchold GM, Greenbaum MP, Roy A, Burns KH, Zhu H, Han DY, Harris RA, Coarfa C, Gunaratne PH, et al. 2009. GASZ is essential for male meiosis and suppression of retrotransposon expression in the male germline. *PLoS Genet* **5:** e1000635.

Mahowald AP. 1968. Polar granules of *Drosophila*. II. Ultrastructural changes during early embryogenesis. *J Exp Zool* **167:** 237–261.

Mahowald AP. 1971. Polar granules of *Drosophila*. 3. The continuity of polar granules during the life cycle of *Drosophila*. *J Exp Zool* **176:** 329–343.

Mahowald AP. 2001. Assembly of the *Drosophila* germ plasm. *Int Rev Cytol* **203:** 187–213.

Maniar JM, Fire AZ. 2011. EGO-1, a C. elegans RdRP, modulates gene expression via production of mRNA-templated short antisense RNAs. *Curr Biol* **21:** 449–459.

Markussen FH, Michon AM, Breitwieser W, Ephrussi A. 1995. Translational control of oskar generates short OSK, the isoform that induces pole plasma assembly. *Development* **121:** 3723–3732.

Marlow FL, Mullins MC. 2008. Bucky ball functions in Balbiani body assembly and animal-vegetal polarity in the oocyte and follicle cell layer in zebrafish. *Dev Biol* **321:** 40–50.

Marston DJ, Goldstein B. 2006. Symmetry breaking in *C. elegans*: Another gift from the sperm. *Dev Cell* **11:** 273–274.

Megosh HB, Cox DN, Campbell C, Lin H. 2006. The role of PIWI and the miRNA machinery in *Drosophila* germline determination. *Curr Biol* **16:** 1884–1894.

Meikar O, Da Ros M, Liljenback H, Toppari J, Kotaja N. 2010. Accumulation of piRNAs in the chromatoid bodies purified by a novel isolation protocol. *Exp Cell Res* **316:** 1567–1575.

Mosquera L, Forristall C, Zhou Y, King ML. 1993. A mRNA localized to the vegetal cortex of Xenopus oocytes encodes a protein with a nanos-like zinc finger domain. *Development* **117:** 377–386.

Moussa F, Oko R, Hermo L. 1994. The immunolocalization of small nuclear ribonucleoprotein particles in testicular cells during the cycle of the seminiferous epithelium of the adult rat. *Cell Tissue Res* **278:** 363–378.

Murchison EP, Stein P, Xuan Z, Pan H, Zhang MQ, Schultz RM, Hannon GJ. 2007. Critical roles for Dicer in the female germline. *Genes Dev* **21:** 682–693.

Nakamura A, Amikura R, Hanyu K, Kobayashi S. 2001. Me31B silences translation of oocyte-localizing RNAs through the formation of cytoplasmic RNP complex during *Drosophila* oogenesis. *Development* **128:** 3233–3242.

Nakamura A, Sato K, Hanyu-Nakamura K. 2004. *Drosophila cup* is an eIF4E binding protein that associates with Bruno and regulates *oskar* mRNA translation in oogenesis. *Dev Cell* **6:** 69–78.

Navarro RE, Shim EY, Kohara Y, Singson A, Blackwell TK. 2001. *cgh-1*, a conserved predicted RNA helicase required for gametogenesis and protection from physiological germline apoptosis in *C elegans*. *Development* **128:** 3221–3232.

Nguyen Chi M, Chalmel F, Agius E, Vanzo N, Khabar KS, Jegou B, Morello D. 2009. Temporally regulated traffic of HuR and its associated ARE-containing mRNAs from the chromatoid body to polysomes during mouse spermatogenesis. *PLoS One* **4:** e4900.

Nishida KM, Saito K, Mori T, Kawamura Y, Nagami-Okada T, Inagaki S, Siomi H, Siomi MC. 2007. Gene silencing mechanisms mediated by Aubergine piRNA complexes in *Drosophila* male gonad. *RNA* **13:** 1911–1922.

Nishida KM, Okada TN, Kawamura T, Mituyama T, Kawamura Y, Inagaki S, Huang H, Chen D, Kodama T, Siomi H, et al. 2009. Functional involvement of Tudor and dPRMT5 in the piRNA processing pathway in *Drosophila* germlines. *EMBO J* **28:** 3820–3831.

Noble SL, Allen BL, Goh LK, Nordick K, Evans TC. 2008. Maternal mRNAs are regulated by diverse P body-related mRNP granules during early *Caenorhabditis elegans* development. *J Cell Biol* **182:** 559–572.

Ohara T, Sakaguchi Y, Suzuki T, Ueda H, Miyauchi K, Suzuki T. 2007. The 3′ termini of mouse Piwi-interacting RNAs are 2′-O-methylated. *Nat Struct Mol Biol* **14:** 349–350.

Oko R, Korley R, Murray MT, Hecht NB, Hermo L. 1996. Germ cell-specific DNA and RNA binding proteins p48/52 are expressed at specific stages of male germ cell development and are present in the chromatoid body. *Mol Reprod Dev* **44:** 1–13.

Ollinger R, Reichmann J, Adams IR. 2010. Meiosis and retrotransposon silencing during germ cell development in mice. *Differentiation* **79:** 147–158.

Pan J, Goodheart M, Chuma S, Nakatsuji N, Page DC, Wang PJ. 2005. RNF17, a component of the mammalian germ cell nuage, is essential for spermiogenesis. *Development* **132:** 4029–1039.

Paniagua R, Nistal M, Amat P, Rodriguez MC. 1985. Presence of ribonucleoproteins and basic proteins in the nuage and intermitochondrial bars of human spermatogonia. *J Anat* **143:** 201–206.

Parvinen M. 2005. The chromatoid body in spermatogenesis. *Int J Androl* **28:** 189–201.

Patil VS, Kai T. 2010. Repression of retroelements in *Drosophila* germline via piRNA pathway by the Tudor domain protein Tejas. *Curr Biol* **8:** 724–730.

Patterson JR, Wood MP, Schisa JA. 2011. Assembly of RNP granules in stressed and aging oocytes requires nucleoporins and is coordinated with nuclear membrane blebbing. *Dev Biol* **15:** 173–185.

Pek JW, Kai T. 2011a. Non-coding RNAs enter mitosis: Functions, conservation and implications. *Cell Div* **6:** 6.

Pek JW, Kai T. 2011b. A role for vasa in regulating mitotic chromosome condensation in *Drosophila*. *Curr Biol* **21:** 39–44.

Pepling ME, Wilhelm JE, O'Hara AL, Gephardt GW, Spradling AC. 2007. Mouse oocytes within germ cell cysts and primordial follicles contain a Balbiani body. *Proc Natl Acad Sci* **104:** 187–192.

Petrella LN, Wang W, Spike CA, Rechtsteiner A, Reinke V, Strome S. 2011. synMuv B proteins antagonize germline fate in the intestine and ensure *C. elegans* survival. *Development* **138:** 1069–1079.

Pitt JN, Schisa JA, Priess JR. 2000. P granules in the germ cells of *Caenorhabditis elegans* adults are associated with clusters of nuclear pores and contain RNA. *Dev Biol* **219:** 315–333.

Rangan P, DeGennaro M, Jaime-Bustamante K, Coux RX, Martinho RG, Lehmann R. 2009. Temporal and spatial control of germ-plasm RNAs. *Curr Biol* **19:** 72–77.

Reuter M, Chuma S, Tanaka T, Franz T, Stark A, Pillai RS. 2009. Loss of the Mili-interacting Tudor domain-containing protein-1 activates transposons and alters the Mili-associated small RNA profile. *Nat Struct Mol Biol* **16:** 639–646.

Ritter R. 1890. Die Entwicklung der Geschlechtsorgane und des Darmes bei Chiromomus. *Zeit furr Wiss Zool* Bd **50:** 408–427.

Rongo C, Lehmann R. 1996. Regulated synthesis, transport and assembly of the *Drosophila* germ plasm. *Trends Genet* **12:** 102–109.

Rongo C, Broihier HT, Moore L, Van Doren M, Forbes A, Lehmann R. 1997. Germ plasm assembly and germ cell migration in *Drosophila. Cold Spring Harb Symp Quant Biol* **62:** 1–11.

Roper K. 2007. Rtnl1 is enriched in a specialized germline ER that associates with ribonucleoprotein granule components. *J Cell Sci* **120:** 1081–1092.

Roucou X. 2009. Prion protein and RNA: A view from the cytoplasm. *Front Biosci* **14:** 5157–5164.

Rouelle-Rossier VB, Biggiogera M, Fakan S. 1993. Ultrastructural detection of calcium and magnesium in the chromatoid body of mouse spermatids by electron spectroscopic imaging and electron energy loss spectroscopy. *J Histochem Cytochem* **41:** 1155–1162.

Saito K, Sakaguchi Y, Suzuki T, Suzuki T, Siomi H, Siomi MC. 2007. Pimet, the *Drosophila* homolog of HEN1, mediates 2′-O-methylation of Piwi- interacting RNAs at their 3′ ends. *Genes Dev* **21:** 1603–1608.

Sato K, Sugita T, Kobayashi K, Fujita K, Fujii T, Matsumoto Y, Mikami T, Nishizuka N, Nishizuka S, Shojima K, et al. 2001. Localization of mitochondrial ribosomal RNA on the chromatoid bodies of marine planarian polyclad embryos. *Dev Growth Differ* **43:** 107–114.

Schupbach T, Wieschaus E. 1986. Germline autonomy of maternal-effect mutations altering the embryonic body pattern of *Drosophila. Dev Biol* **113:** 443–448.

Senti KA, Brennecke J. 2010. The piRNA pathway: A fly's perspective on the guardian of the genome. *Trends Genet* **26:** 499–509.

Seydoux G, Braun RE. 2006. Pathway to totipotency: Lessons from germ cells. *Cell* **127:** 891–904.

Sheth U, Pitt J, Dennis S, Priess JR. 2010. Perinuclear P granules are the principal sites of mRNA export in adult *C. elegans* germ cells. *Development* **137:** 1305–1314.

Shibata N, Tsunekawa N, Okamoto-Ito S, Akasu R, Tokumasu A, Noce T. 2004. Mouse RanBPM is a partner gene to a germline specific RNA helicase, mouse vasa homolog protein. *Mol Reprod Dev* **67:** 1–7.

Shoji M, Tanaka T, Hosokawa M, Reuter M, Stark A, Kato Y, Kondoh G, Okawa K, Chujo T, Suzuki T, et al. 2009. The TDRD9-MIWI2 complex is essential for piRNA-mediated retrotransposon silencing in the mouse male germline. *Dev Cell* **17:** 775–787.

Smith LD. 1966. The role of a "germinal plasm" in the formation of primordial germ cells in *Rana pipiens. Dev Biol* **14:** 330–347.

Snee MJ, Macdonald PM. 2004. Live imaging of nuage and polar granules: Evidence against a precursor-product relationship and a novel role for Oskar in stabilization of polar granule components. *J Cell Sci* **117:** 2109–2120.

Snee MJ, Macdonald PM. 2009. Dynamic organization and plasticity of sponge bodies. *Dev Dyn* **238:** 918–930.

Soderstrom KO. 1977. Effect of actinomycin D on the structure of the chromatoid body in the rat spermatids. *Cell Tissue Res* **184:** 411–421.

Soderstrom KO, Parvinen M. 1976. Incorporation of (3H)uridine by the chromatoid body during rat spermatogenesis. *J Cell Biol* **70:** 239–246.

Soper SF, van der Heijden GW, Hardiman TC, Goodheart M, Martin SL, de Boer P, Bortvin A. 2008. Mouse maelstrom, a component of nuage, is essential for spermatogenesis and transposon repression in meiosis. *Dev Cell* **15:** 285–297.

Spiegelman M, Bennett D. 1973. A light- and electron-microscopic study of primordial germ cells in the early mouse embryo. *J Embryol Exp Morphol* **30:** 97–118.

Spike C, Meyer N, Racen E, Orsborn A, Kirchner J, Kuznicki K, Yee C, Bennett K, Strome S. 2008. Genetic analysis of the *Caenorhabditis elegans* GLH family of P-granule proteins. *Genetics* **178:** 1973–1987.

Squirrell JM, Eggers ZT, Luedke N, Saari B, Grimson A, Lyons GE, Anderson P, White JG. 2006. CAR-1, a protein that localizes with the mRNA decapping component DCAP-1, is required for cytokinesis and ER organization in *Caenorhabditis elegans* embryos. *Mol Biol Cell* **17:** 336–344.

Strasser MJ, Mackenzie NC, Dumstrei K, Nakkrasae LI, Stebler J, Raz E. 2008. Control over the morphology and segregation of Zebrafish germ cell granules during embryonic development. *BMC Dev Biol* **8:** 58.

Strome S, Wood WB. 1982. Immunofluorescence visualization of germ-line-specific cytoplasmic granules in embryos, larvae, and adults of *Caenorhabditis elegans. Proc Natl Acad Sci* **79:** 1558–1562.

Strome S, Wood WB. 1983. Generation of asymmetry and segregation of germ-line granules in early *C. elegans* embryos. *Cell* **35:** 15–25.

Strome S, Martin P, Schierenberg E, Paulsen J. 1995. Transformation of the germ line into muscle in mes-1 mutant embryos of *C. elegans. Development* **121:** 2961–2972.

Styhler S, Nakamura A, Swan A, Suter B, Lasko P. 1998. *vasa* is required for GURKEN accumulation in the oocyte, and is involved in oocyte differentiation and germline cyst development. *Development* **125:** 1569–1578.

Subramaniam K, Seydoux G. 1999. *nos-1* and *nos-2*, two genes related to *Drosophila nanos*, regulate primordial germ cell development and survival in *Caenorhabditis elegans. Development* **126:** 4861–4871.

Suzuki A, Tsuda M, Saga Y. 2007. Functional redundancy among Nanos proteins and a distinct role of Nanos2 during male germ cell development. *Development* **134:** 77–83.

Swetloff A, Conne B, Huarte J, Pitetti JL, Nef S, Vassalli JD. 2009. Dcp1-bodies in mouse oocytes. *Mol Biol Cell* **20:** 4951–4961.

Tam OH, Aravin AA, Stein P, Girard A, Murchison EP, Cheloufi S, Hodges E, Anger M, Sachidanandam R, Schultz RM, et al. 2008. Pseudogene-derived small interfering RNAs regulate gene expression in mouse oocytes. *Nature* **453:** 534–538.

Tanaka T, Nakamura A. 2008. The endocytic pathway acts downstream of Oskar in Drosophila germ plasm assembly. *Development* **135:** 1107–1117.

Tanaka SS, Toyooka Y, Akasu R, Katoh-Fukui Y, Nakahara Y, Suzuki R, Yokoyama M, Noce T. 2000. The mouse homolog of *Drosophila* Vasa is required for the development of male germ cells. *Genes Dev* **14:** 841–853.

Tang XM, Lalli MF, Clermont Y. 1982. A cytochemical study of the Golgi apparatus of the spermatid during spermiogenesis in the rat. *Am J Anat* **163:** 283–294.

Terry LJ, Wente SR. 2009. Flexible gates: Dynamic topologies and functions for FG nucleoporins in nucleocytoplasmic transport. *Eukaryot Cell* **8:** 1814–1827.

Theusch EV, Brown KJ, Pelegri F. 2006. Separate pathways of RNA recruitment lead to the compartmentalization of the zebrafish germ plasm. *Dev Biol* **292:** 129–141.

Thomson T, Lin H. 2009. The biogenesis and function of PIWI proteins and piRNAs: Progress and prospect. *Annu Rev Cell Dev Biol* **25:** 355–376.

Thomson T, Liu N, Arkov A, Lehmann R, Lasko P. 2008. Isolation of new polar granule components in *Drosophila* reveals P body and ER associated proteins. *Mech Dev* **125:** 865–873.

Thorne-Tjomsland G, Clermont Y, Hermo L. 1988. Contribution of the Golgi apparatus components to the formation of the acrosomic system and chromatoid body in rat spermatids. *Anat Rec* **221:** 591–598.

Toyooka Y, Tsunekawa N, Takahashi Y, Matsui Y, Satoh M, Noce T. 2000. Expression and intracellular localization of mouse Vasa-homologue protein during germ cell development. *Mech Dev* **93:** 139–149.

Tsai-Morris CH, Sheng Y, Lee E, Lei KJ, Dufau ML. 2004. Gonadotropin-regulated testicular RNA helicase (GRTH/Ddx25) is essential for spermatid development and completion of spermatogenesis. *Proc Natl Acad Sci* **101:** 6373–6378.

Tsuda M, Sasaoka Y, Kiso M, Abe K, Haraguchi S, Kobayashi S, Saga Y. 2003. Conserved role of nanos proteins in germ cell development. *Science* **301:** 1239–1241.

Unhavaithaya Y, Shin TH, Miliaras N, Lee J, Oyama T, Mello CC. 2002. MEP-1 and a homolog of the NURD complex component Mi-2 act together to maintain germline-soma distinctions in C. elegans. *Cell* **111:** 991–1002.

Unhavaithaya Y, Hao Y, Beyret E, Yin H, Kuramochi-Miyagawa S, Nakano T, Lin H. 2009. MILI, a PIWI-interacting RNA-binding protein, is required for germ line stem cell self-renewal and appears to positively regulate translation. *J Biol Chem* **284:** 6507–6519.

Updike DL, Strome S. 2009a. A genomewide rnai screen for genes that affect the stability, distribution and function of P granules in *Caenorhabditis elegans*. *Genetics* **183:** 1397–1419.

Updike DL, Strome S. 2009b. P granule assembly in *C. elegans germ cells*. *J Androl* **31:** 53–60.

Updike DL, Hachey SJ, Kreher J, Strome S. 2011. P granules extend the nuclear pore complex environment in the C. elegans germ line. *J Cell Biol* **192:** 939–948.

Vagin VV, Klenov MS, Kalmykova AI, Stolyarenko AD, Kotelnikov RN, Gvozdev VA. 2004. The RNA interference proteins and *vasa* locus are involved in the silencing of retrotransposons in the female germline of *Drosophila melanogaster*. *RNA Biol* **1:** 54–58.

Vagin VV, Sigova A, Li C, Seitz H, Gvozdev V, Zamore PD. 2006. A distinct small RNA pathway silences selfish genetic elements in the germline. *Science* **313:** 320–324.

Vagin VV, Hannon GJ, Aravin AA. 2009a. Arginine methylation as a molecular signature of the Piwi small RNA pathway. *Cell Cycle* **8:** 4003–4004.

Vagin VV, Wohlschlegel J, Qu J, Jonsson Z, Huang X, Chuma S, Girard A, Sachidanandam R, Hannon GJ, Aravin AA. 2009b. Proteomic analysis of murine Piwi proteins reveals a role for arginine methylation in specifying interaction with Tudor family members. *Genes Dev* **23:** 1749–1762.

van Dijk E, Cougot N, Meyer S, Babajko S, Wahle E, Séraphin B. 2002. Human Dcp2: A catalytically active mRNA decapping enzyme located in specific cytoplasmic structures. *EMBO J* **21:** 6915–6924.

Vanzo NF, Ephrussi A. 2002. Oskar anchoring restricts pole plasm formation to the posterior of the *Drosophila* oocyte. *Development* **129:** 3705–3714.

Vasileva A, Tiedau D, Firooznia A, Müller-Teichert T, Jessberger R. 2009. Tdrd6 is required for spermiogenesis, chromatoid body architecture, and regulation of miRNA expression. *Curr Biol* **19:** 630–639.

Ventela S, Toppari J, Parvinen M. 2003. Intercellular organelle traffic through cytoplasmic bridges in early spermatids of the rat: Mechanisms of haploid gene product sharing. *Mol Biol Cell* **14:** 2768–2780.

Verdel A, Jia S, Gerber S, Sugiyama T, Gygi S, Grewal SI, Moazed D. 2004. RNAi-mediated targeting of heterochromatin by the RITS complex. *Science* **303:** 672–676.

Voronina E, Seydoux G. 2010. The *C. elegans* homolog of nucleoporin Nup98 is required for the integrity and function of germline P granules. *Development* **137:** 1441–1450.

Vourekas A, Kirino Y, Mourelatos Z. 2010. Elective affinities: A Tudor-Aubergine tale of germline partnership. *Genes Dev* **24:** 1963–1966.

Walt H, Armbruster BL. 1984. Actin and RNA are components of the chromatoid bodies in spermatids of the rat. *Cell Tissue Res* **236:** 487–490.

Wang C, Lehmann R. 1991. Nanos is the localized posterior determinant in *Drosophila*. *Cell* **66:** 637–647.

Wang C, Dickinson LK, Lehmann R. 1994. Genetics of nanos localization in *Drosophila*. *Dev Dyn* **199:** 103–115.

Wang D, Kennedy S, Conte D Jr, Kim JK, Gabel HW, Kamath RS, Mello CC, Ruvkun G. 2005. Somatic misexpression of germline P granules and enhanced RNA interference in retinoblastoma pathway mutants. *Nature* **436:** 593–597.

Wang Y, Zayas RM, Guo T, Newmark PA. 2007. Nanos function is essential for development and regeneration of planarian germ cells. *Proc Natl Acad Sci* **104:** 5901–5906.

Wang J, Saxe JP, Tanaka T, Chuma S, Lin H. 2009. Mili interacts with tudor domain-containing protein 1 in regulating spermatogenesis. *Curr Biol* **19:** 640–644.

Watanabe T, Totoki Y, Toyoda A, Kaneda M, Kuramochi-Miyagawa S, Obata Y, Chiba H, Kohara Y, Kono T, Nakano T, et al. 2008. Endogenous siRNAs from naturally formed dsRNAs regulate transcripts in mouse oocytes. *Nature* **453:** 539–543.

Watanabe T, Chuma S, Yamamoto Y, Kuramochi-Miyagawa S, Totoki Y, Toyoda A, Hoki Y, Fujiyama A, Shibata T, Sado T, et al. 2011. MITOPLD is a mitochondrial protein essential for nuage formation and piRNA biogenesis in the mouse germline. *Dev Cell* **20:** 364–375.

Werner G, Werner K. 1995. Immunocytochemical localization of histone H4 in the chromatoid body of rat spermatids. *J Submicrosc Cytol Pathol* **27:** 325–330.

Wilhelm JE, Mansfield J, Hom-Booher N, Wang S, Turck CW, Hazelrigg T, Vale RD. 2000. Isolation of a ribonucleoprotein complex involved in mRNA localization in *Drosophila* oocytes. *J Cell Biol* **148:** 427–440.

Wilhelm JE, Hilton M, Amos Q, Henzel WJ. 2003. Cup is an eIF4E binding protein required for both the translational repression of *oskar* and the recruitment of Barentsz. *J Cell Biol* **163:** 1197–1204.

Wilhelm JE, Buszczak M, Sayles S. 2005. Efficient protein trafficking requires *trailer hitch*, a component of a ribonucleoprotein complex localized to the ER in *Drosophila. Dev Cell* **9:** 675–685.

Wilk K, Bilinski S, Dougherty MT, Kloc M. 2005. Delivery of germinal granules and localized RNAs via the messenger transport organizer pathway to the vegetal cortex of *Xenopus* oocytes occurs through directional expansion of the mitochondrial cloud. *Int J Dev Biol* **49:** 17–21.

Wilsch-Brauninger M, Schwarz H, Nusslein-Volhard C. 1997. A sponge-like structure involved in the association and transport of maternal products during *Drosophila* oogenesis. *J Cell Biol* **139:** 817–829.

Yabuta Y, Ohta H, Abe T, Kurimoto K, Chuma S, Saitou M. 2011. TDRD5 is required for retrotransposon silencing, chromatoid body assembly, and spermiogenesis in mice. *J Cell Biol* **192:** 781–795.

Yajima M, Wessel GM. 2011. The DEAD-box RNA helicase Vasa functions in embryonic mitotic progression in the sea urchin. *Development* **138:** 2217–2222.

Yano T, Lopez de Quinto S, Matsui Y, Shevchenko A, Shevchenko A, Ephrussi A. 2004. Hrp48, a *Drosophila* hnRNPA/B homolog, binds and regulates translation of oskar mRNA. *Dev Cell* **6:** 637–648.

Zaessinger S, Busseau I, Simonelig M. 2006. Oskar allows nanos mRNA translation in *Drosophila* embryos by preventing its deadenylation by Smaug/CCR4. *Development* **133:** 4573–4583.

Zhang Y, Yan L, Zhou Z, Yang P, Tian E, Zhang K, Zhao Y, Li Z, Song B, Han J, et al. 2009. SEPA-1 mediates the specific recognition and degradation of P granule components by autophagy in *C. elegans. Cell* **136:** 308–321.

Zhou Y, King ML. 2004. Sending RNAs into the future: RNA localization and germ cell fate. *IUBMB Life* **56:** 19–27.

Germ Cell Intercellular Bridges

Michael P. Greenbaum[1], Tokuko Iwamori[2], Gregory M. Buchold[2,3], and Martin M. Matzuk[2,3,4]

[1]Department of Radiation Oncology, Baylor College of Medicine, Houston, Texas 77030

[2]Vanderbilt University Medical Center, Nashville, Tennessee 37232; Departments of Pathology and Immunology, Baylor College of Medicine, Houston, Texas 77030

[3]Molecular and Cellular Biology, Baylor College of Medicine, Houston, Texas 77030

[4]Molecular and Human Genetics, Baylor College of Medicine, Houston, Texas 77030

Correspondence: mmatzuk@tmc.bcm.edu

Stable intercellular bridges are a conserved feature of gametogenesis in multicellular animals observed more than 100 years ago, but their function was unknown. Many of the components necessary for this structure have been identified through the study of cytokinesis in *Drosophila*; however, mammalian intercellular bridges have distinct properties from those of insects. Mammalian germ cell intercellular bridges are composed of general cytokinesis components with additional germ cell–specific factors including TEX14. TEX14 is an inactive kinase essential for the maintenance of stable intercellular bridges in gametes of both sexes but whose loss specifically impairs male meiosis. TEX14 acts to impede the terminal steps of abscission by competing for essential component CEP55, blocking its interaction in nongerm cells with ALIX and TSG101. Additionally, TEX14-interacting protein RBM44, whose localization in stabile intercellular bridges is limited to pachytene and secondary spermatocytes, may participate in processes such as RNA transport but is nonessential to the maintenance of intercellular bridge stability.

Intercellular communication is an essential function for all multicellular organisms. Almost every process in the human body, including reproduction, depends on intercellular communication. All of the common methods of intercellular communication found in humans can also be found in the reproductive system. Some examples include endocrine signaling along the hypothalamic-pituitary-gonadal axis, paracrine and autocrine signaling from Sertoli cells (Saez et al. 1987), and gap junction communication. Germ cells also use an additional, unique method of intercellular communication in the form of large, 0.5–3-μm, stable intercellular bridges, sometimes called cytoplasmic bridges. The scale of intercellular bridges makes them unlike any other form of intercellular communication. They join the cytoplasm of neighboring germ cells, forming syncytium, and are large enough to allow the passage of mitochondria, let alone molecular signals. Intercellular bridges are conserved more than a billion years of evolution from insects to humans. They are formed during cell division in both male and female gametogenesis. Although they have been studied in

Cite this article as *Cold Spring Harb Perspect Biol* doi: 10.1101/cshperspect.a005850

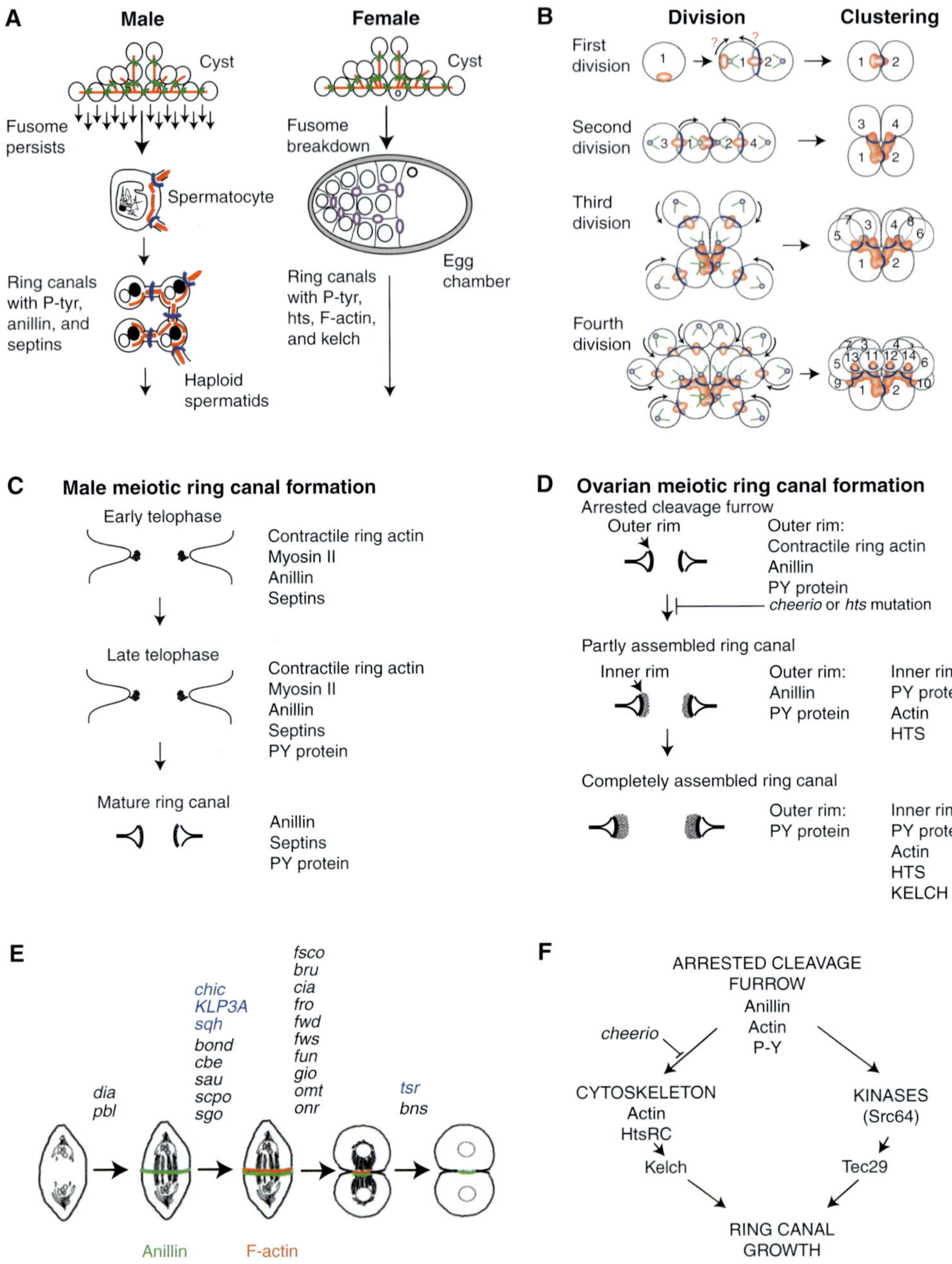

Figure 1. Intercellular bridge formation in *Drosophila*. Both male and female germ cells initially go through four divisions to form a 16-cell cyst (*A*,*B*). In males, meiotic divisions increase the number of germ cells in the cyst to 64. These cells remain connected through the intercellular bridges by the fusome (*A*). In females, the fusome initially connects all 16 cells through intercellular bridges, with two cells containing the greater fraction of the fusions (*B*, cells 1 and 2). The fusome then breaks down after oocyte selection. (*A*, Adapted from Hime et al. 1996; reprinted with permission from The Company of Biologists © 1996. *B*, Adapted from Huynh and St Johnston 2004; reprinted with permission from Elsevier © 2004.) Both male (*C*) and female (*D*) *Drosophila* intercellular bridge formation begins in both sexes with an increase in phosphotyrosine epitopes, but subsequent processes are distinct. (*See facing page for legend.*)

detail by microscopy for more than half a century, we have only recently begun to learn how mammalian germ cell intercellular bridges form and function at the molecular level. Inroads have been made to determine the role of this unique mode of intercellular communication. Many molecules and organelles have been observed to pass through intercellular bridges, but their role in fertility is not fully understood. Although the bridges themselves have been conserved, the way they form and their function appears to vary between gender and species.

This article will review our current understanding of how intercellular bridges form and the role they play in gametogenesis. Examples will primarily be drawn from mammalian and *Drosophila* models where the most study has been completed. We will first provide a general description of somatic cell cytokinesis, because modification of this process forms intercellular bridges. Next, we will review *Drosophila melanogaster* ring canals, the name given to intercellular bridges in that species. We will conclude with what is known about formation and function of mammalian intercellular bridges.

INTERCELLULAR BRIDGES (RING CANALS) IN *DROSOPHILA MELANOGASTER*

Syncytial development of germ cell cysts is conserved from fruit flies to mammals in both males (Burgos and Fawcett 1955; Fawcett et al. 1959; Dym and Fawcett 1971; Moens and Hugenholtz 1975; Huckins and Oakberg 1978; Weber and Russell 1987; Ren and Russell 1991; Ventela et al. 2003) and females (Gondos et al. 1971; Pepling and Spradling 1998, 2001; Pepling et al. 1999). *Drosophila* male and female gametes develop clonally (Rasmussen 1973; Spradling et al. 2001; Gonzalez-Reyes 2003) from a single founder cell, the gonial cell in males and the cystoblast in females. In both genders, the founder cell initially divides four times. "Incomplete" cytokinesis during these divisions results in the formation of a 16-cell cyst (Fig. 1A) (Hime et al. 1996; Huynh and St Johnston 2004). The cells of the cyst are connected by intercellular bridges, also called ring canals in *Drosophila*. The number of intercellular bridges borne by each germ cell in the cyst directly correlates to the round of division from which it was derived. Thus, the founder cell and its first daughter cell will each have four intercellular bridges, their two daughter cells will have three bridges, the next four cells will have two bridges, and the final eight cells have only one bridge, for a total of 15 shared intercellular bridges (Fig. 1B) (intercellular bridges in cysts are in green).

Sixteen cells is the final number of germ cells in the ovarian cyst. Once this number is reached, the cyst migrates posteriorly in the ovary as it matures, and one germ cell will be selected as the oocyte and supported by the other 15 "nurse" cells (Huynh and St Johnston 2004). When male cysts reach 16 cells, two meiotic divisions occur to form a total of 63 shared intercellular bridges. This pattern of division is highly reproducible and, at least in females, is dependent on a structure called the fusome. By comparison, the number of cells in a mammalian germ cell cyst is much greater and less regulated (see below). Furthermore, no

Figure 1. (*Continued*) Male cleavage furrows lose actin and myosin but retain anillin and septins (*C*), whereas female cleavage furrows form an inner ring with Hts and Kelch and lose anillin (*D*). (*D*, Adapted from Robinson and Cooley 1996; reprinted with permission from Elsevier © 1996.) (*E*) Genes for specific steps in cytokinesis in *Drosophila* spermatocytes are shown. The mutants often prevent the formation of intercellular bridges, demonstrating the close relationship between cytokinesis components and intercellular bridge formation in male *Drosophila*. The gene names, and homologs when known, are provided in the text. (*E*, Adapted from Giansanti et al. 2004; reprinted with permission from American Society for Cell Biology © 2004.) (*F*) Cytoskeletal and non-tyrosine kinase pathways intersect to form female intercellular bridges. Src kinase activates Tec29, which localizes to the forming intercellular bridge. Phosphorylation of cytoskeletal components, such as kelch, is necessary for intercellular bridge formation. It is unknown which kinase is responsible for phosphorylating kelch. (*F*, Adapted from Cooley 1998; reprinted with permission from Cell Press © 1998.)

equivalent of the fusome has been identified in mammals.

The fusome (Fig. 1A,B, red) is a structure containing spectrin and hu-li tai shao (Hts), an adducin homolog (Yue and Spradling 1992) that connects the germ cells of the cyst through their intercellular bridges. The fusome is derived from a spherical structure in the germline stem cell called the spectrosome (Lin et al. 1994; de Cuevas et al. 1996). In females, the cystoblast inherits one third of the fusome from the germline stem cell (Deng and Lin 1997). As the germ cells divide to form the cyst the fusome grows and is distributed, but a larger fraction of the fusome is kept in one of the two cells formed by the first division. This asymmetric division is repeated three times. At the end, the cell with the most fusome material (one of the first two cells) will become the oocyte (de Cuevas and Spradling 1998), and the fusome breaks down (Fig. 1B). *Hts* and *α-spectrin* mutants lack a fusome and fail to specify an oocyte. Additionally, the divisions become randomly oriented and result in a variable number of cells (Yue and Spradling 1992; Lin et al. 1994; de Cuevas et al. 1996). In males, all 16 germ cells of the cyst are equivalent, and the fusome is maintained although meiosis and connects all 64 haploid spermatids (Fig. 1B) (Hime et al. 1996). The fusome plays a role in centrosome inheritance in spermatogenesis. A male sterile allele of *hts* eliminates the fusome. In this mutant dividing spermatogonia all have two centrosomes and a normal mitotic spindle, but spermatocytes have a variable number of centrosomes and defective spindles.

Passage of the fusome through intercellular bridges is a common feature of male and female *Drosophila* gametogenesis, and fusome communication through intercellular bridges is essential for fertility. Although intercellular bridges play a similar role in both genders in this respect, the mechanism of intercellular bridge formation differs in male and female *Drosophila*.

Stable intercellular bridges between male germ cells are composed of proteins involved in cytokinesis (Hime et al. 1996), namely, anillin and three *Drosophila* septins, Peanut (Pnut), Sep1, and Sep2. Septins are required for somatic cell cytokinesis in some, but not all, cell types (Glotzer 2005), and are capable of self-assembly into rings (Kinoshita et al. 2002). Anillin, which localizes early to the cleavage furrow (Field and Alberts 1995), directly interacts with septins (Field and Alberts 1995; Kinoshita et al. 2002), and was initially identified as an actin-binding protein in *Drosophila* embryos (Miller et al. 1989).

The earliest sign of intercellular bridge formation in both male and female *Drosophila* is the appearance of phosphotyrosine epitopes in late telophase (Fig. 1C,D) (Robinson and Cooley 1996). These "rings" of phosphotyrosine epitopes remain in the intercellular bridge even after it matures, growing in diameter with the bridge as the germ cells develop. After the appearance of phosphotyrosine epitopes, myosin II and actin from the contractile ring are lost from the mature intercellular bridge in male *Drosophila* (Fig. 1C). In contrast, actin remains a major component of the mature female intercellular bridge (Fig. 1D), but anillin and septins are no longer present. These are not the only compositional differences between male and female *Drosophila* intercellular bridges. Hts and kelch, essential proteins in the female intercellular bridge, are absent from male bridges. In fact, only one protein has been identified as a component of both male and female intercellular bridges, Pavarotti/MKLP1 (Carmena et al. 1998; Minestrini et al. 2002). Interestingly, although male germ cell intercellular bridges have different components from female germ cell intercellular bridges, they may be similar to intercellular bridges that have been reported connecting somatic cells of the follicular epithelium in ovary, imaginal disc cells, and larval brain in *Drosophila* (Kramerova and Kramerov 1999). There is also a single report of somatic intercellular bridges seen by electron microscopy in mammals, but the authors caution that it may be the result of a tissue preparation artifact (Witkin et al. 1995).

Brill et al. (2000) used *Drosophila* spermatocytes to screen for cytokinesis-defective mutations, because mutants with failure of cytokinesis in meiosis are easily identified by their sterility and multinucleated spermatids.

Nineteen genes were identified, including 16 novel loci and three known cytokinesis genes. The mammalian homolog is given, if known. However, detailed mapping of the mutated genes was not performed, so the actual gene products were not all identified. The mutants blocked cytokinesis at several stages (Fig. 1B) (Brill et al. 2000). Two mutants, *diaphanous* (*dia*, *Diaphanous* homolog) and *pebble* (*pbl*, *ECT2/RhoGEF* homolog), prevent formation of the actin and anillin rings in early telophase. Five of the mutants, *scapolo* (*scpo*), *celibe* (*cbe*), *james bond* (*bond*), *sauron* (*sau*), and *smeagol* (*sgo*), affect the central spindle and the F-actin ring, but do not prevent anillin accumulation at the equator. Ten genes, *four score* (*fsco*), *brunelleschi* (*bru*), *ciambella* (*cia*), *four way stop* (*fws*, *Cog5* homolog), *funnel cakes* (*fun*), *omelette* (*omt*), *onion rings* (*onr*), *four wheel drive* (*fwd*, a phosphatidylinositol 4-kinase), *and giotto* (*gio*), are primarily required for constriction of the F-actin ring. Lastly, *bird's nest soup* (*bns*) was required for actin ring disassembly. It is unclear if any of these genes are specifically involved in intercellular bridge formation. The effect on spermatogonial intercellular bridges in these mutants was not reported, but it can be inferred that the phenotypes were specific for spermatocytes. The fact that the known genes, *dia, pbl, fwd,* and *fws*, have previously been implicated in cytokinesis further supports the model that male *Drosophila* intercellular bridges are composed of cytokinesis proteins.

Female *Drosophila* intercellular bridges are required even after the oocyte is specified. Intercellular bridges connecting the 15 nurse cells to the oocyte grow dramatically in size from <1 μm to ~ 10 μm as the egg chamber matures (Cooley 1998). The asymmetric division of the fusome required for oocyte specification also causes microtubules to become polarized with their minus ends directed at the oocyte (Grieder et al. 2000). These modifications allow the ring canals to transfer several cytoplasmic components from the nurse cells to the oocyte, increasing its size. For example, oocyte-specific proteins and mRNAs (e.g., *Bicoid D* [Mische et al. 2007] and *oskar* [Huynh and St Johnston 2004]) are concentrated in the oocyte. The centrosomes from the nurse cells are inactivated and transferred to the oocyte (Bolivar et al. 2001), and mitochondria are also concentrated in the oocyte (Cox and Spradling 2003).

As in males, the first evidence of female *Drosophila* intercellular bridge formation is an increase in phosphotyrosine epitopes (Fig. 1D) (Robinson and Cooley 1996). The phosphotyrosines are seen as 0.5–1-μm rings in the stalled cleavage furrows at the end of the third mitosis (Robinson et al. 1994). The development of female intercellular bridges is known to require the convergence of cytoskeletal and kinase signaling pathways not regularly involved in cytokinesis (Fig. 1F) (Cooley 1998). First, consider the cytoskeletal pathway (Fig. 1F). By the fourth division hu-li tai shao (Hts) (Robinson et al. 1994) and additional filamentous actin (Tilney et al. 1996) localize to the inner rim of the intercellular bridge (Fig. 1D). Over the next few hours Kelch is added to the inner rim (Robinson et al. 1994), while anillin is removed from the outer rim of the intercellular bridge, and the bridge's diameter increases to 3–4 μm (Robinson and Cooley 1997). At this point, the bridges are considered "mature," but will continue to increase in diameter. Incorporation of Hts, Kelch, and filamentous actin all depend on *cheerio* (the mammalian *filamin* homolog), an actin cross-linking protein. Other proteins involved in actin dynamics such as the Arp2/3 complex (Hudson and Cooley 2002), Cortactin (Somogyi and Rorth 2004), and *Akap200* (Jackson and Berg 2002) have also been shown to regulate the intercellular bridge cytoskeleton.

Kelch is able to dimerize through its BTB domain and bind to actin with its Kelch repeat domain (Robinson and Cooley 1997). It may therefore serve as an actin cross-linking protein in the growing intercellular bridge (Robinson and Cooley 1997). *Kelch* mutant females are sterile, but otherwise the flies are completely normal (Robinson et al. 1994). Loss of Kelch causes disorganization of the actin filaments of the intercellular bridge preventing cytoplasm transport from the nurse cells to the oocyte (Robinson et al. 1994). The *kelch* gene produces a single transcript separated into two open reading frames by a stop codon. Although both the truncated, ORF1, and full-length versions of the protein are expressed by tRNA suppression of the stop codon, and both proteins are in the intercellular bridge. Oddly, ORF1 is sufficient to fully rescue the *kelch* intercellular bridge phenotype but full length is not.

Kelch is one protein in which the kinase regulatory pathway converges with the cytoskeletal pathway (Fig. 1F). The ability for Kelch to organize actin is regulated by the *Drosophila* Src nonreceptor tyrosine kinase homolog Src64. Src64 is does not effect the localization of Hts of Kelch, but its kinase activity is required for expansion of the intercellular bridge diameter (Dodson et al. 1998). Src64 signaling is required for phosphorylation of Kelch at tyrosine 627 (Kelso et al. 2002). Using site-directed mutagenesis, KelchY627A is able to rescue the actin disorganization phenotype of *kelch* mutant intercellular bridges, but the bridges fail to expand, phenocopying the *src64* mutant phenotype. Actin monomer exchange is also greatly reduced with KelchY627A. These results suggest that a major function of Src64 signaling is negative regulation of actin cross-linking by Kelch (Kelso et al. 2002).

Although Src64 is required for the phosphorylation of Kelch, it may not be the kinase that actually phosphorylates the protein. Mutations of another nonreceptor tyrosine kinase, Tec29, phenocopies the Src64 intercellular bridge phenotype (Roulier et al. 1998). Tec29 localizes to the intercellular bridge. Although Src64 does not localize to the intercellular bridge, it interacts with Tec29, and Src64 activity is required for Tec29 localization (Roulier et al. 1998; Lu 2004; Lu et al. 2004). The loss of either kinase results in a dramatic decrease in phosphotyrosine epitopes in the intercellular bridge. Although Tec29 may be the kinase that regulates Kelch, the only function currently known for Tec29 kinase activity is the creation of phosphotyrosine binding sites on unidentified intercellular bridge substrates for its own SH2 domain (Lu et al. 2004).

This overview of male and female *Drosophila* intercellular bridges shows that, although essential for fertility in both genders, they form by very different mechanisms. Mammalian intercellular bridges were also initially thought to have an essential role in fertility for both genders. The next section will review the current understanding of mammalian germ cell cysts.

MAMMALIAN INTERCELLULAR BRIDGES

Until 1955, the term "intercellular bridge" referred to short processes that appeared to connect squamous epithelial cells (Fawcett 1961). Electron microscopy has since shown that these structures are actually desmosomes, not true cytoplasmic connections between cells. As reviewed by Dym and Fawcett (1971), as early as 1865 LaValette St. George teased apart fresh testis and noticed that cells were able to divide, yet remain connected in a chain. Later observations by Sertoli (1877) (Fig. 2A) and von Ebner (1888) (Fig. 2B) described spermatogonia connected by processes. It was not until 1955 when electron microscopy studies examining the fine structure of cat spermatids defined the intercellular bridge as a stable cytoplasmic channel connecting cells (Fawcett et al. 1959). Shortly after, intercellular bridges were found in many animals including Hydra, fruit fly, opossum, pigeon, rat, hamster, guinea pig, rabbit, monkey, and man (Fawcett et al. 1959).

Intercellular bridges have become the basis for describing early stages of spermatogenesis. A widely used model of spermatogenesis is based on the work of Huckins (1971) and Oakberg (1971). Male germ cells start as diploid spermatogonia, proceed through meiosis as spermatocytes, and complete development as haploid spermatids. The complete process involves nine to 11 spermatogonial divisions (Huckins 1971; de Rooij and Russell 2000; Chiarini-Garcia and Russell 2002) plus two meiotic divisions. Each division results in the formation of an intercellular bridge (Weber and Russell 1987), and spermatogonia are initially classified by their intercellular bridges. The nomenclature for spermatogonial development is as follows:

$$A_s \rightarrow A_{pr} \rightarrow A_{al}(\text{multiple rounds}) \rightarrow A_1 \rightarrow A_2 \rightarrow A_3 \rightarrow A_4 \rightarrow \text{In} \rightarrow \text{B}.$$

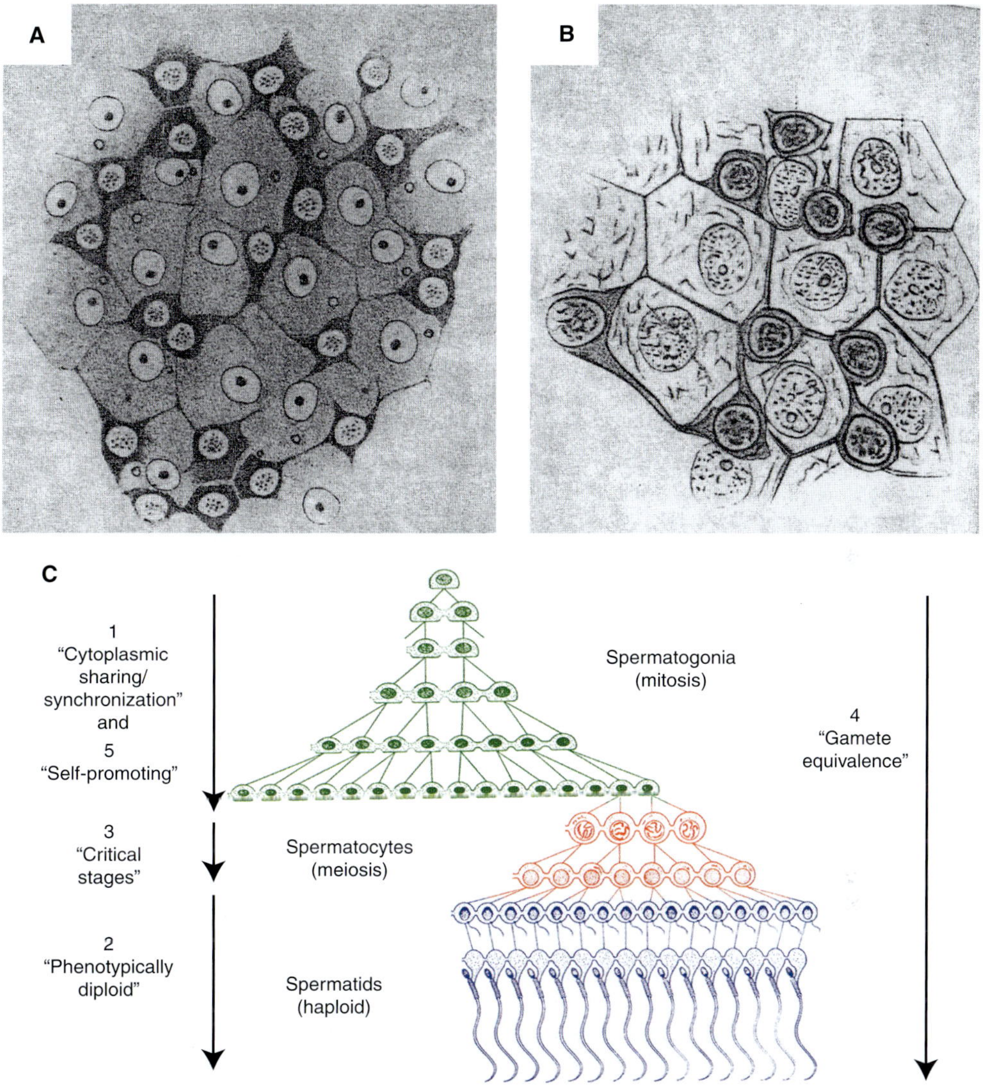

Figure 2. Mammalian intracellular bridges. Nineteenth-century sketches of intercellular bridges. Observations by Sertoli (1877) (*A*) and von Ebner (1888) (*B*) show darker staining spermatogonia connected by intercellular bridges. The lightly staining cells are Sertoli cells. (*C*) Overview of proposed models for intercellular bridge function. Each model specifies a particular period when intercellular bridges are necessary. The "cytoplasmic sharing" and "self-promoting" models involve all dividing cells, spermatogonia, and spermatocytes. The "phenotypically diploid" model involves haploid postmeiotic cells. The "critical stage" model addresses entry into meiosis. The "gamete equivalence" could involve any stage of spermatogenesis. (*C*, Adapted from Dym and Fawcett 1971; reprinted with permission from Society for the Study of Reproduction © 1971.)

The A type single (A_s) spermatogonium is the stem cell for spermatogenesis and is the only dividing male germ cell without intercellular bridges. A single mouse testis is thought to contain about 35,000 A_s spermatogonia (Tegelenbosch and de Rooij 1993). The A_s cell differentiates by dividing into A type paired (A_{pr}) spermatogonia. A_{pr} spermatogonia divide to form a chain of four A type aligned (A_{al}) spermatogonia, and A_{al} spermatogonia can go through multiple rounds of division before differentiating into A_1 spermatogonia (Huckins

1971; de Rooij and Russell 2000). The remaining stages through intermediate and B type spermatogonia link mitotic division with progression to the next stage. This process theoretically leads to syncytia joining thousands of cells (Moens and Hugenholtz 1975; de Rooij and Russell 2000). In practice, up to 650 cells have been observed in a single syncytium (Ren and Russell 1991).

The different stages of spermatogonia can be defined by a skilled examiner by their appearance and morphology in light (Huckins 1971; Chiarini-Garcia et al. 2001) and electron microscopy (Chiarini-Garcia and Russell 2002). Early stages of spermatogonia, such as A_s, are most easily identified in whole mounted tubules. The examiner focuses on the basal plane, where the least differentiated cells lie (Clermont and Bustos-Obregon 1968). A_s cells are identified when no other A type spermatogonia is within 25μm of the cell. A_s, A_{pr}, and A_{al} spermatogonia can also be identified with the molecular markers such as promyelocytic leukemia zinc finger (PLZF), a protein required for spermatogonial stem cell self-renewal (Buaas et al. 2004).

The morphology of intercellular bridges has been studied exhaustively by microscopy (e.g., Burgos and Fawcett 1955; Fawcett et al. 1959; Dym and Fawcett 1971; Moens and Hugenholtz 1975; Huckins and Oakberg 1978; Weber and Russell 1987; Ren and Russell 1991; Ventela et al. 2003). The diameter of the bridge ranges in size starting at under 1 μm in spermatogonia, 1–1.3 μm in intermediate and B type spermatogonia, 1.4–1.7 μm in spermatocytes, 1.8 μm in step 1 spermatids, 3 μm in step 18 spermatids, and then shrinking slightly to 2 μm in step 19 spermatids (Weber and Russell 1987). By electron microscopy, a prominent electron dense "bridge density" ranging from 30–60 nm in thickness lines the walls of the intercellular bridge. In cytokinesis, the bridge density begins to form on the inner edge of the cleavage furrow during ingression. A midbody with normal appearance forms and then breaks down with the central spindle, leaving the intercellular bridge and the lining bridge density (Weber and Russell 1987).

Several roles for mammalian intercellular bridges in spermatogenesis have been proposed (Fig. 2C) (Guo and Zheng 2004). Two hypotheses have the most support. The first hypothesis is that intercellular bridges permit "cytoplasmic sharing" of essential signals for the synchronous cell divisions seen in longitudinal segments of seminiferous tubules (Fawcett et al. 1959; Huckins and Oakberg 1978; Ren and Russell 1991). In this untested model, synchronization is proposed to play an essential role in progression through spermatogenesis. A second hypothesis points out that intercellular bridges may be necessary for haploid germ cells to remain "phenotypically diploid" after meiosis (Fawcett et al. 1959; Erickson 1973; Braun et al. 1989). The most extreme example of this would be for genes with only one copy (e.g., genes on the sex chromosomes). Supporting this second hypothesis, mRNA expressed from a single copy transgene encoding human growth hormone (hGH) under the control of the protamine 1 regulatory sequences was observed to distribute evenly in haploid spermatids (Braun et al. 1989). The protein product was also found equally distributed among the spermatids. In a nontransgenic example, the testis brain RNA-binding protein (TB-RBP) functions as an RNA-binding protein in the testis and binds to *Akap4* mRNA. *Akap4* is an X-linked gene essential for sperm motility and fertility (Miki et al. 2002). *Akap4* mRNA forms a complex with TB-RBP and the Ter ATPase and passes through intercellular bridges between spermatids (Morales et al. 2002). Additionally, because of the large size of intercellular bridges, even organelles (Ventela et al. 2003) have been observed to move between haploid spermatids.

A second hypothesis proposes that intercellular communication between germ cells occurs at "critical stages," such as coordinating the entry into meiosis (Stanley et al. 1972; Robinson and Cooley 1996). In this model, the event is coordinated by the "rapid" transfer of an unidentified signal/substance through the syncytium. A third hypothesis is that intercellular bridges create "gamete equivalence" and would thus result in a more uniform population of sperm which, overall, would be of higher quality (Guo and Zheng 2004). A final hypothesis is that intercellular bridges allow for the elimination of cells with "self-promoting genes" (LeGrand 2001). In this model, a mutation exists which would give a selective advantage to the mutated gamete but would be detrimental to future progeny. For example, the cell could have uncontrolled and greatly increased cell proliferation and have a mutation in a gene required for apoptosis. Normally functioning cells in the syncytium detect something is wrong through the intercellular bridges and the problem is eliminated by apoptosis of the entire cyst. Of course, because all the cells are clonally related, the mutation would have to be acquired somehow after the initial A_s cell division.

To test these hypotheses requires manipulation of the intercellular bridge. The molecular mechanism of mammalian intercellular bridge formation is a recent area of study. The protein TEX14 is expressed specifically in germ cells and localizes to the intercellular bridge in mice (Greenbaum et al. 2006) and humans (Fig. 3A) (Greenbaum et al. 2007). The protein contains three amino-terminal ankyrin repeats, centrally is a dead-kinase domain, and the remainder contains coiled-coil motifs. Interestingly, when compared by BLAST analysis to the *Drosophila* genome, the dead-kinase domain bears more resemblance to *src64* and *tec29* (see above) than any other kinases.

Knockout of mouse *Tex14* eliminates the formation of intercellular bridges (Greenbaum et al. 2006) and results in male infertility. It was found that without intercellular bridges spermatogenesis fails to complete meiosis. In fact, molecular studies showed that only a very small number of spermatocytes make it to pachytene. Importantly, this result proved that intercellular bridges are not required for the transit amplification of germ cells from the A_s stem cells spermatogonia through the initiation of meiosis.

Although it is clear completion of meiosis during spermatogenesis requires intercellular bridges, the exact role of the intercellular bridge remains unknown. Many of the proposed roles listed above may be relevant. Perhaps intercellular bridges do allow passage of a signal for entering the "critical stage" of meiosis through an unknown mechanism. Interestingly, mammalian somatic cells die through a p53-dependent mechanism if they fail to complete cytokinesis (Postiglione et al. 2009). For germ cells, incomplete cytokinesis is the norm and it is completion that results in increased apoptosis (Greenbaum et al. 2006). The one function intercellular bridges are known to play, passage of RNA material in haploid cells, could not be evaluated because no germ cells progressed through meiosis. Thus, intercellular bridges are required for spermatogenesis but their essential role remains undiscovered.

Embryonic mammalian male (Gondos and Hobel 1971; Fukuda et al. 1975) and female

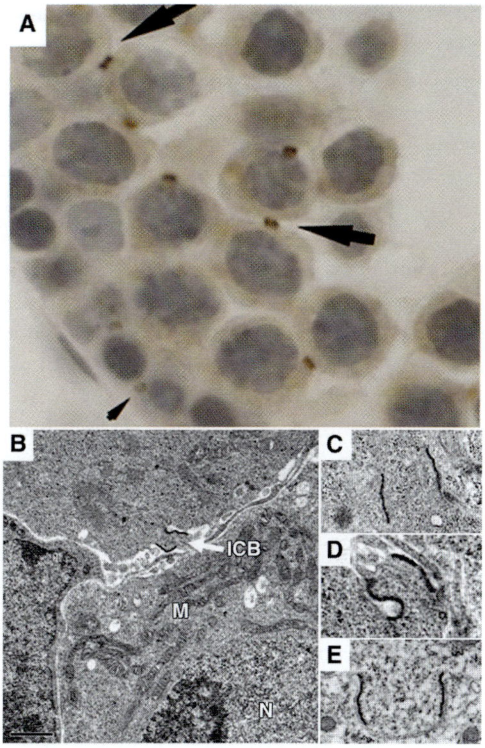

Figure 3. TEX14 is a critical component of mammalian intracellular bridges. (*A*) Immunohistochemical staining of control testes using anti-TEX14 antibody labels ring-shaped structures (arrows) that are absent from TEX14 null germ cells. (*B–E*) Electron microscopy reveals electron dense material lining cytoplasmic channels between newborn ovarian germ cells. No intercellular bridges were found in knockout ovaries. ICB, intercellular bridge; M, mitochondria; N, nucleus.

(Gondos et al. 1971; Pepling and Spradling 1998, 2001; Pepling et al. 1999) germ cells are also connected by intercellular bridges. Mitotic division of germ cells occurs for a finite period during embryogenesis in mammals. Germ cell division in the embryonic ovary is synchronized, presumably through intercellular bridges (Pepling and Spradling 1998). In female mice, germ cells enter meiosis around E14.5 (Borum 1961; Pepling and Spradling 1998). No divisions occur beyond meiosis, so no new intercellular bridges form beyond embryogenesis. Female germ cell cysts break down over the second through fourth day after birth (Pepling and Spradling 2001) leaving single oocytes to form primary follicles. Similar to *Drosophila*, intercellular bridges have been proposed to allow "nurse cells" to promote oocyte development (Pepling and Spradling 2001). Supporting this idea in mice, mitochondria have been shown to reorganize and increase in number around the time of birth (Pepling et al. 1999; Pepling and Spradling 2001). In males, newborn mouse prospermatogonia migrate and arrive at the seminiferous tubule basement membrane within the first few postnatal days (McGuinness and Orth 1992). Mitosis resumes at postnatal day three when mitotic figures are seen in 10% of the prospermatogonia before reaching the basement membrane (Nebel et al. 1961; McGuinness and Orth 1992).

The *Tex14* knockout mice were used to evaluate the role of embryonic intercellular bridges. Tex14 was found to localize to embryonic bridges and intercellular bridges seemed to be absent in *Tex14* knockouts by electron microscopy (Fig. 3B). Still, a second method of confirmation was desired to show the bridges were truly not present. Immunofluorescence is the best way to look at a large amount of tissue for intercellular bridges, but there were no other known embryonic bridge components. A biochemical approach was taken to find more components of the intercellular bridge (Greenbaum et al. 2007). Testis bridges were used because they were more abundant and easier to obtain. Several new components were identified, and the order in which they joined and left the intercellular bridge could be followed in newborn testis. Using one of these new intercellular bridge markers it was confirmed by immunofluorescence that the embryonic intercellular bridges were gone.

Surprisingly, *Tex14* knockout females remained fertile without intercellular bridges! This was not the case in *Drosophila* in which intercellular bridges play a well-described critical role in fertility (see above). Apparently, despite a billion years of conservation, female intercellular bridges are not essential. That is not to say they do not play an important function. It is obvious that selective pressures exist in nature that are absent in a controlled laboratory environment. However, there is a possibility that the female bridges exist as a vestigial structure purely because of the essential role intercellular bridges play in spermatogenesis.

A SPECIAL FORM OF CYTOKINESIS

A typical description of cytokinesis is presented as, "Cytokinesis is the final process in cell division through which one cell is separated into two daughter cells." Clearly, this is not the best definition for a section whose central topic is intercellular bridges, the stable cytoplasmic connection joining the daughter cells. Nonetheless, it is important to describe the basic principles of cytokinesis, as intercellular bridge formation is a modification of how the vast majority of cells divide. Although the most dramatic event of cytokinesis in mammalian cells is ingression of the cleavage furrow, this single event is not the whole picture. The process of cytokinesis can be divided into at least three broad stages as follows: (1) selecting the site of cell division (Burgess and Chang 2005), (2) furrow ingression and formation of a contractile ring (Matsumura 2005; Wang 2005), and (3) abscission of the midbody (von Dassow and Bement 2005). Although the end result of this process, cell separation, is the same in all eukaryotic organisms, there are several variations on the theme (Guertin et al. 2002), with plants even replacing the processes of furrow formation and ingression with a cytoskeletal/vesicular framework called the phragmoplast (Otegui et al. 2005). For our purpose, the

 Cite this article as *Cold Spring Harb Perspect Biol* doi: 10.1101/cshperspect.a005850

following section, when possible, will review the mechanism of cytokinesis as it relates to mammalian cells. Furthermore, although early events in cytokinesis will be addressed briefly, the greatest focus will be on the late events of cytokinesis, as these directly relate to intercellular bridge formation.

Early in cytokinesis, it is necessary to select the site of cell division. In animal cells, the region where the cleavage furrow will form is specified by the microtubules of the mitotic apparatus at the onset of anaphase. This linkage to the mitotic apparatus provides the necessary coordination of chromosomal segregation with cell division. The cleavage site forms at the former metaphase plate, bisecting the two centrosomes. Microtubule depolymerization studies and manipulation of the spindle apparatus (Rappaport 1986; Rieder et al. 1997) show that the mitotic spindle somehow signals to the cortex to determine the region where the cleavage furrow will form. Despite more than a hundred years of study pioneered by Ray Rappaport (Canman and Wells 2004), most aspects of the signaling process remain unknown (Burgess and Chang 2005). Both the signal and the region of the mitotic apparatus that emits the signal remain to be identified. Although many models have been proposed (Canman and Wells 2004; Burgess and Chang 2005), they each have drawbacks. The mechanism of cleavage furrow site selection seems distinct from intercellular bridge formation because it is an early event in cytokinesis that occurs normally in mammalian germ cells. For this reason, these models will not be discussed in further detail.

Furrow ingression follows site selection. The predominant model for the process of furrow formation is the contractile ring hypothesis (Satterwhite and Pollard 1992). This process is very similar in yeast and animal cells, which both divide using an actomyosin contractile ring. Actin and myosin II assemble in a ring around the perimeter of the putative cleavage site. Like a contracting myofiber, antiparallel actin filaments slide together through the action of myosin II resulting in contraction of the ring. The process often draws comparison to a "purse string" as it squeezes the plasma membrane (Wang 2005). The contraction of the equatorial membrane generates enough force to bend a flexible microneedle (Rappaport 1967). As in cleavage site selection (above), the mitotic spindle (now called the central spindle or spindle midzone during furrowing) is generally accepted to regulate cleavage furrowing (Matsumura 2005).

Several evolutionarily conserved complexes essential for cytokinesis have been identified in the central spindle (Glotzer 2005). These are the Aurora B-INCENP-CSC1-Survivin complex, the PRC1-KIF4 complex, the ECT2/RhoGEF (Glotzer 2005), and the mitotic kinesin-like protein 1 (MKLP1)-Male germ cell Rac GTPase-activating protein (MgcRacGAP) complex (also called centralspindlin) (Mishima et al. 2002). Loss of any of these complexes causes cytokinesis failure. It should be pointed out that the name MgcRacGAP (Male germ cell Rac GTPase-activating protein) is somewhat misleading. Although originally described as having predominantly male germ cell expression (Toure et al. 1998), like MKLP1, it is ubiquitously expressed and required for cytokinesis in all cells. Accordingly, MgcRacGAP homozygous gene trap mice die during preimplantation development with multinucleated blastomeres and a dramatic decrease in cell number at day E3.5 because of cytokinesis failure (Van de Putte et al. 2001).

The completion of furrow ingression forms a transient intercellular bridge called the midbody, which connects the daughter cells at the end of cytokinesis. Germ cells form intercellular bridges by altering the fate of the midbody. Rather than discard the transient intercellular bridge by abscission, it is modified into a permanent structure. The composition of the midbody matrix contains several of the conserved complexes mentioned previously (Glotzer 2005). In particular, the centralspindlin complex (Mishima et al. 2002), MKLP1 (Matuliene and Kuriyama 2004), and MgcRacGAP (Arar et al. 1999), is essential for forming the midbody matrix. RNA interference (RNAi) against MKLP1 in mammalian cells prevents formation of a normal midbody matrix resulting in unstable midbodies and failure of cytokinesis (Matuliene and Kuriyama 2004). MKLP1 in the midbody exists in a complex with 1-to-1 stoichiometry with MgcRacGAP, and this complex, centralspindlin, has been reconstituted and shown to bind microtubules in vitro (Mishima et al. 2002). It also indirectly recruits components needed for abscission.

Abscission is the final process in cytokinesis by which the midbody is removed to achieve truly individual cells. Compared to cleavage furrow ingression, which takes ~20 min, abscission is a relatively slow process, requiring 1–3 h to complete (von Dassow and Bement 2005). At least two processes are necessary to achieve abscission. First, the organization of the matrix in the middle of the midbody forms a central ring. By light microscopy, this structure is called the Flemming body, a slightly bulging region right in the middle of the midbody. The formation of the midbody ring is followed by membrane fusion events to cleave the midbody (von Dassow and Bement 2005). Many of the proteins required in these processes have been identified (Gromley et al. 2003, 2005; Low et al. 2003).

There are several vesicle-associated proteins whose essential role in abscission has been identified. The first identified proteins of this class were syntaxin 2 and endobrevin (Low et al. 2003), members of the soluble NSF attachment receptor (SNARE) complex required for SNARE mediated membrane fusion events (Hanson et al. 1997). Syntaxin 2 and endobrevin colocalize laterally to the midbody matrix. RNAi showed that both proteins are essential for abscission of the midbody in HeLa cells (Low et al. 2003).

The identification of other trafficking proteins came, surprisingly, from studying centrosomal proteins. Centriolin, initially known only as a centrosome component, was found to localize to the midbody of HeLa cells (Gromley et al. 2003). RNAi of centriolin did not lead to a microtubule defect but resulted in a novel cytokinesis phenotype (Gromley et al. 2003). Rather that form multinucleated cells, as seen in previous cytokinesis defects, loss of centriolin caused long, thin intercellular bridges to remain connecting the daughter cells. Further studies showed that the SNARE proteins, syntaxin 2, endobrevin, and snapin, require centriolin for localization (Gromley et al. 2005). The exocyst complex (Hsu et al. 2004) is required for vesicle mediated secretion and also plays a role in abscission. Centriolin is also required for localization of the exocyst complex to the midbody. Exocyst complex proteins sec3, sec5, sec8, sec15, and exo84 failed to localize to the midbody in centriolin RNAi knockdown experiments (Gromley et al. 2005). Another centrosomal protein, centrosome protein 55 kDa (CEP55), was found to localize to the midbody matrix in somatic cells (Fabbro et al. 2005; Martinez-Garay et al. 2006). CEP55 was found to bind to the centralspindlin complex (Zhao et al. 2006). Deficiency of CEP55 leads to arrest at the midbody stage in somatic cells and formation of multinucleated cells (Carlton and Martin-Serrano 2007; Morita et al. 2007). In addition, knockdown of the direct downstream interacting partners of CEP55 such as ESCORT-1 (endosomal sorting complex required for transport-1) or ALIX (ALG-2 interacting protein X, also known as programmed cell death 6 interacting protein or PCD6IP), leads to a similar phenotype (Carlton and Martin-Serrano 2007; Morita et al. 2007). These studies suggest that these interactions are essential for somatic cell abscission (Carlton and Martin-Serrano 2007; Morita et al. 2007; Carlton et al. 2008).

FORMATION OF THE STABLE INTERCELLULAR BRIDGE

Knockout of mouse *Tex14* showed that the formation of stable intercellular bridges is essential for spermatogenesis and fertility (Greenbaum et al. 2006). However, until quite recently, it remained unclear how TEX14 participated in stable intercellular bridge formation to prevent abscission and the completion of cytokinesis in male germ cells.

CEP55 was identified as a component of stable intercellular bridges, and is perfectly colocalized with TEX14 as ring-shaped intercellular bridge structures in germ cells throughout the seminiferous tubules at all stages of spermatogenesis. The bridge diameter expands from the juvenile to the adult stage, suggesting that additional factors such as CEP55 and TEX14 are added to the stable intercellular bridge during spermatogenesis. In the female, CEP55 is expressed and colocalized with TEX14 in the embryonic 18.5-day-old mouse ovary.

In somatic cells, abscission requires CEP55, ALIX, and TSG101 (a component of the ESCRT complex). CEP55 is recruited from the centrosome to the midbody and interacts with MKLP1 (Fabbro et al. 2005; Martinez-Garay et al. 2006; Zhao et al. 2006). Furthermore, CEP55 forms homodimers through coiled-coil interactions, and subsequently, ALIX and TSG101 are recruited to the midbody by binding the "hinge" region of CEP55 (Fig. 4, lower panel) (Martinez-Garay et al. 2006; Carlton and Martin-Serrano 2007; Morita et al. 2007). Glycine (G)-proline (P)-proline (P)-X-X-X-tyrosine (Y) (GPPX3Y) motifs in ALIX and TSG101 are critical for this interaction with the "hinge" region of CEP55 (Morita et al. 2007; Lee et al. 2008). CEP55 was also identified as a TEX14 interactor (Iwamori et al. 2010). Nineteen TEX14 orthologs contain a GPPX3Y, which mimic the ALIX and TSG101 interactions with CEP55 (Iwamori et al. 2010; Wang et al. 2001; Wu et al. 2003). Although there is significant divergence of these orthologs in their carboxy-terminal regions compared to the

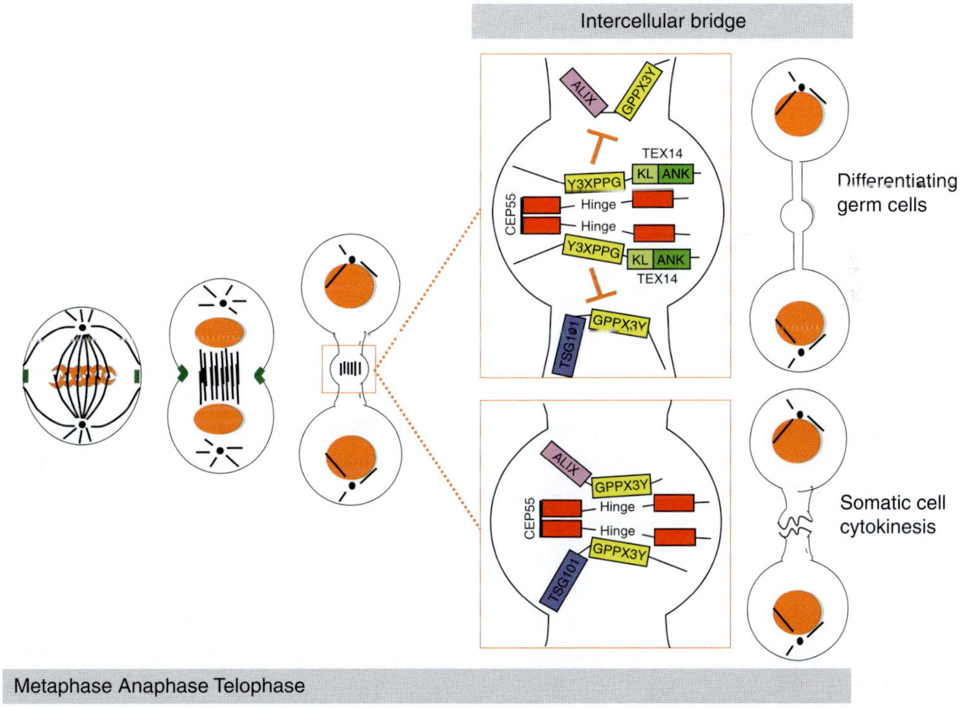

Figure 4. Stable intercellular bridge formation requires TEX14 disruption of interactions between CEP55 and other cytokinesis components. During mitosis, the contractile ring (green) is established adjacent to the metaphase plate. At anaphase, spindle microtubules assume a bipolar orientation (central spindle). In telophase, somatic cells (*left* lower box), a complex is assembled at the midbody (red open box) composed of ALIX (blue box) and TSG101 (pink box) bound to the hinge region of CEP55 (red box) through their GPPX3Y domains, resulting in abcission. In contrast, in differentiating germ cells the hinge region of CEP55 instead interacts with TEX14 and forms a stable ring-like structure, potentially excluding ALIX and TSG101 from the midbody and preventing abscission. KL, kinase-like domain; ANK, ankyrin domain.

ankyrin repeats (ANK) and kinase-like (KL) domains in their amino termini, 17 of the 19 proteins share a similar GPPX3Y motif in their carboxy termini, with the exceptions being the *Xenopus laevis* and dolphin orthologs. In addition, several additional amino acids are conserved between TEX14 and other GPPX3Y-containing proteins including an alanine upstream of the GPPX3Y motif (also present in human ALIX and TSG101), serine/threonine/alanine within the motif, and two prolines downstream (one also present in TSG101). We hypothesize the GPPX3Y motif of TEX14 binds strongly to CEP55 to block similar GPPX3Y motifs of ALIX and TSG101 preventing their interaction with CEP55 and localization to the midbody (Fig. 4, upper panel). In particular, the first proline and the tyrosine of the GPPX3Y motif were sufficient for this portion of TEX14 to interact with CEP55. The exogenous expression of TEX14 in cell culture precludes ALIX from the midbody and produces interconnected somatic cells. By binding endogenous CEP55 and blocking ALIX and TSG101 from interacting with CEP55, TEX14 inhibits the completion of cytokinesis and stabilizes the transient intercellular bridge. This shifts the role of CEP55 from a midbody organizer that recruits proteins for abscission to the midbody to a major component in the stable intercellular bridge. TEX14:CEP55 complexes are critical for the formation of stable intercellular bridges.

In general, abscission can be described as MKLP1 → CEP55 → ALIX/TSG101 → → → abscission; however, in differentiating male (and presumably female) germ cells, the steps involved appear to be MKLP1 → TEX14 → CEP55 → intercellular bridge. The fact that TEX14 appears early in the process of cytokinesis bound to MKLP1 and filling the intercellular bridge (Greenbaum et al. 2007) may allow TEX14 to compete for CEP55 more effectively. In testes of TEX14 null mice, CEP55 localizes to the midbody by binding MKLP1 and, in the absence of local TEX14, its interaction with ALIX and TSG101 is permitted. Complete cytokinesis occurs, disrupting the transient intercellular bridge, and resulting in an eventual failure of spermatogenesis during meiotic prophase (Greenbaum et al. 2006).

OTHER MAMMALIAN INTERCELLULAR BRIDGE FUNCTIONS

Recently our group identified a second testis-abundant component of the mammalian intercellular bridge, RBM44 (RNA binding motif 44), through its association with TEX14 (Iwamori et al. 2011). RBM44 is not, however, a known component of the midbody, but RBM44 orthologs are present in placental and marsupial mammals, birds, and fish. Distinct from TEX14, RBM44 does not localize to all intercellular bridges, but only to those in meiotic germ cells. Based on the presence of an RRM (RNA recognition motif) domain in RBM44 and robust immunostaining of spermatocyte cytoplasm in addition to the bridge, the protein may serve to transport mRNA, including across intercellular bridges. Furthermore, targeted disruption of *Rbm44* results in an unexpected increase in sperm number without the loss of intercellular bridges. Therefore, the mechanistic role of RBM44, unlike the CEP55-TEX14 interaction, is not to physically stabilize the bridge but may be to eliminate select meiotic germ cells, perhaps as a quality checkpoint. Because physiologic apoptosis of spermatocytes predominantly occurs during the meiotic divisions, potentially by activation of the spindle checkpoint, it is notable that the peak of RBM44 staining occurs during stage XII of spermatogenesis when the meiotic divisions occur. Alternatively, the lack of infertility in RBM44 null mice may be due at least partially from redundancy with the more than 100 RRM domain-containing proteins; however, these possess little similarity to RBM44 outside the RRM domain. The RRM domain of RBM22 has the highest identity to that of RBM44, with those of RBMX, RBMXL2, RBMXRT, RBMY1A1, RBM4, RBM4B, RBM14, RBM15B, BRUNOL6, and DAZAP1 only slightly less identical. Tissue microarray data suggest that some of this group may be testis-expressed (e.g., RBM4B, RBM14) or testis-specific (e.g., RBMXL2) (UCSC genes http://genome.ucsc.edu/). RBM22 mRNA appears to increase over ovarian follicular development, and RBM4 and RBM14 are expressed in male germ cells, with distinct kinetics (Mammalian Reproductive Genetics database http://mrg.genetics.washington.edu/). Intriguingly, RBM22 interacts with ALG-2 (Montaville et al. 2006), a binding partner of midbody protein ALIX, so TEX14 interactions with RRM domain-containing proteins could act to redundantly block ALIX entry into the midbody by a CEP55-independent mechanism.

CONCLUDING REMARKS

Additional questions remain regarding mammalian stable intercellular bridges. TEX14 knockout studies strongly support the "critical stage" hypothesis (Fig. 2C), but do not exclude the "synchronization" or "phenotypically diploid" hypotheses. The particular functions contributed by intercellular bridges essential to completion of meiotic prophase in spermatocytes have yet to be elucidated. Intercellular bridges are not essential to ovarian folliculogenesis, but whether they contribute to oocyte function or act redundantly with oocyte-specific pathways is unclear. To ensure that germ cell intercellular bridges remain stable in the face of environmental insults, additional protein interactions may be required in addition to the TEX14:CEP55 interaction, such as between TEX14 and other known or novel bridge components. Exclusion of abscission components from the midbody by the TEX14: CEP55 interaction may also indirectly affect other abscission events, such as those involving the ESCRT complex. Because TEX14 exerts such a large role in the formation of stable intercellular bridges and interacts with general abscission components, we believe that characterization of the key molecular mechanisms underpinning this germ cell–specific variation on abscission will facilitate drug therapy targeted at disrupting stabile intercellular bridges in germ cells for male contraception. Conversely, a small molecule mimic of TEX14 might be applicable as a cancer therapeutic targeted at blocking abscission in continuously proliferating cancer cells.

ACKNOWLEDGMENTS

These studies were supported in part by the Eunice Kennedy Shriver NICHD/NIH through cooperative agreements U54-HD07495 as part of the Specialized Cooperative Centers Program in Reproduction and Infertility Research, R01-HD057880, and U01-HD60496 as part of the Cooperative Program in Male Contraception (to M.M.M.) and T32HD07165 (to G.M.B).

REFERENCES

Arar C, Ott MO, Toure A, Gacon G. 1999. Structure and expression of murine mgcRacGAP: Its developmental regulation suggests a role for the Rac/MgcRacGAP signalling pathway in neurogenesis. *Biochem J* 343 Pt **1:** 225–230.

Bolivar J, Huynh JR, Lopez-Schier H, Gonzalez C, St Johnston D, Gonzalez-Reyes A. 2001. Centrosome migration into the *Drosophila* oocyte is independent of BicD and egl, and of the organisation of the microtubule cytoskeleton. *Development* **128:** 1889–1897.

Borum K. 1961. Oogenesis in the mouse. A study of the meiotic prophase. *Exp Cell Res* **24:** 495–507.

Braun RE, Behringer RR, Peschon JJ, Brinster RL, Palmiter RD. 1989. Genetically haploid spermatids are phenotypically diploid. *Nature* **337:** 373–376.

Brill JA, Hime GR, Scharer-Schuksz M, Fuller MT. 2000. A phospholipid kinase regulates actin organization and intercellular bridge formation during germline cytokinesis. *Development* **127:** 3855–3864.

Buaas FW, Kirsh AL, Sharma M, McLean DJ, Morris JL, Griswold MD, de Rooij DG, Braun RE. 2004. Plzf is required in adult male germ cells for stem cell self-renewal. *Nat Genet* **36:** 647–652.

Burgess DR, Chang F. 2005. Site selection for the cleavage furrow at cytokinesis. *Trends Cell Biol* **15:** 156–162.

Burgos MH, Fawcett DW. 1955. Studies on the fine structure of the mammalian testis. I. Differentiation of the spermatids in the cat (*Felis domestica*). *J Biophys Biochem Cytol* **1:** 287–300.

Canman JC, Wells WA. 2004. Rappaport furrows on our minds: The ASCB Cytokinesis Meeting Burlington, VT July 22–25, 2004. *J Cell Biol* **166:** 943–948.

Carlton JG, Martin-Serrano J. 2007. Parallels between cytokinesis and retroviral budding: A role for the ESCRT machinery. *Science* **316:** 1908–1912.

Carlton JG, Agromayor M, Martin-Serrano J. 2008. Differential requirements for Alix and ESCRT-III in cytokinesis and HIV-1 release. *Proc Natl Acad Sci* **105:** 10541–10546.

Carmena M, Riparbelli MG, Minestrini G, Tavares AM, Adams R, Callaini G, Glover DM. 1998. *Drosophila* polo kinase is required for cytokinesis. *J Cell Biol* **143:** 659–671.

Chiarini-Garcia H, Russell LD. 2002. Characterization of mouse spermatogonia by transmission electron microscopy. *Reproduction* **123:** 567–577.

Chiarini-Garcia H, Hornick JR, Griswold MD, Russell LD. 2001. Distribution of type A spermatogonia in the mouse is not random. *Biol Reprod* **65:** 1179–1185.

Clermont Y, Bustos-Obregon E. 1968. Re-examination of spermatogonial renewal in the rat by means of seminiferous tubules mounted "in toto". *Am J Anat* **122:** 237–247.

Cooley L. 1998. *Drosophila* ring canal growth requires Src and Tec kinases. *Cell* **93:** 913–915.

Cox RT, Spradling AC. 2003. A Balbiani body and the fusome mediate mitochondrial inheritance during *Drosophila* oogenesis. *Development* **130:** 1579–1590.

de Cuevas M, Spradling AC. 1998. Morphogenesis of the *Drosophila* fusome and its implications for oocyte specification. *Development* **125:** 2781–2789.

de Cuevas M, Lee JK, Spradling AC. 1996. α-spectrin is required for germline cell division and differentiation in the *Drosophila* ovary. *Development* **122:** 3959–3968.

de Rooij DG, Russell LD. 2000. All you wanted to know about spermatogonia but were afraid to ask. *J Androl* **21:** 776–798.

Deng W, Lin H. 1997. Spectrosomes and fusomes anchor mitotic spindles during asymmetric germ cell divisions and facilitate the formation of a polarized microtubule array for oocyte specification in *Drosophila*. *Dev Biol* **189:** 79–94.

Dodson GS, Guarnieri DJ, Simon MA. 1998. Src64 is required for ovarian ring canal morphogenesis during *Drosophila* oogenesis. *Development* **125:** 2883–2892.

Dym M, Fawcett DW. 1971. Further observations on the numbers of spermatogonia, spermatocytes, and spermatids connected by intercellular bridges in the mammalian testis. *Biol Reprod* **4:** 195–215.

Erickson RP. 1973. Haploid gene expresion versus meiotic drive: The relevance of intercellular bridges during spermatogenesis. *Nat New Biol* **243:** 210–212.

Fabbro M, Zhou BB, Takahashi M, Sarcevic B, Lal P, Graham ME, Gabrielli BG, Robinson PJ, Nigg EA, Ono Y, et al. 2005. Cdk1/Erk2- and Plk1-dependent phosphorylation of a centrosome protein, Cep55, is required for its recruitment to midbody and cytokinesis. *Dev Cell* **9:** 477–488.

Fawcett DW. 1961. Intercellular bridges. *Exp Cell Res* Suppl **8:** 174–187.

Fawcett DW, Ito S, Slautterback D. 1959. The occurrence of intercellular bridges in groups of cells exhibiting synchronous differentiation. *J Biophys Biochem Cytol* **5:** 453–460.

Field CM, Alberts BM. 1995. Anillin, a contractile ring protein that cycles from the nucleus to the cell cortex. *J Cell Biol* **131:** 165–178.

Fukuda T, Hedinger C, Groscurth P. 1975. Ultrastructure of developing germ cells in the fetal human testis. *Cell Tissue Res* **161:** 55–70.

Giansanti MG, Farkas RM, Bonaccorsi S, Lindsley DL, Wakimoto BT, Fuller MT, Gatti M. 2004. Genetic dissection of meiotic cytokinesis in *Drosophila* males. *Mol Biol Cell* **15:** 2509–2522.

Glotzer M. 2005. The molecular requirements for cytokinesis. *Science* **307:** 1735–1739.

Gondos B, Hobel CJ. 1971. Ultrastructure of germ cell development in the human fetal testis. *Z Zellforsch Mikrosk Anat* **119:** 1–20.

Gondos B, Bhiraleus P, Hobel CJ. 1971. Ultrastructural observations on germ cells in human fetal ovaries. *Am J Obstet Gynecol* **110:** 644–652.

Gonzalez-Reyes A. 2003. Stem cells, niches and cadherins: A view from *Drosophila*. *J Cell Sci* **116:** 949–954.

Greenbaum MP, Yan W, Wu MH, Lin YN, Agno JE, Sharma M, Braun RE, Rajkovic A, Matzuk MM. 2006. TEX14 is essential for intercellular bridges and fertility in male mice. *Proc Natl Acad Sci* **103:** 4982–4987.

Greenbaum MP, Ma L, Matzuk MM. 2007. Conversion of midbodies into germ cell intercellular bridges. *Developmental Biology* **305:** 389–396.

Grieder NC, de Cuevas M, Spradling AC. 2000. The fusome organizes the microtubule network during oocyte differentiation in *Drosophila*. *Development* **127:** 4253–4264.

Gromley A, Jurczyk A, Sillibourne J, Halilovic E, Mogensen M, Groisman I, Blomberg M, Doxsey S. 2003. A novel human protein of the maternal centriole is required for the final stages of cytokinesis and entry into S phase. *J Cell Biol* **161:** 535–545.

Gromley A, Yeaman C, Rosa J, Redick S, Chen CT, Mirabelle S, Guha M, Sillibourne J, Doxsey SJ. 2005. Centriolin anchoring of exocyst and SNARE complexes at the midbody is required for secretory-vesicle-mediated abscission. *Cell* **123:** 75–87.

Guertin DA, Trautmann S, McCollum D. 2002. Cytokinesis in eukaryotes. *Microbiol Mol Biol Rev* **66:** 155–178.

Guo GQ, Zheng GC. 2004. Hypotheses for the functions of intercellular bridges in male germ cell development and its cellular mechanisms. *J Theor Biol* **229:** 139–146.

Hanson PI, Heuser JE, Jahn R. 1997. Neurotransmitter release—four years of SNARE complexes. *Curr Opin Neurobiol* **7:** 310–315.

Hime GR, Brill JA, Fuller MT. 1996. Assembly of ring canals in the male germ line from structural components of the contractile ring. *J Cell Sci* **109:** 2779–2788.

Hsu SC, TerBush D, Abraham M, Guo W. 2004. The exocyst complex in polarized exocytosis. *Int Rev Cytol* **233:** 243–265.

Huckins C. 1971. The spermatogonial stem cell population in adult rats. I. Their morphology, proliferation and maturation. *Anat Rec* **169:** 533–557.

Huckins C, Oakberg EF. 1978. Morphological and quantitative analysis of spermatogonia in mouse testes using whole mounted seminiferous tubules, I. The normal testes. *Anat Rec* **192:** 519–528.

Hudson AM, Cooley L. 2002. A subset of dynamic actin rearrangements in *Drosophila* requires the Arp2/3 complex. *J Cell Biol* **156:** 677–687.

Huynh JR, St Johnston D. 2004. The origin of asymmetry: Early polarisation of the *Drosophila* germline cyst and oocyte. *Curr Biol* **14:** R438–449.

Iwamori T, Iwamori N, Ma L, Edson MA, Greenbaum MP, Matzuk MM. 2010. TEX14 interacts with CEP55 to block cell abscission. *Mol Cell Biol* **30:** 2280–2292.

Iwamori T, Lin YN, Ma L, Iwamori N, Matzuk MM. 2011. Identification and characterization of RBM44 as a novel intercellular bridge protein. *PLoS One* **6:** e17066.

Jackson SM, Berg CA. 2002. An A-kinase anchoring protein is required for protein kinase A regulatory subunit localization and morphology of actin structures during oogenesis in *Drosophila*. *Development* **129:** 4423–4433.

Kelso RJ, Hudson AM, Cooley L. 2002. *Drosophila Kelch* regulates actin organization via Src64-dependent tyrosine phosphorylation. *J Cell Biol* **156:** 703–713.

Kinoshita M, Field CM, Coughlin ML, Straight AF, Mitchison TJ. 2002. Self- and actin-templated assembly of Mammalian septins. *Dev Cell* **3:** 791–802.

Kramerova IA, Kramerov AA. 1999. Mucinoprotein is a universal constituent of stable intercellular bridges in *Drosophila melanogaster* germ line and somatic cells. *Dev Dyn* **216:** 349–360.

Lee HH, Elia N, Ghirlando R, Lippincott-Schwartz J, Hurley JH. 2008. Midbody targeting of the ESCRT machinery by a noncanonical coiled coil in CEP55. *Science* **322:** 576–580.

LeGrand EK. 2001. Genetic conflict and apoptosis. *Perspect Biol Med* **44:** 509–521.

Lin H, Yue L, Spradling AC. 1994. The *Drosophila* fusome, a germline-specific organelle, contains membrane skeletal proteins and functions in cyst formation. *Development* **120:** 947–956.

Low SH, Li X, Miura M, Kudo N, Quinones B, Weimbs T. 2003. Syntaxin 2 and endobrevin are required for the terminal step of cytokinesis in mammalian cells. *Dev Cell* **4:** 753–759.

Lu N, Guarnieri DJ, Simon MA. 2004. Localization of Tec29 to ring canals is mediated by Src64 and PtdIns(3,4,5)P3-dependent mechanisms. *EMBO J* **23:** 1089–1100.

Martinez-Garay I, Rustom A, Gerdes HH, Kutsche K. 2006. The novel centrosomal associated protein CEP55 is present in the spindle midzone and the midbody. *Genomics* **87:** 243–253.

Matsumura F. 2005. Regulation of myosin II during cytokinesis in higher eukaryotes. *Trends Cell Biol* **15:** 371–377.

Matuliene J, Kuriyama R. 2004. Role of the midbody matrix in cytokinesis: RNAi and genetic rescue analysis of the mammalian motor protein CHO1. *Mol Biol Cell* **15:** 3083–3094.

McGuinness MP, Orth JM. 1992. Reinitiation of gonocyte mitosis and movement of gonocytes to the basement membrane in testes of newborn rats in vivo and in vitro. *Anat Rec* **233:** 527–537.

Miki K, Willis WD, Brown PR, Goulding EH, Fulcher KD, Eddy EM. 2002. Targeted disruption of the *Akap4* gene causes defects in sperm flagellum and motility. *Dev Biol* **248:** 331–342.

Miller KG, Field CM, Alberts BM. 1989. Actin-binding proteins from *Drosophila* embryos: A complex network of interacting proteins detected by F-actin affinity chromatography. *J Cell Biol* **109:** 2963–2975.

Minestrini G, Mathe E, Glover DM. 2002. Domains of the Pavarotti kinesin-like protein that direct its subcellular distribution: Effects of mislocalisation on the tubulin and actin cytoskeleton during *Drosophila* oogenesis. *J Cell Sci* **115:** 725–736.

Mische S, Li M, Serr M, Hays TS. 2007. Direct observation of regulated ribonucleoprotein transport across the nurse cell/oocyte boundary. *Mol Biol Cell* **18:** 2254–2263.

Mishima M, Kaitna S, Glotzer M. 2002. Central spindle assembly and cytokinesis require a kinesin-like protein/RhoGAP complex with microtubule bundling activity. *Dev Cell* **2:** 41–54.

Montaville P, Dai Y, Cheung CY, Giller K, Becker S, Michalak M, Webb SE, Miller AL, Krebs J. 2006. Nuclear translocation of the calcium-binding protein ALG-2 induced by the RNA-binding protein RBM22. *Biochim Biophys Acta* **1763:** 1335–43.

Moens PB, Hugenholtz AD. 1975. The arrangement of germ cells in the rat seminiferous tubule: An electron-microscope study. *J Cell Sci* **19:** 487–507.

Morales CR, Lefrancois S, Chennathukuzhi V, El-Alfy M, Wu X, Yang J, Gerton GL, Hecht NB. 2002. A TB-RBP and Ter ATPase complex accompanies specific mRNAs from nuclei through the nuclear pores and into intercellular bridges in mouse male germ cells. *Dev Biol* **246:** 480–494.

Morita E, Sandrin V, Chung HY, Morham SG, Gygi SP, Rodesch CK, Sundquist WI. 2007. Human ESCRT and ALIX proteins interact with proteins of the midbody and function in cytokinesis. *EMBO J* **26:** 4215–4227.

Nebel BR, Amarose AP, Hacket EM. 1961. Calendar of gametogenic development in the prepuberal male mouse. *Science* **134:** 832–833.

Oakberg EF. 1971. Spermatogonial stem-cell renewal in the mouse. *Anat Rec* **169:** 515–531.

Otegui MS, Verbrugghe KJ, Skop AR. 2005. Midbodies and phragmoplasts: Analogous structures involved in cytokinesis. *Trends Cell Biol* **15:** 404–413.

Pepling ME, Spradling AC. 1998. Female mouse germ cells form synchronously dividing cysts. *Development* **125:** 3323–3328.

Pepling ME, Spradling AC. 2001. Mouse ovarian germ cell cysts undergo programmed breakdown to form primordial follicles. *Dev Biol* **234:** 339–351.

Pepling ME, de Cuevas M, Spradling AC. 1999. Germline cysts: A conserved phase of germ cell development? *Trends Cell Biol* **9:** 257–262.

Postiglione I, Chiaviello A, Palumbo G. 2009. Twilight effects of low doses of ionizing radiation on cellular systems: A bird's eye view on current concepts and research. *Med Oncol* **27:** 495–509.

Rappaport R. 1967. Cell division: Direct measurement of maximum tension exerted by furrow of echinoderm eggs. *Science* **156:** 1241–1243.

Rappaport R. 1986. Establishment of the mechanism of cytokinesis in animal cells. *Int Rev Cytol* **105:** 245–281.

Rasmussen SW. 1973. Ultrastructural studies of spermatogenesis in *Drosophila melanogaster* Meigen. *Z Zellforsch Mikrosk Anat* **140:** 125–144.

Ren HP, Russell LD. 1991. Clonal development of interconnected germ cells in the rat and its relationship to the segmental and subsegmental organization of spermatogenesis. *Am J Anat* **192:** 121–128.

Rieder CL, Khodjakov A, Paliulis LV, Fortier TM, Cole RW, Sluder G. 1997. Mitosis in vertebrate somatic cells with two spindles: Implications for the metaphase/anaphase transition checkpoint and cleavage. *Proc Natl Acad Sci* **94:** 5107–5112.

Robinson DN, Cooley L. 1996. Stable intercellular bridges in development: The cytoskeleton lining the tunnel. *Trends Cell Biol* **6:** 474–479.

Robinson DN, Cooley L. 1997. *Drosophila* kelch is an oligomeric ring canal actin organizer. *J Cell Biol* **138:** 799–810.

Robinson DN, Cant K, Cooley L. 1994. Morphogenesis of *Drosophila* ovarian ring canals. *Development* **120:** 2015–2025.

Roulier EM, Panzer S, Beckendorf SK. 1998. The Tec29 tyrosine kinase is required during *Drosophila* embryogenesis and interacts with Src64 in ring canal development. *Mol Cell* **1:** 819–829.

Saez JM, Perrard-Sapori MH, Chatelain PG, Tabone E, Rivarola MA. 1987. Paracrine regulation of testicular function. *J Steroid Biochem* **27:** 317–329.

Satterwhite LL, Pollard TD. 1992. Cytokinesis. *Curr Opin Cell Biol* **4:** 43–52.

Sertoli E. 1877. Sulla struttura dei canalicoli seminiferi dei testicoli studiata in rapport allo sviluppo dei nemaspermi. *Arch Sci Med* **2:** 107–146, 267–295.

Somogyi K, Rorth P. 2004. Cortactin modulates cell migration and ring canal morphogenesis during *Drosophila* oogenesis. *Mech Dev* **121:** 57–64.

Spradling A, Drummond-Barbosa D, Kai T. 2001. Stem cells find their niche. *Nature* **414:** 98–104.

Stanley HP, Bowman JT, Romrell LJ, Reed SC, Wilkinson RF. 1972. Fine structure of normal spermatid differentiation in *Drosophila melanogaster*. *J Ultrastruct Res* **41:** 433–466.

Tegelenbosch RA, de Rooij DG. 1993. A quantitative study of spermatogonial multiplication and stem cell renewal in the C3H/101 F1 hybrid mouse. *Mutat Res* **290:** 193–200.

Tilney LG, Tilney MS, Guild GM. 1996. Formation of actin filament bundles in the ring canals of developing *Drosophila* follicles. *J Cell Biol* **133:** 61–74.

Toure A, Dorseuil O, Morin L, Timmons P, Jegou B, Reibel L, Gacon G. 1998. MgcRacGAP, a new human GTPase-activating protein for Rac and Cdc42 similar to *Drosophila* rotundRacGAP gene product, is expressed in male germ cells. *J Biol Chem* **273:** 6019–6023.

Van de Putte T, Zwijsen A, Lonnoy O, Rybin V, Cozijnsen M, Francis A, Baekelandt V, Kozak CA, Zerial M, Huylebroeck D. 2001. Mice with a homozygous gene trap vector insertion in mgcRacGAP die during preimplantation development. *Mech Dev* **102:** 33–44.

Ventela S, Toppari J, Parvinen M. 2003. Intercellular organelle traffic through cytoplasmic bridges in early spermatids of the rat: Mechanisms of haploid gene product sharing. *Mol Biol Cell* **14:** 2768–2780.

von Dassow G, Bement WM. 2005. A ring-like template for abscission. *Dev Cell* **9:** 578–580.

von Ebner V. 1888. Zur Spermatogenese bei den Saugethieren. *Arch Mikrosk Anat* **31:** 236–292.

Wang YL. 2005. The mechanism of cortical ingression during early cytokinesis: thinking beyond the contractile ring hypothesis. *Trends Cell Biol* **15:** 581–588.

Wang PJ, McCarrey JR, Yang F, Page DC. 2001. An abundance of X-linked genes expressed in spermatogonia. *Nat Genet* **27:** 422–426.

Weber JE, Russell LD. 1987. A study of intercellular bridges during spermatogenesis in the rat. *Am J Anat* **180:** 1–24.

Witkin JW, O'Sullivan H, Silverman AJ. 1995. Novel associations among gonadotropin-releasing hormone neurons. *Endocrinology* **136:** 4323–4330.

Wu MH, Rajkovic A, Burns KH, Yan W, Lin YN, Matzuk MM. 2003. Sequence and expression of testis-expressed gene 14 (Tex14): A gene encoding a protein kinase preferentially expressed during spermatogenesis. *Gene Expr Patterns* **3:** 231–236.

Yue L, Spradling AC. 1992. *hu-li tai shao*, a gene required for ring canal formation during *Drosophila* oogenesis, encodes a homolog of adducin. *Genes Dev* **6:** 2443–2454.

Zhao WM, Seki A, Fang G. 2006. Cep55, a microtubule-bundling protein, associates with centralspindlin to control the midbody integrity and cell abscission during cytokinesis. *Mol Biol Cell* **17:** 3881–3896.

Small Noncoding RNAs in the Germline

Jonathan P. Saxe and Haifan Lin

Yale Stem Cell Center and Department of Cell Biology, Yale University School of Medicine, New Haven, Connecticut 06511

Correspondence: haifan.lin@yale.edu

Small noncoding RNAs have emerged as potent regulators of gene expression, especially in the germline. We review the biogenesis and regulatory function of three major small noncoding RNA pathways in the germline: The small interfering RNA (siRNA) pathway that leads to the degradation of target mRNAs, the microRNA (miRNA) pathway that mostly represses the translation of target mRNAs, and the newly discovered Piwi-interacting RNA (piRNA) pathway that appears to have diverse functions in epigenetic programming, transposon silencing, and the regulation of mRNA translation and stability. The siRNA and miRNA pathways are present in the germline as well as many somatic tissues, whereas the piRNA pathway is predominantly confined to the germline. Investigation of the three small RNA pathways has started to reveal a new dimension of gene regulation with defining roles in germline specification and development.

It is an emerging theme among diverse higher eukaryotes that a large portion of the genome is actively transcribed. Many of these transcripts, however, do not encode proteins, and are thus called "noncoding RNAs (ncRNAs)." Noncoding RNAs range in size from 19–21 base pairs (bp) up to tens of kilobases in length and play crucial roles in organismal development. The regulatory role of noncoding RNAs is especially prominent within the germline, where they are required for mRNA splicing (snRNPs), protein synthesis (tRNAs and ribosomal RNAs), RNA maturation and modification (snoRNAs), maintenance of telomere ends (*terc*), and epigenetic regulation of chromosomes such as X inactivation (long noncoding RNAs).

Among ncRNAs, there are three major types of small ncRNAs: small interfering RNAs (siRNAs, 21–23 bp), microRNAs (miRNAs, 21–23 bp), and the germline-enriched piwi-interacting RNAs (piRNAs, mostly 24–34 bp). A common characteristic of these ncRNAs is that they bind to proteins of the Piwi/Argonaute family. These proteins are evolutionarily conserved, and characterized by the presence of the PAZ (Piwi-Argonaute-Zwille) and PIWI domains that are essential for their function. This family can be subdivided into two distinct subfamilies: The Argonaute (Ago) subfamily, which bind siRNAs and miRNAs and are ubiquitously present in somatic and germline cells, and the Piwi subfamily that binds piRNAs and is highly enriched in the germline. Here, we review small noncoding RNAs, with an emphasis on germline-enriched piRNAs.

Cite this article as *Cold Spring Harb Perspect Biol* doi: 10.1101/cshperspect.a002717

AN OVERVIEW OF THE ARGONAUTE/PIWI PROTEIN FAMILY

Ago proteins are broadly expressed in most eukaryotes, whereas Piwi proteins are found only in sexually reproducing organisms, namely animals and ciliates (Fig. 1) (Beyret and Lin 2008). Ago/Piwi proteins contain a central PAZ domain of 100 residues that forms a pocket to bind the 3′ end of small RNAs. Mutational analysis of this domain, however, has suggested that it is not critical for RNA binding, which is instead mediated by the Piwi domain (Wang et al. 2008). The Piwi domain is a 350-residue carboxy-terminal domain involved in RNA binding and slicing. Site-directed mutagenesis has highlighted several critical residues that bind RNA using a combination of hydrogen bonding and base-pair stacking. The Piwi domain lies to one side of the ribonucleic acid complex to cut RNA bound within the domain (Song et al. 2004). Consistently, many Ago and Piwi proteins have been shown to possess slicing activity.

THE siRNA AND miRNA PATHWAYS

Characterization and Functions of siRNAs

RNA interference (RNAi), or posttranscriptional gene silencing (PTGS), has been characterized in many species. It was first observed in animals in *Caenorhabditis elegans* (Fire et al. 1998), in which endogenous transcripts could be effectively degraded by feeding worms with sequence-specific double-stranded RNA (dsRNA) (Fire et al. 1998). Sequence-specific degradation was also observed in *Drosophila* embryo extracts (Tuschl et al. 1999) and requires processing of dsRNA strands into 19–21-bp siRNA duplexes (Zamore et al. 2000; Elbashir et al. 2001b). Nearly identical processing occurs in mammalian cells (Billy et al. 2001; Elbashir et al. 2001a). The RNAse III-like enzyme Dicer produces siRNAs from dsRNAs in worms (Grishok et al. 2001), flies (Bernstein et al. 2001; Hutvagner et al. 2001), and mammals (Bernstein et al. 2001; Billy et al. 2001), which are then integrated into Ago2-containing silencing complexes that mediate degradation of complementary transcripts. The biogenesis and function of the siRNA pathway is summarized in Figure 2A. This selective degradation of specific RNA messages defends against viral infection, represses transposition, and regulates endogenous gene expression.

Eight recent simultaneous reports detailed a novel class of siRNAs, termed endogenous siRNAs (endo-siRNAs), in *Drosophila* embryos and mouse oocytes. *Drosophila* endo-siRNAs are 21 bp in length, exhibit little strand or

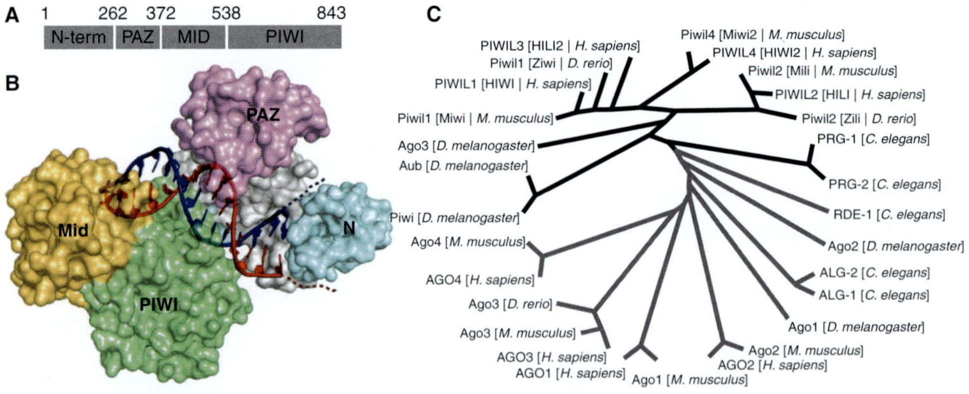

Figure 1. The PIWI/Ago family of proteins. (*A*) Schematic representation of the amino-terminal (N-term), PAZ, Mid, and PIWI domains of *Drosophila* Piwi. (*B*) Crystal structure of *Thermus thermophilus* Ago bound to a 21-bp guide RNA (red) and a 19-bp target RNA (blue). Figure created from data from Wang et al. (2009). (*C*) Radial cladogram of the Piwi (black branches) and Ago (gray branches) subfamilies. Human, mouse, zebrafish, and selected *Caenorhabditis elegans* Piwi and Ago proteins are shown.

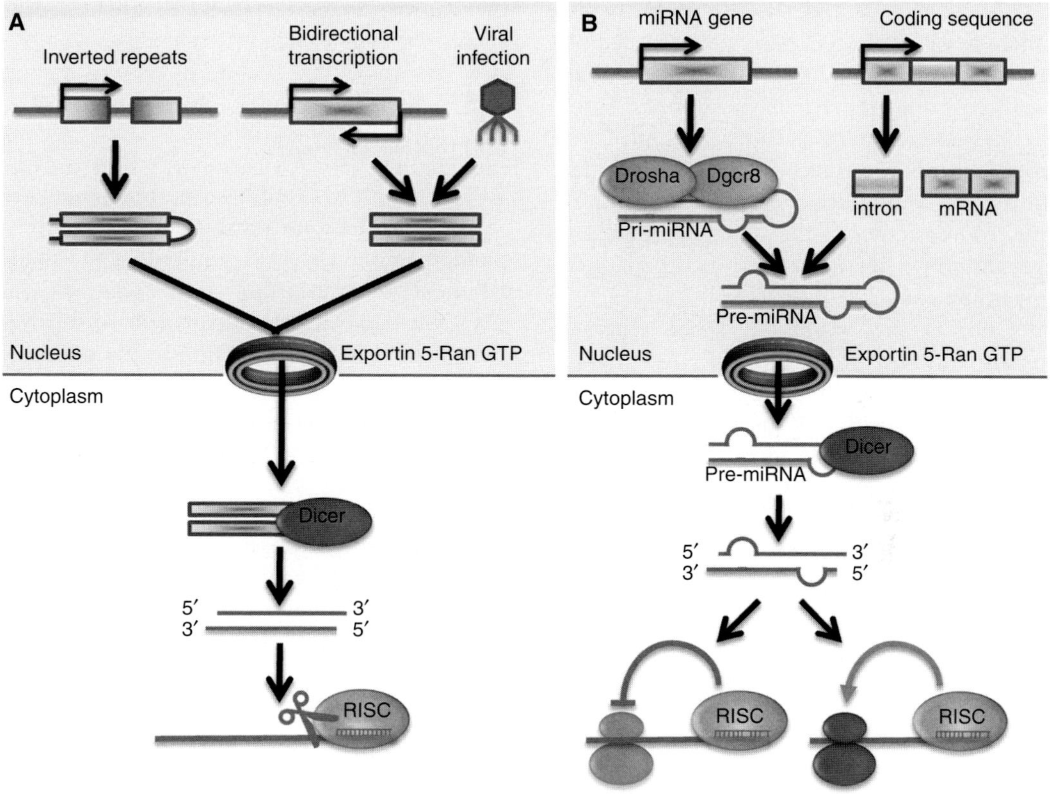

Figure 2. The siRNA and miRNA pathways. (*A*) Endogenous siRNAs are produced from inverted repeats or bidirectional transcription. The long dsRNA precursors are transported into the cytoplasm by Exportin5. Exogenous siRNAs are introduced by viral infection or transfection and are localized to the cytoplasm. In both cases, long dsRNAs are processed by Dicer to form mature siRNAs that are then incorporated into RNA-induced silencing complexes (RISCs) which mediate target cleavage. (*B*) miRNAs are produced by transcription of pri-miRNA precursors from miRNA genes, which are processed to pre-miRNAs by the microprocessor containing Drosha and Dgcr8. Alternatively, pre-miRNAs are also produced by splicing of intronic sequences containing miRNA sequences. Pre-miRNAs are exported from the nucleus by Exportin and processed into mature miRNAs by Dicer. miRNAs are then incorporated into RISCs that can mediate translational silencing or activation.

sequence bias, and arise mainly from repetitive genomic loci, suggesting a role for endo-siRNAs in defense against transposons. Consistently, there is a large increase in the activity of a subset of transposable elements in *ago2* and *dicer2* null flies (Chung et al. 2008; Czech et al. 2008; Ghildiyal et al. 2008; Kawamura et al. 2008; Okamura et al. 2008a,b). Profiling the small RNA population of mouse oocytes identified a large population of functional endo-siRNAs corresponding to transposon sequences, suggesting a conserved role for this pathway in genome defense (Golden et al. 2008; Tam et al. 2008; Watanabe et al. 2008).

Characterization and Function of miRNAs

miRNAs are produced through two distinct mechanisms (Fig. 2B). First, miRNAs are derived from primary transcripts, termed pri-miRNAs. Pri-miRNA genes are frequently arranged in tandem and several mature miRNAs can be derived from a single pri-miRNA. Pri-miRNAs are processed by a protein complex containing Drosha and DGCR8 to form pre-miRNAs of ~70–100 base pairs in length. Pre-miRNAs adopt a hairpin structure and are exported from the nucleus via exportin 5 and Ran GTPase (Kim et al. 2009). Second, "mirtrons," miRNA sequences contained within introns of coding sequences, are released by spliceosome activity rather than Drosha cleavage and then fall into the miRNA pathway (Okamura et al. 2007; Ruby et al. 2007).

Once in the cytoplasm, Dicer generates mature miRNAs from pre-miRNAs. Mature miRNAs are 19–23-bp RNA duplexes containing a two-nucleotide 3′ overhang. Although either strand is capable of mediating translational silencing, only the strand whose 5′ end is more loosely bound to its complement is then incorporated into RNA-induced silencing complexes (RISCs) (Elbashir et al. 2001c; Martinez et al. 2002; Schwarz et al. 2003).

miRNAs generally regulate their targets through translational repression. The first two known RNAs, *let-7* and *let-4*, were identified in genetic screens for heterochronic genes in *C. elegans*. These two genes produce short, ~22-nt transcripts with no protein-coding potential and bind to 3′UTRs of specific mRNAs, inhibiting their translation into protein. Additional miRNAs were soon discovered in *C. elegans* (Lau et al. 2001; Lee and Ambros 2001), *Drosophila*, and human cell lines (Lagos-Quintana et al. 2001). To date, more than 600 miRNAs have been identified in mice and humans (miRBase http://microrna.sanger.ac.uk/cgi-bin/sequences/browse.pl. Sanger Institute Wellcome Trust).

miRNAs function in a variety of developmental, physiological, and pathophysiological processes. Single miRNAs can direct hematopoetic and neural stem/progenitor cells toward a specific cell fate (Chen et al. 2004; Cheng et al. 2009) and maintain pluripotency of embryonic stem cells (for review see Gangaraju and Lin 2009). miRNAs often display exquisite developmental, temporal, and tissue-specific expression, to the extent that it is now possible to accurately classify human cancers on the basis of their miRNA expression (Lu et al. 2005). This powerful diagnostic approach could prove to be extremely useful in cancer therapy.

Germline Functions of the siRNA and miRNA Pathways

siRNA and miRNA pathway components have important functions in the germline. In *Drosophila*, maternal Dicer1 is required for the establishment of primordial germ cells (PGCs) (Megosh et al. 2006). Impairment of miRNA biogenesis by deletion of Dicer1 or Loquacious results in defects in maintenance of germline stem cells (GSCs) (Forstemann et al. 2005; Park et al. 2007); clonal analysis showed that this was a cell-autonomous defect in GSC proliferation (Hatfield et al. 2005). This function is evolutionarily conserved, as mutations in Dicer1 in *C. elegans* likewise results in defects in germline formation and sterility (Knight and Bass 2001). Further studies have highlighted roles for specific miRNAs in these processes in *Drosophila* (Iovino et al. 2009; Yang et al. 2009).

Small RNAs are crucial to ensuring genomic integrity in the germline. Endo-siRNAs play critical roles in genome surveillance to keep transposons in check within the female germline, ensuring that they are degraded in an Ago2- and Dicer2-dependent manner (Czech et al. 2008). Indeed, recent studies have highlighted the predominance of endo-siRNAs, rather than miRNAs, as the key class of small ncRNAs in oocytes (Suh et al. 2010). A distinct small RNA pathway, utilizing the Piwi subfamily of the Argonaute/Piwi family of proteins, is also necessary for maintaining genomic integrity.

PIWI PROTEINS AND piRNAs

The Piwi subfamily of proteins functions specifically within the germline. *Drosophila* PIWI, the founding member the PIWI/AGO family, was discovered in a p-element insertion screen for genes that disrupt GSC function (Lin and Spradling 1997). *Piwi* mutant flies are sterile (Cox et al. 1998) because of defects in self-renewal of GSCs (Cox et al. 1998, 2000).

Germline Functions of PIWI Proteins

PIWI proteins are mainly expressed in the germline, although several somatic functions have been determined. Within the germline, PIWI proteins are involved in germline determination and gametogenesis.

Formation of the Germline

In *Drosophila*, loss of maternal *piwi* transcript results in loss of PGC formation. Furthermore, overexpression of increasing amounts of maternal *piwi* leads to a linear increase in the number of new PCGs formed (Megosh et al. 2006). Piwi interacts with the miRNA pathway components Dicer1 and Fmrp (Megosh et al. 2006) which play key roles in PGC specification, implying that Piwi could act to control miRNA-mediated mechanisms of PGC formation.

The Male Germline

Drosophila Piwi proteins function within the male germline. Piwi was discovered in a screen for mutants that disrupt GSC self-renewal in the testis. The Piwi GSC phenotype can be enhanced and suppressed by different haploinsufficient mutations, suggesting a dose-dependent effect of *piwi* in regulating GSC function (Smulders-Srinivasan and Lin 2003). The *Drosophila* Piwi protein Ago3 is also required for GSC maintenance, as male 5–7-day-old flies do not contain Vasa-positive germ cells adjacent to the somatic hub cells (Li et al. 2009).

Mice have three Piwi homologs: Miwi, Mili, and Miwi2, which are all required for male fertility. Mili is expressed in PGCs at approximately embryonic day 12.5 (E12.5) of development and in postnatal GSCs, spermatogonia, spermatocytes, and round spermatids (Kuramochi-Miyagawa et al. 2004; Wang et al. 2009). Miwi is expressed in postnatal late primary spermatocytes through elongating spermatids (Deng and Lin 2002). In contrast, Miwi2 is expressed in gonocytes, from E15.5 when they are established through ~3 dpp when they become activated to form GSCs (Aravin et al. 2008). Miwi2 is also expressed in Sertoli cells, although this is not required for fertility (Carmell et al. 2007).

Each Piwi mutant displays unique defects in spermatogenesis. Miwi mutants are arrested at the round spermatid phase, a phenotype strongly resembling that observed in *crem* knockout mice (Sassone-Corsi 2000). *Miwi* mutants lose expression of *act* (activator of Crem in testis) and a range of Crem target genes, which Miwi directly binds and stabilizes. The primary defect in *mili*-null mice occurs within GSCs, which fail to undergo self-renewing divisions (Unhavaithaya et al. 2009). The few GSC progeny can continue to differentiation, but arrest as pachytene spermatocytes and eventually undergo apoptosis, resulting in full male sterility (Kuramochi-Miyagawa et al. 2004). Miwi2-mutant animals also display defects during meiosis, with arrest in leptotene spermatocytes and mutant spermatogonia undergoing massive apoptosis (Carmell et al. 2007). These data suggest that Miwi2 sets up an epigenetic program, which is required for GSC maintenance even after its own expression has vanished, and that lack of Miwi2 activates a DNA damage checkpoint. Consistently, mutation of *Drosophila* Atr and/or Chk2 is sufficient to overcome some (but not all) defects caused by mutations in the piRNA pathway components *armitage* or *aubergine* (Klattenhoff et al. 2007).

The Female Germline

Drosophila Piwi proteins are essential for oogenesis and female fertility. Piwi is required for GSC self-renewal and is expressed in both germline cells and supporting somatic tissue in the *Drosophila* ovary (Cox et al. 1998). *Piwi* mutants also display altered number of nurse cells and abnormal egg chamber polarity (Cox et al. 1998). Clonal analyses showed that somatic Piwi expression regulates GSC number and division (Cox et al. 1998, 2000; Szakmary et al. 2005), whereas germline expression of Piwi promotes GSC division in a cell-autonomous manner (Cox et al. 2000). This implies that Piwi utilizes distinct regulatory mechanisms to respond to and provide appropriate signals for GSC maintenance. *Ago3* mutants display female sterility, with mutants laying many fewer eggs than their wild-type counterparts (Li et al. 2009). In contrast to the effects of Piwi family mutations on *Drosophila* oogenesis, there are no known mutations of these proteins that affect fertility in female mice or humans.

Identification and Characterization of piRNAs

piRNAs in Mammals

Several groups simultaneously identified piRNAs from total testicular RNA and Piwi protein immunoprecipitates (Aravin et al. 2006; Girard et al. 2006; Grivna et al. 2006a; Lau et al. 2006; Watanabe et al. 2006). Radiolabeling of mouse testicular RNA revealed an abundant population of 28–32-nt small RNAs that were associated with MIWI and absent in *miwi*$^{-/-}$ mice. Subsequent cloning revealed 40 distinct RNAs (Grivna et al. 2006a). Five percent of this set overlapped with another set of 367 26–30-nt small RNAs isolated in an independent cloning effort (Watanabe et al. 2006). As their expression is dependent on Piwi proteins, these small RNAs were termed piwi-interacting RNAs, or piRNAs.

piRNAs are a highly complex family of small ncRNAs with more than 60,000 distinct species. Deep sequencing of RNAs immunoprecipitated with Miwi (Girard et al. 2006) and Mili (Aravin et al. 2006) revealed more than 50,000 and 15,000 additional piRNAs, respectively. Mili piRNAs are 26–28 nt, whereas Miwi piRNAs are slightly larger ($\sim$30 nt). Both sets are biased for a 5′ uracil (94.2% and 88%). Mapping of these sequences confirmed that piRNAs are highly clustered in distinct genomic loci. Approximately 16% of piRNA sequences mapped to repetitive elements, and none to telomeres (Girard et al. 2006). Clusters ranged in size from $\sim$1 to 127 kb and displayed interesting strand characteristics. In some clusters piRNAs map exclusively to a single strand; in other clusters piRNAs may be processed from two nonoverlapping, antisense transcripts separated by a short, piRNA-depleted spacer (Aravin et al. 2006), and rat and human clusters are syntenic (Fig. 3).

piRNA expression is developmentally regulated, with different populations arising from distinct clusters in prepachytene cells, including fetal germ cells, as compared to later cell types (Aravin et al. 2007b; Kuramochi-Miyagawa et al. 2008). Accumulation of prepachytene piRNAs is Mili-dependent whereas later populations are more dependent on Miwi.

piRNAs have been identified in other mammalian species, including humans (Aravin et al. 2006; Girard et al. 2006; Lau et al. 2006), rats (Lau et al. 2006), marsupials (Devor et al. 2008), and platypus (Murchison et al. 2008). Similar to mouse piRNAs, these populations all are larger than siRNAs and miRNAs, cluster in discrete genomic loci, have a strong preference for uracil at their 5′ ends, and contain a phosphate group on their 5′ ends and a 2-O-methyl group on their 3′ ends.

piRNAs in Insects

The first piRNAs were identified in divergent species and characterized in *Drosophila*. Small RNA cloning efforts in plants (Llave et al. 2002) and trypanosomes (Djikeng et al. 2001) identified a class of 24–26-nt small RNAs, which displayed a 5′U bias and evidence of RNAse III-mediated cleavage. These RNAs corresponded to highly repetitive elements and were termed repeat-associated small RNAs, or rasiRNAs. Subsequent efforts in *Drosophila* identified rasiRNAs that clustered in specific genomic loci and were enriched in embryos and testes (Aravin et al. 2003). The Piwi protein Aub was found to be required for expression of *Suppressor of Stellate (Su(Ste))* rasiRNAs (Aravin et al. 2004), which could be coimmunoprecipitated with Piwi and Aub but not with Ago1 (Vagin et al. 2006). Deep sequencing of small RNAs associated with Piwi (Brennecke et al. 2007; Saito et al. 2006; Yin and Lin 2007), Aub (Brennecke et al. 2007), and Ago3 (Brennecke et al. 2007; Gunawardane et al. 2007) revealed a number of characteristics of *Drosophila* piRNAs.

First, distinct piRNA populations associated with each Piwi protein. Piwi-associated piRNAs were larger on average than Aub piRNAs, which were larger than Ago3 piRNAs. Second, Gunawardane et al. (2007) identified a pattern in which Piwi- and Aub-associated piRNAs

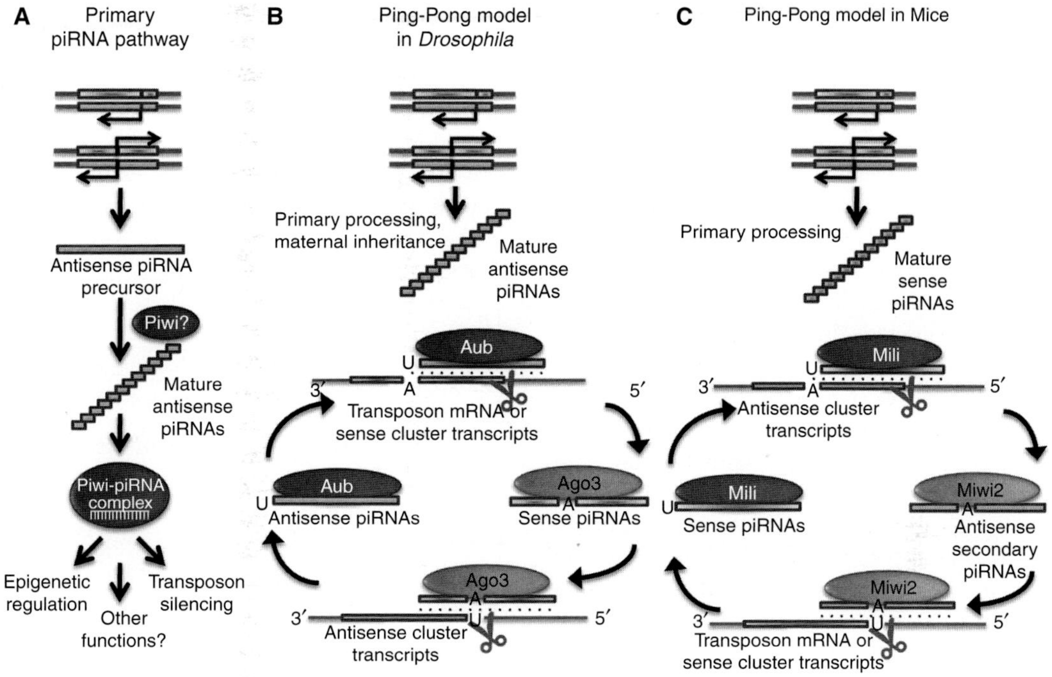

Figure 3. Biogenesis of piRNAs. (*A*) In somatic cells of the *Drosophila* ovary, antisense piRNA precursors are produced from uni- and bidirectional piRNA clusters and processed into mature antisense piRNAs through an undefined mechanism. These piRNAs are them bound by Piwi and exert divergent effects including epigenetic regulation, transposon silencing, and others. A similar biogenesis mechanism utilizing Miwi and Mili may function in pachytene mouse testes. (*B*) Within the *Drosophila* germline, Aub and Ago3 function in a piRNA amplification loop. Antisense piRNAs are maternally inherited and/or produced from uni- and bidirectional piRNA clusters and bind to Aub. Aub-piRNA complexes catalyze degradation of transposon mRNAs and sense cluster transcripts, producing sense piRNAs that are then bound by Ago3. Ago3-piRNA complexes then produce additional antisense piRNAs through degradation of antisense cluster transcripts. (*C*) A similar mechanism functions during generation of prepachytene mouse piRNAs. Mili binds sense piRNAs, likely produced from sense piRNA precursors. The Mili-piRNA complex mediates degradation of antisense cluster transcripts, and the resulting antisense piRNAs are bound by Miwi2. Miwi2-piRNA complexes then mediate destruction of transposon mRNAs, generating additional sense piRNAs that are incorporated into Mili-piRNA complexes.

contain a 5′U at the first nucleotide position and Ago3-associated piRNAs have an adenine at their 10th nucleotide position, with a 10-nucleotide complementarity between Ago3- and Aub-associated piRNAs. Furthermore, Ago3-associated sequences contain a high level of transposon sense strand bias, whereas Piwi- and Aub-associated piRNAs contain a high level of bias for antisense strands (Brennecke et al. 2007). These data are the "signature" of the ping-pong model of piRNA generation (see below). Third, *Drosophila* piRNAs map to specific genomic clusters. Most of these clusters are transposon-rich, unlike in mammals where few clusters correspond to transposonic regions. Concordantly, the proportion of piRNAs mapping to repeat-associated sequences is considerably higher in *Drosophila* than in mice (~50% compared to ~16%) (Aravin et al. 2006; Brennecke et al. 2007).

piRNAs in Other Organisms

piRNAs have been identified in a number of other organisms, including zebrafish, planaria, silkworm, and *C. elegans*. Zebrafish have two Piwi-like proteins, Ziwi and Zili, which function in germ cell maintenance and differentiation, meiosis, and transposon silencing. Zebrafish piRNAs are present in both male and female gametes and display the piRNA characteristics described above (Houwing et al. 2008, 2007), as do those isolated from the planaria *S. mediterranea* (Friedlander et al. 2009), which contains two piRNA-binding Piwi proteins (Palakodeti et al. 2008).

C. elegans possesses several diverse families of small RNAs that bind specific Piwi/Ago proteins. These include the 21U-RNAs, 22G-RNAs, 26G-RNAs, and distinct categories of primary and secondary siRNAs (Claycomb et al. 2009; Han et al. 2009; Ruby et al. 2006). 21U-RNAs are likely the piRNAs of nematodes, and were discovered through deep sequencing of RNAs with a 5′ monophosphate and 3′ hydroxyl group. 21U-RNAs invariantly contain a 5′U and are 21 bp in length, are not derived from a dsRNA precursor, are largely depleted from coding sequences, and are developmentally regulated (Ruby et al. 2006). Subsequent work showed that 21U-RNAs complex with PRG-1, one of two *C. elegans* Piwi proteins, to regulate spermatogenesis (Batista et al. 2008; Wang and Reinke 2008).

piRNA Biogenesis

The piRNA Precursor

piRNAs are probably processed from a single-stranded RNA precursor. Individual piRNAs lay end-to-end or are slightly overlapping, suggesting that they are not nonspecific breakdown products (Aravin et al. 2006). piRNAs exhibit extreme strand bias, with little mixing of sense and antisense sequences in a particular cluster. This in turn implies that piRNAs in a specific cluster are derived from a single transcript. Consistently, long RNA transcripts corresponding to piRNA clusters have been identified (Lau et al. 2006). Interestingly, these precursors are probably not polyadenylated (Ro et al. 2007).

Recent findings have suggested possible mechanisms that may regulate the transcription of specific precursors. In *Drosophila*, a member of the Heterochromatin Protein 1 (HP1) subfamily of chromo box proteins, Rhino, physically associates with DNA at the 1/42AB piRNA cluster and is required for transcription of precursors from this region. *Rhino* mutations result in transposon activation and female sterility (Volpe et al. 2001; Klattenhoff et al. 2009). Its presence may displace other HP proteins, preventing them from repressing expression at these loci (Klattenhoff et al. 2009).

Processing of the Precursor

There is evidence indicating that piRNAs are produced from their precursors by the cleavage activity of Piwi proteins rather than Dicer. Mutation of *miwi* and *mili* severely reduces levels of mature piRNAs but not miRNAs or Dicer (Grivna et al. 2006a, 2006b; Aravin et al. 2007b), although it is not known if precursors are properly transcribed. In *Drosophila*, piRNAs are expressed normally in Dicer 1 and Dicer 2 mutants (Vagin et al. 2006). Piwi, Ago3, and Aub all contain slicer activity in vitro, arguing that they are capable of generating mature piRNAs from potential precursors (Gunawardane et al. 2007; Saito et al. 2006). The *Drosophila* nucleases *squash* and *zucchini* play roles in ping-pong dependent and independent processing of piRNAs, respectively (Pane et al. 2007; Malone et al. 2009; Saito et al. 2009), although how they fit into these pathways is unknown.

The Ping-Pong Model: Posttranscriptional Amplification

The ping-pong "signature" is particularly prevalent among Aub- and Ago3-piRNAs. Piwi-associated piRNAs, although maintaining an antisense bias, do not display evidence of ping-pong-based biogenesis (Malone et al. 2009). Many piRNA clusters correspond to various transposonic "graveyards," which are transcribed as a precursor in an antisense orientation. This precursor is then processed into mature piRNAs by Aub (and possibly Piwi), either directly or in a processing complex of unknown composition. These mature piRNAs contain complementarity to active transposons, leading to their targeting and cleavage that produces sense piRNAs. These sense piRNAs are complementary to the antisense precursors, leading to their cleavage and feedforward production of more antisense piRNAs. Initial antisense piRNAs may be maternally contributed (Lin 2007; Brennecke et al. 2008), although the chicken-and-egg problem of how the first piRNAs are derived has yet to be fully resolved.

There is considerable genetic evidence in favor of ping-pong-based amplification for a subset of piRNAs. First, *aubergine* mutants display a shift in the overall size of the piRNA population to that of Piwi piRNAs and no longer have a ping-pong signature in their germline piRNAs (Malone et al. 2009). Second, *ago3* heterozygotes contained only ~40% of the sense piRNAs as compared to wild-type controls, whereas antisense piRNAs were not affected to the same degree. This suggests that Ago3 is limiting, and likely required, for the production of sense piRNAs. Third, antisense piRNAs were severely depleted in mutants containing no viable *ago3*, implying that Ago3 acts to stabilize and/or amplify this population (Li et al. 2009). Interestingly, antisense Piwi-bound piRNAs were only slightly affected in *ago3* heterozygotes versus mutants, suggesting that Piwi-piRNAs are mostly generated by a non-ping-pong mechanism.

Prepachytene piRNAs in mice also display a ping-pong signature. In Mili and Miwi2 immunoprecipitates from male 16.5-dpc gonads, 46% and 76% of piRNAs corresponded to transposon sequences, contained a 10 bp-overlap in their 5′ ends and an adenine at their 10th position (Aravin et al. 2007b). Mili preferentially interacts with sense sequences whereas Miwi2-associated piRNAs tend to have an antisense bias. piRNA precursors may be processed by Mili, generating primary piRNAs, which then direct cleavage of antisense transcripts in Miwi2 complexes to generate secondary piRNAs. Note that this is the opposite of what is proposed in *Drosophila*. Consistently, *mili* mutants have no Miwi2-associated piRNAs (Aravin et al. 2008).

Ping-Pong-Independent piRNA Generation

The ping-pong mechanism is highly conserved evolutionarily, functioning in organisms as diverse as *Nematostella* (sea anenome) and *Amphimedon* (sponge) (Grimson et al. 2008). However, the ping-pong mechanism is not the lone, or possibly even primary, mechanism of piRNA biogenesis. The profiling of piRNAs in *Drosophila* revealed that some piRNAs are mapped to 26 specific transposons and are predominantly antisense, even in the absence of *ago3*. They mainly bound Piwi rather than Aub or Ago3, and mapped predominantly to one particular locus termed *flamenco* in which transposons are arranged in an antisense manner. *Piwi* is required for silencing transposons within the *flamenco* locus (Sarot et al. 2004), including *gypsy* which is active within somatic cells of the ovary. Thus a distinct, Piwi-dependent piRNA generation pathway operates within somatic cells (and, to a lesser extent, within germ cells as well) (Li et al. 2009; Malone et al. 2009).

Ping-pong-independent processing of the *flamenco* locus is reminiscent of mouse pachytene piRNA clusters. Only ~17% of piRNAs from these clusters correspond to transposons, whereas repetitive sequences comprise ~50% of the cloned prepachytene piRNAs. Even in *Drosophila*, only ~50% of total piRNAs potentially target transposons. In sum, these data show that the piRNA pathway utilizes multiple biogenic mechanisms.

Regulatory Functions of Piwi Proteins and piRNAs

Piwis and piRNAs as Epigenetic Regulators

Piwi proteins function in epigenetic regulation. In *Drosophila*, Piwi and Aub regulate position-effect variegation (PEV) (Pal-Bhadra et al. 2004), in which a euchromatic gene is silenced when localized in the vicinity of a heterochromatic region (Dorer and Henikoff 1994; Martin-Morris et al. 1997). An artificial example of PEV is observed by tandem insertion of the *white* gene, leading to transcriptional silencing of the transgene and red eye color. Piwi and Aub are required for heterochromatin organization (Pal-Bhadra et al. 2004). Furthermore, HP1a and Piwi physically interact in vitro and in vivo. Piwi directly interacts with chromosomes and is localized to several heterochromatic regions. As PEV is lost when Piwi and HP1a cannot physically interact (Brower-Toland et al. 2007), it is evident that Piwi and Hp1a interact to achieve epigenetic regulation. In this pathway, it is possible that the Piwi-piRNA complex guides HP1a and other epigenetic factors to their target sequences in the genome (Lin and Yin 2008).

Support for this hypothesis comes from a recent study in which Piwi and a specific piRNA were found to play an epigenetic activation role specifically at the piRNA target site (Yin and Lin 2007). This study focused on a genomic sequence in the subtelomeric region of the right arm of the third chromosome (3R-TAS) that corresponds uniquely to a piRNA called the 3R-TAS1 piRNA. 3R-TAS1 chromatin is bound by Piwi and is partially euchromatic. However, in *piwi* mutants it became heterochromatic, indicating the role of Piwi in promoting the euchromatic state of 3R-TAS1. In support of this, the expression of the 3R-TAS1 piRNA is lost in *piwi* mutant flies. In addition, the expression of a protein reporter gene inserted immediately downstream from the 3R-TAS1-corresponding site is also silenced. Furthermore, this epigenetic change is sharply confined within the vicinity of the 3R-TAS1 DNA sequence, suggesting that the piRNA renders Piwi-specific binding to its complementary 3R-TAS sequence. Interestingly, insertion of a P-element 128 bp downstream from the 3R-TAS1 sequence rescued the 3R-TAS1 piRNA expression in the mutant background, restored the euchromatic character of 3R-TAS1 DNA, and partially rescued effects of *piwi* mutation of germline stem cell maintenance (Yin and Lin 2007). These results correlate the expression of 3R-TAS1 piRNA to the epigenetic state of its target genomic sequence.

The recruitment of HP1a by Piwi-piRNA complexes, however, cannot easily explain the opposite effect of Piwi at different sites of the genome because HP1a is mainly known to promote heterochromatin formation. Other proteins might be recruited by Piwi-piRNA complexes to different sites. For example, the HP-like protein Rhino is specifically expressed in the female germline (Volpe et al. 2001; Vermaak et al. 2005). *Rhino* mutants specifically overexpress RNAs of transposons present in double-stranded piRNA clusters, where Rhino is physically present in a wild-type background (Klattenhoff et al. 2009). These results suggest different HP proteins may dictate a permissive vs. nonpermissive chromatin environment and could explain how Piwi could be important for both heterochromatin formation and epigenetic activation (Yin and Lin 2007).

Epigenetic functions of Piwi proteins are conserved in mouse testes, where they are required for establishment of DNA methylation patterns. Mice mutant for *dnmt3l* (DNA methyltransferase 3L, which is required for de novo DNA methylation) display a phenotype similar to *mili* mutants, including loss of transposon suppression and an increase in transposon-derived piRNAs (Bourc'his and Bestor 2004; Aravin et al. 2008). *Mili* and *miwi2* mutant animals show greatly reduced levels of CpG methylation in PGCs, with concordant increases in transposon expression (Kuramochi-Miyagawa et al. 2008). These studies also suggest that specific epigenetic mechanisms mediated by Piwi proteins in mammalian systems are at least partially different from that in *Drosophila*.

Regulation of Transposon Activity Depends on Piwi Protein Localization

Piwi proteins are well characterized for their function in silencing transposition, which has been extensively reviewed (Aravin et al. 2007a; Thomson and Lin 2009). Here, we cover the latest developments in Piwi/piRNA-mediated transposon silencing, and focus on how a piRNA-mediated scanning and silencing mechanism might operate. Clues to this question arise from work utilizing *rhino* mutants. In these flies, Piwi mostly retains nuclear expression whereas Ago3 and Aub are mislocalized (Klattenhoff et al. 2009). In a wild-type background, these two proteins are localized to a perinuclear structure called the nuage. The function of the nuage has not conclusively been shown, but proper nuage formation is apparently critical for transposon silencing.

Mutations in genes that disrupt nuage formation, such as *rhino*, *maelstrom*, and *krimper*, invariably result in elevated levels of transposon transcripts and defects in piRNA production (Findley et al. 2003; Lim and Kai 2007; Klattenhoff et al. 2009). Additionally, localization of Aub and Ago3 to nuage is interdependent (Li et al. 2009), and mutation of either gene also leads to defects in nuage formation and piRNA biogenesis. Thus, nuage assembly and piRNA-mediated silencing appear to be coregulated.

These observations, coupled with the fact that expression of protein-coding transcripts containing transposons in their introns does not seem to be affected by piRNA pathway mutations, led (Klattenhoff et al. 2009) to propose that mature RNA transcripts are exported from the nucleus and are scanned for homology to transposons in the nuage. Protein-coding transcripts pass through, as any transposons containing intronic sequences have been removed by splicing. Transposon transcripts and piRNA precursors, however, are recognized and destroyed by piRNAs in an Ago3/Aub-mediated ping-pong cycle. An alternative explanation may lie in the nature of the primary transcript itself; mRNAs contain polyadenylated 3′ ends, whereas primary piRNA precursors apparently lack this modification (Ro et al. 2007). This may form the basis for recognition of these sequences to recruit the appropriate scanning apparatus. This general model of nuage-based scanning could potentially function in mammalian testes as well.

Proper localization of piRNA pathway components is critical for piRNA-mediated transposon silencing in mouse testis. In adult testes, Miwi and Mili are both localized to the chromatoid body, a mammalian version of the nuage (Kotaja and Sassone-Corsi 2007; see also Voronina et al. 2011). In gonocytes, Mili is located to a specialized form of nuage called the intermitochondrial cement, where it forms granules termed "pi-bodies" (Aravin et al. 2009). Within pi-bodies, Mili interacts with the Tudor domain-containing protein Tdrd1, whose localization and expression depends on Mili (Wang et al. 2009). Deletion of Tdrd1 results in male sterility (Chuma et al. 2006) with severe disruption of the ping-pong cycle. In addition, Miwi2-associated antisense piRNAs corresponding to transposonic sequences are lost, and cellular transcripts instead comprise the vast majority of cloned piRNAs (although overall piRNA biogenesis is intact (Reuter et al. 2009; Vagin et al. 2009; Wang et al. 2009). Several other Tudor domain-containing proteins also interact with Mili, Miwi, and Miwi2, including Tdrd2, Tdrd6, Tdrd7, and Tdrd9 (Hosokawa et al. 2007; Chen et al. 2009; Vagin et al. 2009). Deletion of Tdrd6 and Tdrd9 also result in male sterility (Shoji et al. 2009; Vasileva et al. 2009). Tdrd9 specifically interacts with Miwi2 in gonocyte-specific structures termed "piP-bodies" (Aravin et al. 2009; Shoji et al. 2009). Mael is another component of piP-bodies and is required for Miwi2 and Tdrd9 localization. Interestingly, pi-body components Mili and Tdrd1 are required for proper localization of the piP-body components Miwi2, Tdrd9, and Mael, although the converse is not true (Aravin et al. 2008, 2009). Deletion of Mael and Tdrd9 also lead to male sterility and transposon derepression (Soper et al. 2008; Shoji et al. 2009), and piRNAs from Tdrd9 mutants are highly enriched for sense-strand LINE-element sequences (Shoji et al. 2009). Thus, pi- and piP-bodies may act as a sort of "base station" in which the core piRNA machinery interacts with nuage components such as Tdrd proteins and MVH. These proteins may recruit specific transcripts and/or RNA populations to play roles in piRNA-mediated processes and to ensure correct targeting of transposons for degradation. Despite this, it remains unknown how these distinct granules coordinate their activities to accomplish their functions.

piP bodies may function to coordinate piRNA biogenesis and translational repression. piP bodies are a variant of P-bodies, which act in mRNA storage, decay, and translational regulation (Eulalio et al. 2007), and contain several P-body components such as GW182, DDX6, and XRN1 (Aravin et al. 2009; Shoji et al. 2009). Thus, piP bodies lie at the intersection of the piRNA pathway and mRNA processing, and may act in translational regulation.

The Role of Piwi Proteins and piRNAs in Translation and mRNA Turnover

Mammalian Piwi proteins are predominantly cytoplasmic and cofractionate with polysomes, suggesting a potential role in translational regulation (Grivna et al. 2006a,b; Wang et al. 2009). Miwi associates with mRNAs and piRNAs in polysomes, suggesting that Miwi-bound piRNAs may regulate translation or stability of specific transcripts (Grivna et al. 2006b). The control of Piwi proteins toward the mRNA stability is evident by the fact that Miwi-bound mRNAs are abrogated in *miwi*-null animals (Deng and Lin 2002). Furthermore, the following observations indicate that Piwi proteins may be involved in translational regulation. First, Miwi associates with the cap-binding complex involved in translational initiation, and also with Dicer (Grivna et al. 2006b). *Miwi* mutant animals display abnormal biogenesis of a subset of miRNAs, suggesting that Miwi may regulate translation through miRNA-mediated mechanisms as well as directly through its interactions with mRNA cap-binding proteins. Second, *mili* mutants display defects in global protein translation. Isolated 7 dpp seminiferous tubules, which contain only GSCs, display greatly reduced protein synthesis in *mili* mutants despite containing normal mRNA levels (Unhavaithaya et al. 2009). At the molecular level, Mili interacts with the translation initiation factor eIF3a and with the eIF4E- and eIF4G-containing m7G cap-binding complex. Thus, Mili may be required for various aspects of translation, and defects in these aspects may lead to the defects in GSC self-renewal observed in *mili*-null animals (Unhavaithaya et al. 2009).

Some piRNAs derived from mRNAs may function in translational repression. Recent studies have shown that these piRNAs are disproportionately derived from the 3′UTRs of specific classes of genes (Robine et al. 2009; Saito et al. 2009). These piRNAs are sense-oriented, are not generated through a ping-pong mechanism, and are present in *Drosophila*, *Xenopus*, and mouse (Robine et al. 2009), even though how transcripts are selected for processing is unclear. The 3′UTR of *traffic jam* (*tj*) generates a large amount of piRNAs (compared to other 3′UTRs). Clonal analysis in ovary follicles showed that *piwi* mutant follicles have elevated levels of TJ (Robine et al. 2009). Interestingly, *tj* positively regulates Piwi expression in somatic gonadal tissue, and may negatively regulate Piwi within the germline itself (Saito et al. 2009). As ~35% of small RNAs from 10 dpp testes correspond to 3′UTRs (Robine et al. 2009), it is tempting to speculate that this is a general mechanism for translational repression; however, this remains speculation, and this mechanism may be specific to the *tj* mRNA.

CONCLUDING REMARKS

As reviewed above, three families of small non-coding RNAs have emerged as key regulators of cellular and developmental processes, with their function highlighted in the germline. Among them, miRNAs control various aspects of germline development, from germline specification to GSC self-renewal and gametogenesis. Endo-siRNAs represent a novel, exciting class of small RNAs also with critical roles within the germline. They are known to protect and maintain genomic integrity from the assault of mobile genetic elements such as transposons, a role they share with the germline-specific piRNAs. However, their role in regulating gene expression awaits further exploration. By comparison, the understanding of the piRNA pathway is still at the beginning stage, with fundamental biological questions remaining unanswered. For example, how are piRNAs generated? What exact roles do these molecules play in epigenetic and translational regulation? How do these mechanisms contribute to biological functions such as GSC maintenance, transposon silencing, and meiosis? The coming years will doubtlessly provide exciting answers to these questions and many more.

REFERENCES

Aravin AA, Lagos-Quintana M, Yalcin A, Zavolan M, Marks D, Snyder B, Gaasterland T, Meyer J, Tuschl T. 2003. The small RNA profile during *Drosophila melanogaster* development. *Dev Cell* **5:** 337–350.

Aravin AA, Klenov MS, Vagin VV, Bantignies F, Cavalli G, Gvozdev VA. 2004. Dissection of a natural RNA silencing process in the *Drosophila melanogaster* germ line. *Mol Cell Biol* **24:** 6742–6750.

Aravin A, Gaidatzis D, Pfeffer S, Lagos-Quintana M, Landgraf P, Iovino N, Morris P, Brownstein MJ, Kuramochi-Miyagawa S, Nakano T, et al. 2006. A novel class of small RNAs bind to MILI protein in mouse testes. *Nature* **442:** 203–207.

Aravin AA, Hannon GJ, Brennecke J. 2007a. The Piwi-piRNA pathway provides an adaptive defense in the transposon arms race. *Science* **318:** 761–764.

Aravin AA, Sachidanandam R, Girard A, Fejes-Toth K, Hannon GJ. 2007b. Developmentally regulated piRNA clusters implicate MILI in transposon control. *Science* **316:** 744–747.

Aravin AA, Sachidanandam R, Bourc'his D, Schaefer C, Pezic D, Toth KF, Bestor T, Hannon GJ. 2008. A piRNA pathway primed by individual transposons is linked to de novo DNA methylation in mice. *Mol Cell* **31:** 785–799.

Aravin AA, van der Heijden GW, Castaneda J, Vagin VV, Hannon GJ, Bortvin A. 2009. Cytoplasmic compartmentalization of the fetal piRNA pathway in mice. *PLoS Genet* **5:** e1000764.

Batista PJ, Ruby JG, Claycomb JM, Chiang R, Fahlgren N, Kasschau KD, Chaves DA, Gu W, Vasale JJ, Duan S, et al. 2008. PRG-1 and 21U-RNAs interact to form the piRNA complex required for fertility in *C. elegans*. *Mol Cell* **31:** 67–78.

Bernstein E, Caudy AA, Hammond SM, Hannon GJ. 2001. Role for a bidentate ribonuclease in the initiation step of RNA interference. *Nature* **409:** 363–366.

Beyret E, Lin H. 2008. Piwi-interacting RNAs (piRNAs). In *MicroRNAs from basic science to disease biology* (ed. Appasani K), pp. 489–496. Cambridge University Press, Cambridge.

Billy E, Brondani V, Zhang H, Muller U, Filipowicz W. 2001. Specific interference with gene expression induced by long, double-stranded RNA in mouse embryonal teratocarcinoma cell lines. *Proc Natl Acad Sci* **98:** 14428–14433.

Bourc'his D, Bestor TH. 2004. Meiotic catastrophe and retrotransposon reactivation in male germ cells lacking Dnmt3L. *Nature* **431:** 96–99.

Brennecke J, Aravin AA, Stark A, Dus M, Kellis M, Sachidanandam R, Hannon GJ. 2007. Discrete small RNA-generating loci as master regulators of transposon activity in *Drosophila*. *Cell* **128:** 1089–1103.

Brennecke J, Malone CD, Aravin AA, Sachidanandam R, Stark A, Hannon GJ. 2008. An epigenetic role for maternally inherited piRNAs in transposon silencing. *Science* **322:** 1387–1392.

Brower-Toland B, Findley SD, Jiang L, Liu L, Yin H, Dus M, Zhou P, Elgin SC, Lin H. 2007. *Drosophila* PIWI associates with chromatin and interacts directly with HP1a. *Genes Dev* **21:** 2300–2311.

Carmell MA, Girard A, van de Kant HJ, Bourc'his D, Bestor TH, de Rooij DG, Hannon GJ. 2007. MIWI2 is essential for spermatogenesis and repression of transposons in the mouse male germline. *Dev Cell* **12:** 503–514.

Chen CZ, Li L, Lodish HF, Bartel DP. 2004. MicroRNAs modulate hematopoietic lineage differentiation. *Science* **303:** 83–86.

Chen C, Jin J, James DA, Adams-Cioaba MA, Park JG, Guo Y, Tenaglia E, Xu C, Gish G, Min J, et al. 2009. Mouse Piwi interactome identifies binding mechanism of Tdrkh Tudor domain to arginine methylated Miwi. *Proc Natl Acad Sci* **106:** 20336–20341.

Cheng LC, Pastrana E, Tavazoie M, Doetsch F. 2009. miR-124 regulates adult neurogenesis in the subventricular zone stem cell niche. *Nat Neurosci* **12:** 399–408.

Chuma S, Hosokawa M, Kitamura K, Kasai S, Fujioka M, Hiyoshi M, Takamune K, Noce T, Nakatsuji N. 2006. *Tdrd1/Mtr-1*, a *tudor*-related gene, is essential for male germ-cell differentiation and nuage/germinal granule formation in mice. *Proc Natl Acad Sci* **103:** 15894–15899.

Chung WJ, Okamura K, Martin R, Lai EC. 2008. Endogenous RNA interference provides a somatic defense against *Drosophila* transposons. *Curr Biol* **18:** 795–802.

Claycomb JM, Batista PJ, Pang KM, Gu W, Vasale JJ, van Wolfswinkel JC, Chaves DA, Shirayama M, Mitani S, Ketting RF, et al. 2009. The Argonaute CSR-1 and its 22G-RNA cofactors are required for holocentric chromosome segregation. *Cell* **139:** 123–134.

Cox DN, Chao A, Baker J, Chang L, Qiao D, Lin H. 1998. A novel class of evolutionarily conserved genes defined by piwi are essential for stem cell self-renewal. *Genes Dev* **12:** 3715–3727.

Cox DN, Chao A, Lin H. 2000. piwi encodes a nucleoplasmic factor whose activity modulates the number and division rate of germline stem cells. *Development* **127:** 503–514.

Czech B, Malone CD, Zhou R, Stark A, Schlingeheyde C, Dus M, Perrimon N, Kellis M, Wohlschlegel JA, Sachidanandam R, et al. 2008. An endogenous small interfering RNA pathway in *Drosophila*. *Nature* **453:** 798–802.

Deng W, Lin H. 2002. *miwi*, a murine homolog of *piwi*, encodes a cytoplasmic protein essential for spermatogenesis. *Dev Cell* **2:** 819–830.

Devor EJ, Huang L, Samollow PB. 2008. PiRNA-like RNAs in the marsupial *Monodelphis domestica* identify transcription clusters and likely marsupial transposon targets. *Mamm Genome* **19:** 581–586.

Djikeng A, Shi H, Tschudi C, Ullu E. 2001. RNA interference in *Trypanosoma brucei*: Cloning of small interfering RNAs provides evidence for retroposon-derived 24-26-nucleotide RNAs. *RNA* **7:** 1522–1530.

Dorer DR, Henikoff S. 1994. Expansions of transgene repeats cause heterochromatin formation and gene silencing in *Drosophila*. *Cell* **77:** 993–1002.

Elbashir SM, Harborth J, Lendeckel W, Yalcin A, Weber K, Tuschl T. 2001a. Duplexes of 21-nucleotide RNAs mediate RNA interference in cultured mammalian cells. *Nature* **411:** 494–498.

Elbashir SM, Lendeckel W, Tuschl T. 2001b. RNA interference is mediated by 21- and 22-nucleotide RNAs. *Genes Dev* **15:** 188–200.

Elbashir SM, Martinez J, Patkaniowska A, Lendeckel W, Tuschl T. 2001c. Functional anatomy of siRNAs for mediating efficient RNAi in *Drosophila melanogaster* embryo lysate. *EMBO J* **20:** 6877–6888.

Eulalio A, Behm-Ansmant I, Izaurralde E. 2007. P bodies: At the crossroads of post-transcriptional pathways. *Nat Rev Mol Cell Biol* **8:** 9–22.

Findley SD, Tamanaha M, Clegg NJ, Ruohola-Baker H. 2003. *Maelstrom*, a *Drosophila spindle*-class gene, encodes a protein that colocalizes with Vasa and RDE1/AGO1 homolog, Aubergine, in nuage. *Development* **130:** 859–871.

Fire A, Xu S, Montgomery MK, Kostas SA, Driver SE, Mello CC. 1998. Potent and specific genetic interference by double-stranded RNA in *Caenorhabditis elegans*. *Nature* **391:** 806–811.

Forstemann K, Tomari Y, Du T, Vagin VV, Denli AM, Bratu DP, Klattenhoff C, Theurkauf WE, Zamore PD. 2005. Normal microRNA maturation and germ-line stem cell maintenance requires Loquacious, a double-stranded RNA-binding domain protein. *PLoS Biol* **3:** e236.

Friedlander MR, Adamidi C, Han T, Lebedeva S, Isenbarger TA, Hirst M, Marra M, Nusbaum C, Lee WL, Jenkin JC, et al. 2009. High-resolution profiling and discovery of planarian small RNAs. *Proc Natl Acad Sci* **106:** 11546–11551.

Gangaraju VK, Lin H. 2009. MicroRNAs: Key regulators of stem cells. *Nat Rev Mol Cell Biol* **10:** 116–125.

Ghildiyal M, Seitz H, Horwich MD, Li C, Du T, Lee S, Xu J, Kittler EL, Zapp ML, Weng Z, et al. 2008. Endogenous siRNAs derived from transposons and mRNAs in *Drosophila* somatic cells. *Science* **320:** 1077–1081.

Girard A, Sachidanandam R, Hannon GJ, Carmell MA. 2006. A germline-specific class of small RNAs binds mammalian Piwi proteins. *Nature* **442:** 199–202.

Golden DE, Gerbasi VR, Sontheimer EJ. 2008. An inside job for siRNAs. *Mol Cell* **31:** 309–312.

Grimson A, Srivastava M, Fahey B, Woodcroft BJ, Chiang HR, King N, Degnan BM, Rokhsar DS, Bartel DP. 2008. Early origins and evolution of microRNAs and Piwi-interacting RNAs in animals. *Nature* **455:** 1193–1197.

Grishok A, Pasquinelli AE, Conte D, Li N, Parrish S, Ha I, Baillie DL, Fire A, Ruvkun G, Mello CC. 2001. Genes and mechanisms related to RNA interference regulate expression of the small temporal RNAs that control *C. elegans* developmental timing. *Cell* **106:** 23–34.

Grivna ST, Beyret E, Wang Z, Lin H. 2006a. A novel class of small RNAs in mouse spermatogenic cells. *Genes Dev* **20:** 1709–1714.

Grivna ST, Pyhtila B, Lin H. 2006b. MIWI associates with translational machinery and PIWI-interacting RNAs (piRNAs) in regulating spermatogenesis. *Proc Natl Acad Sci* **103:** 13415–13420.

Gunawardane LS, Saito K, Nishida KM, Miyoshi K, Kawamura Y, Nagami T, Siomi H, Siomi MC. 2007. A slicer-mediated mechanism for repeat-associated siRNA 5′ end formation in *Drosophila*. *Science* **315:** 1587–1590.

Han T, Manoharan AP, Harkins TT, Bouffard P, Fitzpatrick C, Chu DS, Thierry-Mieg D, Thierry-Mieg J, Kim JK. 2009. 26G endo-siRNAs regulate spermatogenic and zygotic gene expression in *Caenorhabditis elegans*. *Proc Natl Acad Sci* **106:** 18674–18679.

Hatfield SD, Shcherbata HR, Fischer KA, Nakahara K, Carthew RW, Ruohola-Baker H. 2005. Stem cell division is regulated by the microRNA pathway. *Nature* **435:** 974–978.

Hosokawa M, Shoji M, Kitamura K, Tanaka T, Noce T, Chuma S, Nakatsuji N. 2007. Tudor-related proteins TDRD1/MTR-1, TDRD6 and TDRD7/TRAP: Domain composition, intracellular localization, and function in male germ cells in mice. *Dev Biol* **301:** 38–52.

Houwing S, Kamminga LM, Berezikov E, Cronembold D, Girard A, van den Elst H, Filippov DV, Blaser H, Raz E, Moens CB, et al. 2007. A role for Piwi and piRNAs in germ cell maintenance and transposon silencing in zebrafish. *Cell* **129:** 69–82.

Houwing S, Berezikov E, Ketting RF. 2008. Zili is required for germ cell differentiation and meiosis in zebrafish. *EMBO J* **27:** 2702–2711.

Hutvagner G, McLachlan J, Pasquinelli AE, Balint E, Tuschl T, Zamore PD. 2001. A cellular function for the RNA-interference enzyme Dicer in the maturation of the *let-7* small temporal RNA. *Science* **293:** 834–838.

Iovino N, Pane A, Gaul U. 2009. miR-184 has multiple roles in *Drosophila* female germline development. *Dev Cell* **17:** 123–133.

Kawamura Y, Saito K, Kin T, Ono Y, Asai K, Sunohara T, Okada TN, Siomi MC, Siomi H. 2008. *Drosophila* endogenous small RNAs bind to Argonaute 2 in somatic cells. *Nature* **453:** 793–797.

Kim VN, Han J, Siomi MC. 2009. Biogenesis of small RNAs in animals. *Nat Rev Mol Cell Biol* **10:** 126–139.

Klattenhoff C, Bratu DP, McGinnis-Schultz N, Koppetsch BS, Cook HA, Theurkauf WE. 2007. *Drosophila* rasiRNA pathway mutations disrupt embryonic axis specification through activation of an ATR/Chk2 DNA damage response. *Dev Cell* **12:** 45–55.

Klattenhoff C, Xi H, Li C, Lee S, Xu J, Khurana JS, Zhang F, Schultz N, Koppetsch BS, Nowosielska A, et al. 2009. The *Drosophila* HP1 homolog Rhino is required for transposon silencing and piRNA production by dual-strand clusters. *Cell* **138:** 1137–1149.

Knight SW, Bass BL. 2001. A role for the RNase III enzyme DCR-1 in RNA interference and germ line development in *Caenorhabditis elegans*. *Science* **293:** 2269–2271.

Kotaja N, Sassone-Corsi P. 2007. The chromatoid body: A germ-cell-specific RNA-processing centre. *Nature Reviews Mol Cell Biol* **8:** 85–90.

Kuramochi-Miyagawa S, Kimura T, Ijiri TW, Isobe T, Asada N, Fujita Y, Ikawa M, Iwai N, Okabe M, Deng W, et al. 2004. *Mili*, a mammalian member of *piwi* family gene, is essential for spermatogenesis. *Development* **131:** 839–849.

Kuramochi-Miyagawa S, Watanabe T, Gotoh K, Totoki Y, Toyoda A, Ikawa M, Asada N, Kojima K, Yamaguchi Y, Ijiri TW, et al. 2008. DNA methylation of retrotransposon genes is regulated by Piwi family members MILI and MIWI2 in murine fetal testes. *Genes Dev* **22:** 908–917.

Lagos-Quintana M, Rauhut R, Lendeckel W, Tuschl T. 2001. Identification of novel genes coding for small expressed RNAs. *Science* **294:** 853–858.

Lau NC, Lim LP, Weinstein EG, Bartel DP. 2001. An abundant class of tiny RNAs with probable regulatory roles in *Caenorhabditis elegans*. *Science* **294:** 858–862.

Lau NC, Seto AG, Kim J, Kuramochi-Miyagawa S, Nakano T, Bartel DP, Kingston RE. 2006. Characterization of the piRNA complex from rat testes. *Science* **313:** 363–367.

Lee RC, Ambros V. 2001. An extensive class of small RNAs in *Caenorhabditis elegans. Science* **294:** 862–864.

Li C, Vagin VV, Lee S, Xu J, Ma S, Xi H, Seitz H, Horwich MD, Syrzycka M, Honda BM, et al. 2009. Collapse of germline piRNAs in the absence of Argonaute3 reveals somatic piRNAs in flies. *Cell* **137:** 509–521.

Lim AK, Kai T. 2007. Unique germ-line organelle, nuage, functions to repress selfish genetic elements in *Drosophila melanogaster. Proc Natl Acad Sci***104:** 6714–6719.

Lin H. 2007. piRNAs in the germ line. *Science* **316:** 397.

Lin H, Spradling AC. 1997. A novel group of *pumilio* mutations affects the asymmetric division of germline stem cells in the *Drosophila* ovary. *Development* **124:** 2463–2476.

Lin H, Yin H. 2008. A novel epigenetic mechanism in *Drosophila* somatic cells mediated by Piwi and piRNAs. *Cold Spring Harb Symp Quant Biol* **73:** 273–281.

Llave C, Kasschau KD, Rector MA, Carrington JC. 2002. Endogenous and silencing-associated small RNAs in plants. *Plant Cell* **14:** 1605–1619.

Lu J, Getz G, Miska EA, Alvarez-Saavedra E, Lamb J, Peck D, Sweet-Cordero A, Ebert BL, Mak RH, Ferrando AA, et al. 2005. MicroRNA expression profiles classify human cancers. *Nature* **435:** 834–838.

Malone CD, Brennecke J, Dus M, Stark A, McCombie WR, Sachidanandam R, Hannon GJ. 2009. Specialized piRNA pathways act in germline and somatic tissues of the *Drosophila* ovary. *Cell* **137:** 522–535.

Martin-Morris LE, Csink AK, Dorer DR, Talbert PB, Henikoff S. 1997. Heterochromatic trans-inactivation of *Drosophila* white transgenes. *Genetics* **147:** 671–677.

Martinez J, Patkaniowska A, Urlaub H, Luhrmann R, Tuschl T. 2002. Single-stranded antisense siRNAs guide target RNA cleavage in RNAi. *Cell* **110:** 563–574.

Megosh HB, Cox DN, Campbell C, Lin H. 2006. The role of PIWI and the miRNA machinery in *Drosophila* germline determination. *Curr Biol* **16:** 1884–1894.

miRBase http://microrna.sanger.ac.uk/cgi-bin/sequences/browse.pl. Manchester (UK): Sanger Institute Wellcome Trust, -.c.J., 2009. Available from: http://microrna.sanger.ac.uk.

Murchison EP, Kheradpour P, Sachidanandam R, Smith C, Hodges E, Xuan Z, Kellis M, Grutzner F, Stark A, Hannon GJ. 2008. Conservation of small RNA pathways in platypus. *Genome Res* **18:** 995–1004.

Okamura K, Hagen JW, Duan H, Tyler DM, Lai EC. 2007. The mirtron pathway generates microRNA-class regulatory RNAs in *Drosophila. Cell* **130:** 89–100.

Okamura K, Balla S, Martin R, Liu N, Lai EC. 2008a. Two distinct mechanisms generate endogenous siRNAs from bidirectional transcription in *Drosophila melanogaster. Nat Struct Mol Biol* **15:** 998.

Okamura K, Chung WJ, Ruby JG, Guo H, Bartel DP, Lai EC. 2008b. The *Drosophila* hairpin RNA pathway generates endogenous short interfering RNAs. *Nature* **453:** 803–806.

Palakodeti D, Smielewska M, Lu YC, Yeo GW, Graveley BR. 2008. The PIWI proteins SMEDWI-2 and SMEDWI-3 are required for stem cell function and piRNA expression in planarians. *RNA* **14:** 1174–1186.

Pal-Bhadra M, Leibovitch BA, Gandhi SG, Rao M, Bhadra U, Birchler JA, Elgin SC. 2004. Heterochromatic silencing and HP1 localization in *Drosophila* are dependent on the RNAi machinery. *Science* **303:** 669–672.

Pane A, Wehr K, Schupbach T. 2007. *zucchini* and *squash* encode two putative nucleases required for rasiRNA production in the *Drosophila* germline. *Dev Cell* **12:** 851–862.

Park JK, Liu X, Strauss TJ, McKearin DM, Liu Q. 2007. The miRNA pathway intrinsically controls self-renewal of *Drosophila* germline stem cells. *Curr Biol* **17:** 533–538.

Reuter M, Chuma S, Tanaka T, Franz T, Stark A, Pillai RS. 2009. Loss of the Mili-interacting Tudor domain–containing protein-1 activates transposons and alters the Mili-associated small RNA profile. *Nat Struct Mol Biol* **16:** 639–646.

Ro S, Park C, Song R, Nguyen D, Jin J, Sanders KM, McCarrey JR, Yan W. 2007. Cloning and expression profiling of testis-expressed piRNA-like RNAs. *RNA* **13:** 1693–1702.

Robine N, Lau NC, Balla S, Jin Z, Okamura K, Kuramochi-Miyagawa S, Blower MD, Lai EC. 2009. A broadly conserved pathway generates 3′UTR-directed primary piRNAs. *Curr Biol* **19:** 2066–2076.

Ruby JG, Jan CH, Bartel DP. 2007. Intronic microRNA precursors that bypass Drosha processing. *Nature* **448:** 83–86.

Ruby JG, Jan C, Player C, Axtell MJ, Lee W, Nusbaum C, Ge H, Bartel DP. 2006. Large-scale sequencing reveals 21U-RNAs and additional microRNAs and endogenous siRNAs in *C. elegans. Cell* **127:** 1193–1207.

Saito K, Inagaki S, Mituyama T, Kawamura Y, Ono Y, Sakota E, Kotani H, Asai K, Siomi H, Siomi MC. 2009. A regulatory circuit for *piwi* by the large Maf gene *traffic jam* in *Drosophila. Nature* **461:** 1296–1299.

Sarot E, Payen-Groschene G, Bucheton A, Pelisson A. 2004. Evidence for a *piwi*-dependent RNA silencing of the *gypsy* endogenous retrovirus by the *Drosophila melanogaster flamenco* gene. *Genetics* **166:** 1313–1321.

Sassone-Corsi P. 2000. CREM: A master-switch regulating the balance between differentiation and apoptosis in male germ cells. *Mol Reprod Dev* **56:** 228–229.

Schwarz DS, Hutvagner G, Du T, Xu Z, Aronin N, Zamore PD. 2003. Asymmetry in the assembly of the RNAi enzyme complex. *Cell* **115:** 199–208.

Shoji M, Tanaka T, Hosokawa M, Reuter M, Stark A, Kato Y, Kondoh G, Okawa K, Chujo T, Suzuki T, et al. 2009. The TDRD9-MIWI2 complex is essential for piRNA-mediated retrotransposon silencing in the mouse male germline. *Dev Cell* **17:** 775–787.

Smulders-Srinivasan TK, Lin H. 2003. Screens for *piwi* suppressors in *Drosophila* identify dosage-dependent regulators of germline stem cell division. *Genetics* **165:** 1971–1991.

Song JJ, Smith SK, Hannon GJ, Joshua-Tor L. 2004. Crystal structure of Argonaute and its implications for RISC slicer activity. *Science* **305:** 1434–1437.

Soper SF, van der Heijden GW, Hardiman TC, Goodheart M, Martin SL, de Boer P, Bortvin A. 2008. Mouse Maelstrom, a component of nuage, is essential for spermatogenesis and transposon repression in meiosis. *Dev Cell* **15:** 285–297.

Suh N, Baehner L, Moltzahn F, Melton C, Shenoy A, Chen J, Blelloch R. 2010. MicroRNA function is globally suppressed in mouse oocytes and early embryos. *Curr Biol* **20:** 271–277.

Szakmary A, Cox DN, Wang Z, Lin H. 2005. Regulatory relationship among *piwi*, *pumilio*, and *bag-of-marbles* in *Drosophila* germline stem cell self-renewal and differentiation. *Curr Biol* **15:** 171–178.

Tam OH, Aravin AA, Stein P, Girard A, Murchison EP, Cheloufi S, Hodges E, Anger M, Sachidanandam R, Schultz RM, et al. 2008. Pseudogene-derived small interfering RNAs regulate gene expression in mouse oocytes. *Nature* **453:** 534–538.

Thomson T, Lin H. 2009. The biogenesis and function of PIWI proteins and piRNAs: Progress and prospect. *Ann Rev Cell Dev Biol* **25:** 355–376.

Tuschl T, Zamore PD, Lehmann R, Bartel DP, Sharp PA. 1999. Targeted mRNA degradation by double-stranded RNA in vitro. *Genes Dev* **13:** 3191–3197.

Unhavaithaya Y, Hao Y, Beyret E, Yin H, Kuramochi-Miyagawa S, Nakano T, Lin H. 2009. MILI, a PIWI-interacting RNA-binding protein, is required for germ line stem cell self-renewal and appears to positively regulate translation. *J Biol Chem* **284:** 6507–6519.

Vagin VV, Sigova A, Li C, Seitz H, Gvozdev V, Zamore PD. 2006. A distinct small RNA pathway silences selfish genetic elements in the germline. *Science* **313:** 320–324.

Vagin VV, Wohlschlegel J, Qu J, Jonsson Z, Huang X, Chuma S, Girard A, Sachidanandam R, Hannon GJ, Aravin AA. 2009. Proteomic analysis of murine Piwi proteins reveals a role for arginine methylation in specifying interaction with Tudor family members. *Genes Dev* **23:** 1749–1762.

Vasileva A, Tiedau D, Firooznia A, Muller-Reichert T, Jessberger R. 2009. Tdrd6 is required for spermiogenesis, chromatoid body architecture, and regulation of miRNA expression. *Curr Biol* **19:** 630–639.

Vermaak D, Henikoff S, Malik HS. 2005. Positive selection drives the evolution of *rhino*, a member of the heterochromatin protein 1 family in *Drosophila*. *PLoS Genet* **1:** 96–108.

Volpe AM, Horowitz H, Grafer CM, Jackson SM, Berg CA. 2001. *Drosophila* rhino encodes a female-specific chromo-domain protein that affects chromosome structure and egg polarity. *Genetics* **159:** 1117–1134.

Voronina E, Seydoux G, Sassone-Corsi P, Nagamori I. 2011. RNA granules in germ cells. *Cold Spring Harb Perspect Biol* doi:10.1101/cshperspect.a002774.

Wang G, Reinke V. 2008. A *C. elegans* Piwi, PRG-1, regulates 21U-RNAs during spermatogenesis. *Curr Biol* **18:** 861–867.

Wang Y, Juranek S, Li H, Sheng G, Tuschl T, Patel DJ. 2008. Structure of an argonaute silencing complex with a seed-containing guide DNA and target RNA duplex. *Nature* **456:** 921–926.

Wang J, Saxe JP, Tanaka T, Chuma S, Lin H. 2009. Mili interacts with Tudor domain–containing protein 1 in regulating spermatogenesis. *Curr Biol* **19:** 640–644.

Watanabe T, Takeda A, Tsukiyama T, Mise K, Okuno T, Sasaki H, Minami N, Imai H. 2006. Identification and characterization of two novel classes of small RNAs in the mouse germline: Retrotransposon-derived siRNAs in oocytes and germline small RNAs in testes. *Genes Dev* **20:** 1732–1743.

Watanabe T, Totoki Y, Toyoda A, Kaneda M, Kuramochi-Miyagawa S, Obata Y, Chiba H, Kohara Y, Kono T, Nakano T, et al. 2008. Endogenous siRNAs from naturally formed dsRNAs regulate transcripts in mouse oocytes. *Nature* **453:** 539–543.

Yang Y, Xu S, Xia L, Wang J, Wen S, Jin P, Chen D. 2009. The bantam microRNA is associated with *Drosophila* fragile X mental retardation protein and regulates the fate of germline stem cells. *PLoS Genet* **5:** e1000444.

Yin H, Lin H. 2007. An epigenetic activation role of Piwi and a Piwi-associated piRNA in *Drosophila melanogaster*. *Nature* **450:** 304–308.

Zamore PD, Tuschl T, Sharp PA, Bartel DP. 2000. RNAi: double-stranded RNA directs the ATP-dependent cleavage of mRNA at 21 to 23 nucleotide intervals. *Cell* **101:** 25–33.

Developmental Control of Oocyte Maturation and Egg Activation in Metazoan Models

Jessica R. Von Stetina and Terry L. Orr-Weaver

Whitehead Institute and Department of Biology, Massachusetts Institute of Technology, Cambridge, Massachusetts 02142

Correspondence: weaver@wi.mit.edu

Production of functional eggs requires meiosis to be coordinated with developmental signals. Oocytes arrest in prophase I to permit oocyte differentiation, and in most animals, a second meiotic arrest links completion of meiosis to fertilization. Comparison of oocyte maturation and egg activation between mammals, *Caenorhabditis elegans*, and *Drosophila* reveal conserved signaling pathways and regulatory mechanisms as well as unique adaptations for reproductive strategies. Recent studies in mammals and *C. elegans* show the role of signaling between surrounding somatic cells and the oocyte in maintaining the prophase I arrest and controlling maturation. Proteins that regulate levels of active Cdk1/cyclin B during prophase I arrest have been identified in *Drosophila*. Protein kinases play crucial roles in the transition from meiosis in the oocyte to mitotic embryonic divisions in *C. elegans* and *Drosophila*. Here we will contrast the regulation of key meiotic events in oocytes.

Germ cells exist to produce gametes. Gametogenesis has two crucial events: meiosis and differentiation to produce the specialized features of gametes. In meiosis, two rounds of chromosome segregation occur without an intervening round of DNA replication. The resulting reduction of chromosome number is essential so that fusion of the sperm and egg genomes at fertilization restores a diploid chromosome number. In spermatogenesis, differentiation of sperm follows meiosis. In contrast, oocyte differentiation occurs during meiosis, necessitating arrest and restart of meiosis.

In addition to differences in timing of differentiation relative to meiosis, the distinct differentiated characteristics of sperm and oocytes reflect their unique contributions to fertilization and early embryogenesis. The mature oocyte contains maternal components needed by the fertilized embryos until developmental regulation comes under zygotic control. Although this can be as early as the two-cell stage in mammals, in other animals up to 14 division cycles of early embryogenesis are under maternal control, driven by proteins and mRNAs that must be stockpiled during oogenesis (Tadros et al. 2007). Moreover, patterning of the body axes is laid down in the developing oocyte in some animals (Riechmann and Ephrussi 2001). In contrast to the "moving van" nature of the oocyte, in most organisms the sperm is differentiated as a "race car," with minimal cytoplasm

Cite this article as *Cold Spring Harb Perspect Biol* doi: 10.1101/cshperspect.a005553

and adaptations to be motile and capable of penetrating the egg.

In most animals, oocyte development involves arrest of meiosis at two points, permitting growth, differentiation, and coordination between fertilization and the completion of meiosis (Fig. 1). Differentiation of the oocyte and stockpiling of maternal components occurs during a prolonged arrest in prophase I, the primary arrest point. Oocyte maturation releases this arrest, enabling the oocyte to progress into the meiotic divisions. The secondary arrest point coordinates completion of meiosis with fertilization, and the meiotic stage of this arrest varies between animals (Fig. 1). In most vertebrates, the arrest occurs at metaphase of meiosis II, in insects metaphase of meiosis I, in marine invertebrates such as starfish it is in G1 after the completion of both meiotic divisions, whereas in *C. elegans*, sperm signaling triggers oocyte maturation and there is not a secondary meiotic arrest (Nishiyama and Kishimoto 2010).

Extrinsic cues link developmental input with progression through meiosis in differentiating oocytes. These signals promote the onset, maintenance, and release of the primary arrest as well as the onset and maintenance of the secondary arrest. Long-range hormonal signals affect oocyte maturation and the release of the primary arrest, but local signaling also occurs. Cell–cell interactions between the oocyte and somatic support cells are crucial in regulating oocyte maturation and, in *C. elegans*, signaling from the sperm plays a critical role.

In general, animals make a tremendous investment in oocyte production and they are in limiting supply, making proper regulation of oogenesis vital for reproductive success. Here we discuss recent discoveries on the developmental control of oocyte maturation and egg activation and the extrinsic signals coordinating meiosis with oocyte differentiation. We compare vertebrates, focusing largely on mammals, with the invertebrate models *C. elegans* and *Drosophila* (Table 1). Production of functional oocytes also requires mechanisms to ensure accurate chromosome segregation in the meiotic divisions, but we will not cover this topic here. In addition, regulation of the translation of stockpiled maternal mRNAs is critical, contributing to the developmental timing of both

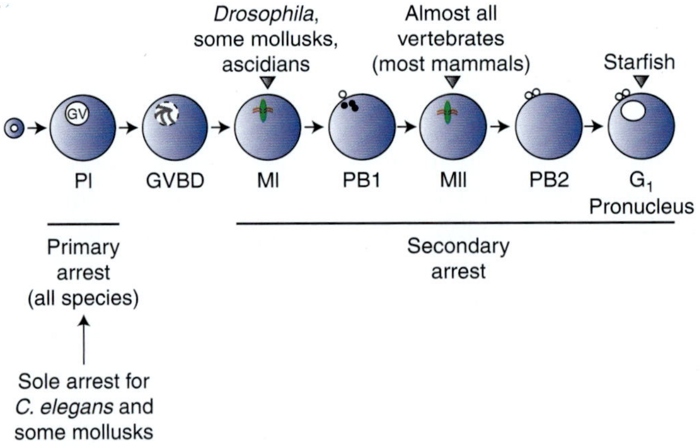

Figure 1. Meiotic arrests during oocyte development. Oocytes from most species undergo a first arrest at prophase I (PI) that is maintained for a few days (*Drosophila*) or for decades (humans). On hormonal or developmental stimulation, oocytes undergo meiotic maturation, release the primary arrest, and enter a second arrest at metaphase I (MI), metaphase II (MII), or postmeiotic G_1 depending on the species. Fertilization triggers release from the second arrest and the completion of meiosis in vertebrates (*Xenopus* and mammals). *Drosophila* releases the secondary arrest in a sperm-independent manner. GV, germinal vesicle; PI, prophase I; GVBD, germinal vesicle breakdown; MI, metaphase I; PB1, polar body 1; MII, metaphase II; PB2, polar body 2. (Adapted from Sagata 1996; reprinted with express permission from Noriyuki Sagata.)

Table 1. Summary of key molecular players and events regulating oocyte meiotic arrest, maturation, and egg activation

Meiotic event	Mouse	*Drosophila*	*Caenorhabditis elegans*
Primary arrest point	Prophase I	Prophase I	Prophase I
Length of prophase I arrest	Months	~1.5 d	~23 min (when ~40 sperm are present) ~ 8–10 h (no sperm present)
Maintenance of prophase I arrest	High cAMP produced by the oocyte; low Cdk1 activity	Low cyclin A; Elgi, Mtrm; inhibition of Polo; inactive Cdk2/Cyclin E	$G\alpha_{o/i}$
Maturation induction signal	LH	Unknown	MSP
Function of surrounding somatic cells for meiotic progression	Granulosa cells → transport of cGMP	Unknown	Gonadal sheath cells → high cAMP
Somatic cell/oocyte communication	Transport of cGMP through gap junctions	Unknown	Transport of unknown molecule through gap junctions
Secondary arrest point	Metaphase II	Metaphase I	None
Secondary arrest mechanism	cMOS; active MAPK; high Cdk1/Cyclin B; Emi2 inhibition of APC	Presence of chiasmata; heterochromatin pairing; chromosome congression to metaphase plate; inactive APC?; high Cdk1/Cyclin B?; SAC	N/A
Egg activation	Requires fertilization	Fertilization-independent	Requires fertilization
Completion of meiosis	Inactivation of Emi2; activation of APC; Ca^{2+} signaling; CamKII	*cortex* form of APC *sarah* (Ca^{2+}?)	EGG4,5 pseudo tyrosine phosphatase; APC activity

LH, luteinizing hormone; MSP, major sperm protein; cGMP, guanosine 3′-5′-cyclic monophosphate; MAPK, mitogen-activated protein kinase; APC, anaphase-promoting complex.

oocyte maturation and egg activation (reviewed in Vardy and Orr-Weaver 2007b; Radford et al. 2008). A detailed review on mammalian oocyte maturation is available (Jaffe and Norris 2010). Two comprehensive reviews on egg activation and release of the secondary meiotic arrest also have been published recently (Horner and Wolfner 2008; Nishiyama and Kishimoto 2010).

MAMMALS

Recent progress in our understanding of the control of meiosis in mammalian oogenesis has highlighted the importance of signaling between the oocyte and surrounding somatic follicle cells. The cell cycle regulators affected by this signaling were identified and their roles in oocyte maturation and maintenance of the secondary arrest largely elucidated from studies using *Xenopus*. In this article, we focus on mammalian oocytes but draw parallels to the regulation identified in *Xenopus*.

Maintenance of the Primary Arrest

The mammalian oocyte enters meiosis during fetal development and arrests at the diplotene stage of prophase I (germinal vesicle [GV] state) from birth to puberty (Fig. 2A) (Hunt and

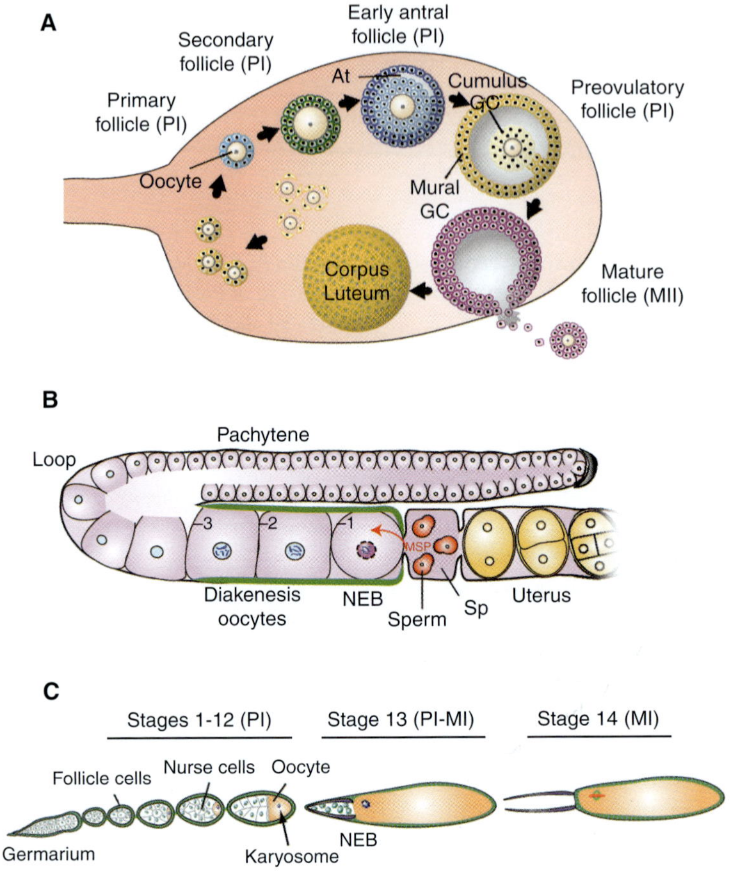

Figure 2. Stages of oocyte development and meiotic progression. (*A*) In mammals, a primary follicle, consisting of a prophase I (PI) arrested primary oocyte surrounded by somatic pregranulosa cells, is generated shortly after birth. Primary oocytes grow, whereas granulosa cells (GC) proliferate to form secondary follicles. As a fluid-filled cavity (antrum, At) begins to form, secondary follicles become early antral follicles. In preovulatory follicles, the fully grown primary oocyte is surrounded by cumulus and mural granulosa cells. After a luteinizing hormone surge, the oocyte undergoes meiotic maturation and produces a secondary oocyte that arrests at metaphase II (MII). During ovulation, the MII oocyte is released into the oviduct where on fertilization, meiosis is resumed and completed. (*B*) In the distal germline of *C. elegans*, syncytial nuclei enter meiosis and are found in pachytene of PI. Around the loop region, these nuclei cellularize to form oocytes that progress to diakenesis I where they arrest. Somatic-derived gonadal sheath cells (green) surround the developing oocytes. In response to sperm and its secreted factor major sperm protein (MSP), the most proximal oocyte (−1) is induced to undergo meiotic maturation (nuclear envelope breakdown [NEB] and rounding up of the cell). The oocyte passes through the spermatheca (Sp) where fertilization occurs, and it then is deposited into the uterus as a one-cell zygote. (*C*) The *Drosophila* oocyte develops within a 16-cell germline cyst surrounded by a monolayer of somatic follicle cells (green). The oocyte enters meiosis in region 2A of the germarium and soon after arrests at PI for most of oogenesis (∼2 d). At stage 13, after a yet unknown developmental or hormonal signal, the oocyte undergoes meiotic resumption and progresses into metaphase I (MI), the secondary arrest point. On ovulation, as the mature stage 14 oocyte travels in the oviduct, rehydration and mechanical pressure trigger the completion of meiosis. (*A*, Adapted from Matzuk and Lamb 2002; reprinted with express permission from Martin Matzuk. *B*, Adapted from Kuwabara 2003; reprinted with express permission from Patty Kuwabara. *C*, Adapted from Xiang et al. 2007; reprinted with express permission from Scott Hawley.)

Hassold 2008). The oocyte remains arrested at this stage for months (mouse) or decades (humans) until a preovulatory hormonal surge (Table 1) (Whitaker 1996; Kishimoto 2003). Because the maturation or M-phase promoting factor (MPF), a complex of the kinase subunit Cdk1 (also known as Cdc2) and the regulatory subunit Cyclin B, triggers the onset of the meiotic divisions, low Cdk1/Cyclin B activity is key in maintaining the prophase I arrest (Sagata 1996). One mechanism to inactivate Cdk1/Cyclin B involves the reduction of Cyclin B levels in GV oocytes by the Anaphase Promoting Complex/Cyclosome (APC/C) and its coactivator Cdh1 (APC^{Cdh1}) (Reis et al. 2006). High levels of the second messenger molecule cyclic adenosine 3′, 5′-monophosphate (cAMP) within the mammalian oocyte also are essential to keep Cdk1/Cyclin B in an inactive state during this primary arrest (Mehlmann 2005b). High cAMP levels promote activation of the protein kinase A (PKA) (Conti et al. 2002; Kovo et al. 2006), which concurrently phosphorylates and activates the kinase Wee1 (a Cdkl inhibitor) (Han et al. 2005) and inactivates the dual-specificity phosphatase Cdc25 (a Cdk1 activator) (Lincoln et al. 2002; Pirino et al. 2009). Thus, cAMP is at the top of a cascade of events leading to the inactivation of Cdk1/Cyclin B.

The inhibitory cAMP pool that is required to maintain meiotic arrest is produced by the oocyte itself (Fig. 3A) (reviewed in Mehlmann 2005b; Jaffe and Norris 2010). Several lines of evidence have shown that endogenous production of cAMP is mediated by the activation of the G protein G_s by the G protein–coupled receptor (GPR3/mouse; GPR12/rat) present on the oocyte plasma membrane and the subsequent stimulation of the adenylyl cyclase (AC), the enzyme responsible for cAMP synthesis (Mehlmann et al. 2002, 2004; Horner et al. 2003; Kalinowski et al. 2004; Freudzon et al. 2005; Hinckley et al. 2005; Mehlmann 2005a; DiLuigi et al. 2008). The crucial role of oocyte pools of cAMP is shown by this source of cAMP being essential for meiotic arrest even when the oocyte is separated from the somatic layer (Vaccari et al. 2008).

In addition to the production of cAMP by the oocyte, somatic-derived inhibitory signals that are transmitted into the oocyte are also required to maintain prophase I arrest (Edry et al. 2006; Jaffe and Norris 2010). Mammalian oocytes develop within a follicle, surrounded by somatic cumulus granulosa cells (inner layer), thus forming an oocyte-cumulus complex (OCC) (Fig. 2A). The OCC connects on one side to somatic mural granulosa cells (outer layer) (Fig. 2A), which are the key source of inhibitory signals (Jaffe and Norris 2010). When this outer layer is removed from the OCC, the oocyte resumes meiosis (Racowsky and Baldwin 1989).

The diffusion of inhibitory signals from the somatic compartment into the oocyte occurs through gap junctions. Consistent with this conclusion, genetic and pharmacological inhibition of gap junctions induce resumption of meiosis (Piontkewitz 1993; Sela-Abramovich et al. 2006; Norris et al. 2008). Guanosine 3′, 5′-cyclic monophosphate (cGMP) has been recently identified as an essential inhibitory signal that diffuses through gap junctions from cumulus granulosa cells into the oocyte (Fig. 3A) (Sela-Abramovich et al. 2008; Norris et al. 2009). cGMP inhibits the oocyte-specific cAMP phosphodiesterase 3A (PDE3A), which metabolizes cAMP, and in turn maintains elevated levels of cAMP (Fig. 3A) (Conti et al. 2002; Hambleton et al. 2005; Norris et al. 2009; Vaccari et al. 2009). These studies highlight the conservation of the role of the gap-junction-mediated communication in the control of oocyte meiosis in both vertebrate and invertebrates species (discussed below).

Recent studies have identified a new pathway involved in the generation of cGMP in cumulus granulosa cells and oocytes. It has been shown that the mural granulosa cells stimulate the generation of cGMP by secreting natriuretic peptide precursor type C (NPPC), which binds to and activates the natriuretic peptide receptor 2 (NPR2), a guanylyl cyclase, expressed in the cumulus granulosa cells (Zhang et al. 2010). In addition to producing its own inhibitory pool of cAMP, the oocyte also induces the expression of NPR2 in the cumulus cells (Zhang et al. 2010).

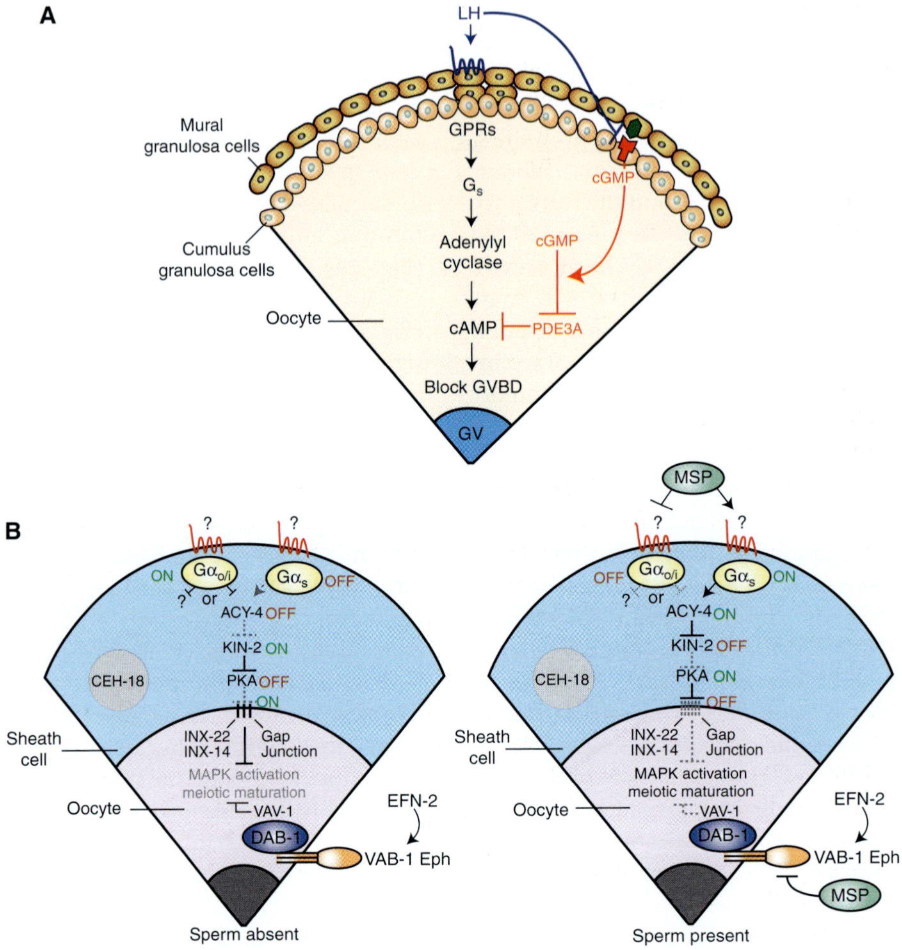

Figure 3. Gap junctional communication between the oocyte and the surrounding somatic cells regulates oocyte meiotic maturation. (*A*) High levels of cyclic adenosine 3′, 5′-monophosphate (cAMP) in the oocyte inhibit meiotic maturation in mammals. The inhibitory cAMP is produced by the oocyte itself via the activation of the GPR3/Gs/adenylyl cyclase pathway. Guanosine 3′, 5′-cyclic monophosphate (cGMP) produced by the cumulus somatic cells flows through gap junctions into the oocyte to inhibit PDE3A, the phosphodiesterase responsible for the hydrolysis of cAMP. The oocyte stimulates the expression of natriuretic peptide receptor 2 (NPR2) (red), a guanylyl cyclase, in the cumulus cells. Mural granulosa cells induce the generation of cGMP by secreting the NPR2 ligand, natriuretic peptide precursor type C (NPPC) (green). Binding of luteinizing hormone (LH) to its G protein-coupled receptor (GPCR) (blue) reverses the inhibition of meiotic maturation by decreasing the synthesis of cGMP in the somatic follicular layer and by blocking its diffusion through gap junctions. (*B*) In *C. elegans*, the $G\alpha_{o/i}$ pathway in gonadal sheath cells (blue) leads to the inactivation of adenylate cyclase 4 (ACY-4) and subsequently protein kinase A (PKA) in the absence of sperm. This inhibition is postulated to stabilize gonadal sheath-to-oocyte gap junctions composed of innexin (Inx) proteins (Inx-22 and Inx-14), and thereby allow the influx of a negative regulatory signal into the oocyte, which blocks MAP kinase (MAPK) activation and meiotic maturation. In parallel, the VAB-1/Ephrin Receptor in the oocyte inhibits MAPK and oocyte maturation. Sperm-derived major sperm protein (MSP) antagonizes both the sheath $G\alpha_{o/i}$ and oocyte VAB-1 signaling pathways, while simultaneously activating the sheath $G\alpha_s$ pathway, resulting in MAPK activation and meiotic maturation. It is hypothesized that the $G\alpha_s$ pathway destabilizes sheath-oocyte gap junctions. The nuclear POU homeodomain protein is required for proper differentiation of gonadal sheath cells. (*B*, Adapted from Govindan et al. 2006 and Sun et al. 2009; reprinted with express permission from Qing-Yuan Sun and David Greenstein, respectively.)

Meiotic Maturation

Activation of Cdk1/Cyclin B is the key molecular event that promotes meiotic maturation and release from the prophase I arrest in most species, including mammals (Sagata 1996). The phosphorylation of downstream targets by active Cdk1/Cyclin B leads to nuclear envelope breakdown (NEB) (also known as germinal vesicle breakdown, GVBD), spindle assembly, and chromosome condensation, the morphological hallmarks of meiotic maturation (Jones 2004). In mammals, the onset of oocyte maturation is controlled by the action of the gonadotrophin hormones. Following puberty, in response to a cyclic preovulatory surge in luteinizing hormone (LH), the prophase I–arrested mammalian oocyte resumes meiosis. This is induced by a drop in intraoocyte cAMP concentration, which leads to the activation of the Cdc25 phosphatase and the relief of Cdk1/Cyclin B inhibition. Because LH receptors are only expressed in the mural granulosa cells and not in the oocyte (Peng et al. 1991), the LH-induced stimulation of meiotic resumption also involves signaling between the oocyte and its somatic support cells.

Reinitiation of meiosis in response to LH involves the reduction of cGMP in the oocyte and consequently of cAMP (Fig. 3A). LH mediates this effect in part by decreasing the concentration of cGMP in the mural granulosa cells through a yet not well-understood mechanism (Norris et al. 2009; Vaccari et al. 2009). The second way by which LH reduces cGMP in the oocyte is by inducing the closure of gap junction communication between the somatic cells (Norris et al. 2008, 2009). Activation of the mitogen-activated protein kinase (MAPK) pathway by LH promotes the phosphorylation of connexin 43 (a major ovarian gap junction protein) and induces a decrease in gap junction permeability between the mural granulosa cells (Sela-Abramovich et al. 2005; Norris et al. 2008). This closure of gap junctions prevents the influx of cGMP from the granulosa cells into the oocyte. The resulting reduction of cGMP by these two mechanisms ultimately leads to the release of the inhibition of PDE3A and the increase in cAMP hydrolysis (Norris et al. 2009).

Current evidence indicates that the LH-induced signaling that mediates meiotic maturation is transmitted from the somatic compartment into the oocyte mostly via the activation of the epidermal growth factor (EGF) receptor pathway (Conti et al. 2006; Panigone et al. 2008). Binding of LH to its G protein–coupled receptor in the granulosa cells (Fig. 3A) leads, in part, to the activation of G_s and the production of cAMP (reviewed in Jaffe and Norris 2010). High cAMP levels in the somatic cells promote PKA activation, and the subsequent synthesis and release of EGF receptor (EGFR) ligands (Panigone et al. 2008). The resulting ligand-dependent transactivation of the EGF receptor is required for the partial activation of MAPK (Panigone et al. 2008), which as discussed above inhibits gap junctions. Activation of the EGFR signaling pathway in response to LH surge also causes a major decrease of cGMP in the granulosa cells (Vaccari et al. 2009).

Regulation of the activity of the APC/C has been shown to be required for the timing of meiotic maturation in mammals. Prior to the onset of meiotic maturation, the phosphatase Cdc14B antagonizes the activity of Cdk1/Cyclin B by activating Cdh1 and promoting Cyclin B degradation by the APC/C (Schindler and Schultz 2009). Consistent with this role, depletion of Cdc14B by RNAi in GV oocytes causes premature meiotic resumption (Schindler and Schultz 2009). On GVBD, the early mitotic inhibitor 1 (Emi1) protein counteracts the function of Cdc14B by inhibiting the APC^{Cdh1} (Marangos et al. 2007). It remains to be determined how Cdc14B and Emi1 are developmentally regulated and inactivated to permit oocyte maturation.

Metaphase II Arrest and Release

A rise in Cdk1/Cyclin B activity is required in mammalian eggs to induce the progression from prophase I to a second arrest at metaphase II, which remains until fertilization (Sagata 1996). The stability of Cdk1/Cyclin B in vertebrates depends on the cytostatic factor

(CSF), an activity in the cytoplasm of mature *Rana pipiens* oocytes that was found to induce metaphase arrest when injected into two-cell embryos (Masui and Markert 1971). Breakthrough discoveries in the *Xenopus laevis* model identified the MOS/MEK1/MAPK/p90^{Rsk} and the early mitotic inhibitor 2 (Emi2) pathways as molecular components of the CSF activity and thus key players involved in the metaphase II arrest (Tunquist and Maller 2003; Nishiyama and Kishimoto 2010). In mammals, MOS/MEK1/MAPK is likely to be required only for maintaining the metaphase II arrest. Mouse oocytes lacking MOS function reach metaphase II arrest but fail to maintain the arrest, as they spontaneously exit meiosis (Colledge et al. 1994; Hashimoto et al. 1994; Verlhac et al. 1996). In contrast, Emi2 plays dual roles in the establishment and maintenance of the second arrest (reviewed in Madgwick and Jones 2007). Both the MOS and Emi2 pathways work together to inhibit the activity of the APC/C and keep Cdk1/Cyclin B active (Madgwick and Jones 2007).

Egg Activation, Release from Metaphase II Arrest, and Entry into Embryogenesis

The oocyte resumes meiosis and becomes competent to begin embryonic development on egg activation. Egg activation is characterized by multiple events including changes in egg coverings to prevent polyspermy, release of the second meiotic arrest and completion of meiosis, posttranscriptional modifications of maternal mRNAs, and cytoskeletal rearrangements (reviewed in Horner and Wolfner 2008). The mammalian egg is activated in a fertilization-dependent manner. The sperm fusion to the egg induces a wave of Ca^{2+} oscillations within the egg that break the Meta II arrest (Jones 2005; Ducibella and Fissore 2008). However, the molecular mechanisms whereby the sperm Ca^{2+} signal induces this response have not been fully defined.

It is known that an elevation of Ca^{2+} at fertilization results in the activation of one downstream effector, calmodulin-dependent protein kinase II (CaMKII) (Tatone et al. 2002; Markoulaki et al. 2004). In mouse eggs, CaMKIIγ is the predominant CaMKII isoform that is essential for egg activation (Backs et al. 2010). CaMKIIγ$^{-/-}$ mice display female-specific sterility because of a failure to decrease MAPK and Cdk1/Cyclin B activities to permit resumption of meiosis (Backs et al. 2010). The direct downstream effectors of CaMKII have not yet been identified. Mechanistically, based on findings in the *Xenopus* system (Liu and Maller 2005; Rauh et al. 2005; Hansen et al. 2006), and some evidence in mammals (Ducibella and Fissore 2008), CaMKII activation likely leads to the degradation of Emi2. After Emi2 is degraded, the APC/C can target Cyclin B and securin (an inhibitor of the separase protease responsible for release of sister chromatid cohesion) for degradation and in turn inactivate Cdk1/Cyclin B to relieve the meiotic block. In addition to CaMKII, calcineurin (calcium and calmodulin-dependent phosphatase) and protein kinase C (PKC) appear to be second effector molecules for metaphase II progression in mouse oocytes, although their role is less clear (Jones 2005).

Fertilization induces completion of meiosis and leads to the restart of the mitotic cycle in the embryo, with the first embryonic cell cycle in the one-cell embryo being mostly driven by maternally provided factors (Kubiak et al. 2008). Zygotic transcription becomes prevalent by the second division and presumably subsequent embryonic cell divisions are under zygotic control.

C. elegans

Regulation of Oocyte Maturation by Sperm in C. elegans

In *C. elegans*, signaling from the somatic cells surrounding the oocyte, in this case the sheath cells of the gonad, is crucial in regulating meiotic maturation (Miller et al. 2003). In this organism, however, maturation is controlled by sperm. Hermaphrodite worms make a limited number of sperm during the initial stages of gametogenesis, before transitioning to oocyte production. These sperm are stored in the spermatheca (Fig. 2B) adjacent to the mature oocytes. Given the finite number of sperm, forcing oocyte maturation to be dependent on the presence of sperm ensures that oocytes are not wasted and each maturing oocyte becomes fertilized. Sperm signaling promotes oocyte maturation and ovulation, releasing the primary meiotic arrest. Fertilization then leads to egg activation and the completion of meiosis (for review, see Yamamoto et al. 2006).

The major sperm protein (MSP) is the ligand that regulates not only oocyte maturation but also oocyte growth and contraction of the sheath cells (Fig. 2B) (Yamamoto et al. 2006). Originally identified as a cytoskeletal protein needed for motility of these amoeboid sperm, MSP appears to be secreted by a vesicle budding process to form a concentration gradient from the spermatheca (Fig. 2B) (Kosinski et al. 2005). MSP can bind to the ephrin receptor VAB-1 on the oocyte (Miller et al. 2003) and recently has been shown to bind to as yet unidentified receptors on the sheath cells (Fig. 3B) (Govindan et al. 2009).

Delineation of the effects of MSP on maturation and the contributions of the oocyte and sheath cells have been facilitated by the possibility of doing mosaic analysis of mutants, RNAi of candidate players, and by the ability to assay MSP effects by direct injection into the gonad. *C. elegans* females, which are mutant hermaphrodites that cannot produce sperm, have been used for these injection assays. MSP induces four events that accompany and facilitate ovulation and fertilization: oocyte maturation, microtubule reorganization in the oocyte undergoing maturation, actin-based cytoplasmic streaming that promotes oocyte growth prior to meiotic arrest, and gonadal sheath cell contractions that promote ovulation (Miller et al. 2001; Harris et al. 2006; Govindan et al. 2009).

MSP acts both on the oocyte and the surrounding sheath cells, and recent studies delineated the consequences of MSP binding to sheath cells (Fig. 3B) (Govindan et al. 2006, 2009). These experiments established that binding of MSP causes the activation of $G\alpha_s$ (*gsa-1*), which in turn activates adenylate cyclase to elevate cAMP levels. Mutation of the *acy-4* adenylate cyclase gene blocks oocyte maturation, and function of the gene is required in the somatic rather than the germline lineage. High overexpression of adenylate cyclase or gain-of-function mutation of $G\alpha_s$ are capable of triggering meiotic maturation even in the absence of sperm, revealing a direct role for $G\alpha_s$ signaling and adenylate cyclase in triggering subsequent maturation steps. Activation of $G\alpha_s$ signaling counteracts inhibitory signals from $G\alpha_{o/i}$ that block oocyte maturation in the absence of MSP. These results imply the presence of a G protein–coupled receptor in sheath cells to which MSP binds, but the receptor has not yet been identified. In summary, $G\alpha_s$-adenylate cyclase mediated signaling functions in the gonadal sheath cells for all characterized MSP-dependent meiotic maturation responses.

MSP signaling promotes oocyte activation by blocking inhibition mediated through signaling between the sheath cells and oocytes (Govindan et al. 2009). This signaling requires the gap junction Innexin (invertebrate homolog of connexin) proteins encoded by *inx-14* and *inx-22* expressed in the germline (Fig. 3B) (Whitten and Miller 2007). Mutation of *inx-22* suppresses the maturation failure of a *gsa-1* mutant, demonstrating the critical inhibitory role of these gap junctions in oocyte maturation. This mutation, however, does not suppress *acy-4* mutants, indicating the existence of other participating pathways.

Although an oocyte receptor for MSP has been identified, the function of MSP signaling within the oocyte in inducing maturation is not clear. Most recent results suggest that it may play a modulatory role in tightly linking oocyte maturation to the presence of sperm. The current data suggest that VAB-1 functions to inhibit meiotic maturation when MSP is absent or limiting and has no essential role in promoting maturation. VAB-1 inhibits meiotic maturation when in, or in transit to, recycling endosomes (Cheng et al. 2008). When MSP is present, the exclusion of VAB-1 from the recycling endosomes requires $G\alpha_s$ and ACY-4 function in the gonadal sheath cells. Recent data suggest that VAB-1 functions in oocyte meiotic maturation by negatively regulating the DAF-18/PTEN pathway (Brisbin et al. 2009). How redistribution of MSP to the plasma membrane permits oocyte maturation remains to be elucidated.

Oocyte maturation is accompanied by activation of MAPK within the oocyte, as assayed by phospho-specific antibodies (Miller et al. 2001). Active phosphorylated MPK-1 MAPK is observed in the one to six oocytes most proximal to the spermatheca (see Fig. 2B) (Lee et al. 2007), and this activation requires MSP and $G\alpha_s$ signaling in the sheath cells, unless the gap junctions are nonfunctional (Whitten and Miller 2007; Govindan et al. 2009). Meiotic maturation, however, is spatially restricted to the -1 oocyte, albeit it is not known how. Presumably active MAPK causes the onset of meiotic chromosome segregation by the conserved pathways defined in vertebrates, although this remains to be shown directly. Several mutants result in MAPK activation even in the absence of sperm, including the transcription factors CEH-18 (a POU homeodomain transcription factor) and EGRH-1 (an EGR family member) (Rose et al. 1997; Clary and Okkema 2010). These transcription factors act in the soma to repress MAPK activation in the oocyte.

Conserved cell cycle regulators such as POLO kinase, which participates in the Cdk1/Cyclin B-Cdc25 autoregulatory loop by promoting the phosphorylation and activation of Cdc25 in many systems (Abrieu et al. 1998; Chase et al. 2000; Roshak et al. 2000; Karaiskou et al. 2004), and Cdk1 also function as positive regulators of meiotic maturation in *C. elegans*. Disruption of POLO-like kinase 1 (PLK-1) or NCC-1/Cdk1, via RNAi, result in defects in nuclear envelope breakdown prior to ovulation (Boxem et al. 1999; Chase et al. 2000). Worm-specific proteins such as the zinc finger domain–containing proteins OMA-1 and OMA-2 also promote meiotic maturation upstream of Cdk1 (Detwiler et al. 2001).

Control of the Transition from Oocyte to Zygote in C. elegans

Release of meiotic arrest in the oocyte by fertilization also ensures proper control of the restart of the mitotic cell cycle in the zygote. During embryogenesis there is a handoff from maternal to zygotic control of development, but the timing of this transition varies markedly between animals. In *C. elegans*, the mechanisms to terminate maternal control begin at oocyte maturation. The MBK-2 kinase, a member of the dual-specificity tyrosine-regulated family (DYRKs), phosphorylates maternal proteins to contribute to the oocyte–zygote transition (Stitzel et al. 2006). For example, phosphorylation of the katanin protein MEI-1 by MBK-2 leads to its degradation, preventing it from severing microtubules after completion of meiosis and thereby possibly disrupting zygotic development. Other proteins phosphorylated by MBK-2 include MEX-2 and MEX-6. MBK-2 phosphorylation of these two proteins primes them for phosphorylation by POLO kinase, permitting them to act to establish anterior-posterior polarity in the early zygote. Phosphorylation of the OMA-1 and OMA-2 proteins by MBK-2 converts their function from that in oocyte maturation to the ability to sequester TAF-4 and block transcription in the germline precursor cells in the embryo (Guven-Ozkan et al. 2008).

The activation of MBK-2 is tightly regulated during oocyte development. Cdk1 phosphorylates MBK-2 during oocyte maturation, making it capable of phosphorylating MEI-1 (Cheng et al. 2009). The mechanism of this activation is unknown, but may involve inactivation or release of an inhibitor. Activity of MBK-2 is blocked, however, until anaphase I by its retention at the oocyte cortex by the EGG-3 protein (Stitzel et al. 2007). In addition, it is bound at the cortex by the EGG-4 and EGG-5 proteins (Cheng et al. 2009; Parry et al. 2009). These proteins have homology to tyrosine phosphatases but appear to lack active catalytic sites. They bind to MBK-2 and have been shown in vitro to block its kinase activity. Ablation of *egg-4* and *egg-5* by RNAi leads to premature phosphorylation of the MEI-1 substrate in oocytes, consistent with EGG-4 and EGG-5 playing crucial roles to restrict MBK-2 kinase activity in vivo. If these genes are ablated or mutated meiosis is abnormal, polar bodies do not form, the eggshell is not properly produced, and polyspermy can occur. The APC, which is active at the metaphase I/anaphase I transition, releases the complex from the cortex, resulting in MBK-2 activation. The APC targets EGG-3 for degradation and possibly EGG-4 and EGG-5 as well.

DROSOPHILA

Prophase I Arrest

The *Drosophila* oocyte, which is differentiated from a 16-cell cyst, enters the early events of prophase I in region 2A in the germarium, and by stage 5 of egg chamber development arrests for the first time in the equivalent of diplotene, with the chromosomes compacted into a karyosome (Fig. 2C) (Page and Hawley 2001; Resnick et al. 2009). The other 15 nurse cell sisters instead enter the endocycle (DNA replication without mitosis) to become highly polyploid (Hong et al. 2003). In *Drosophila*, the p21CIP/p27KIP1/p57KIP2-like Cdk inhibitor, Dacapo (DAP), is required within the oocyte to inhibit the kinase activity of Cdk2/Cyclin E and thus prevent DNA replication or entry into the endocycle (Hong et al. 2003).

Contrary to mammals and *C. elegans*, in *Drosophila* it has not been established whether gap junction–mediated communication between the somatic follicle cells and the oocyte is required to maintain prophase I meiotic arrest. Interestingly, the proteins that compose the gap junction channels, innexins (Inx), are widely expressed during *Drosophila* oogenesis (Stebbings et al. 2002). In fact, in vitro injections with antisera against Inx2 (a gap junction protein between oocyte and follicle cells) into stage 10 follicle-enclosed oocytes significantly affected development past stage 10 by blocking follicle cell differentiation, nurse cell regression, and oocyte growth (Bohrmann and Zimmermann 2008). Those oocytes that bypassed the developmental arrest showed defects in chorion formation; however, meiosis was not analyzed in these oocytes (Bohrmann and Zimmermann 2008). To date, no roles for the *Drosophila* cAMP transduction pathway during meiosis have been established.

Cross talk between the oocyte and the surrounding follicle cells is required for follicle cell fate determination and patterning of the egg and embryo (Ray and Schupbach 1996; Gonzalez-Reyes et al. 1997). Thus, it is possible that this bidirectional communication also regulates meiotic events in the oocyte. In addition, there is signaling between meiotic progression and developmental patterning. Mutants in which meiotic double-strand breaks are not repaired have dorsal/ventral patterning defects in the developing oocyte. These unrepaired breaks are sensed and lead to a failure to translate Gurken, the ligand secreted from the oocyte that directs determination of the adjacent follicle cells (Ghabrial and Schupbach 1999; Abdu et al. 2002).

Meiotic Maturation

The *Drosophila* oocyte undergoes meiotic maturation at stage 13 of oogenesis (Xiang et al. 2007; Von Stetina et al. 2008). Consistent with a potential requirement of active Cdk1/Cyclin B in *Drosophila* during this process, temperature-sensitive mutations in *cdk1* delay the timing of meiotic maturation (Fig. 4) (Von Stetina et al. 2008). In addition, mutants of *twine* (*twe*), the meiotic homolog of the phosphatase Cdc25 (Alphey et al. 1992; Courtot et al. 1992; White-Cooper et al. 1993), also are delayed in the disassembly of the nuclear envelope (Xiang et al. 2007; Von Stetina et al. 2008). Recent discoveries with mutations affecting the developmental timing of maturation identified new regulators of Twine and Polo, confirming the link between maturation and Cdk1/Cyclin B activity.

The *Drosophila* *α-endosulfine* homolog, *endos*, which encodes a conserved phosphoprotein, regulates all aspects of meiotic maturation including the timing of nuclear envelope breakdown (NEB) (Von Stetina et al. 2008). *endos* mutant oocytes display a severe delay in NEB, spindle formation and chromosome congression defects very similar to *twe* mutants. The Endos protein likely exerts its control on meiotic maturation by controlling Cdk1/Cyclin B activity, as it controls the protein stability of

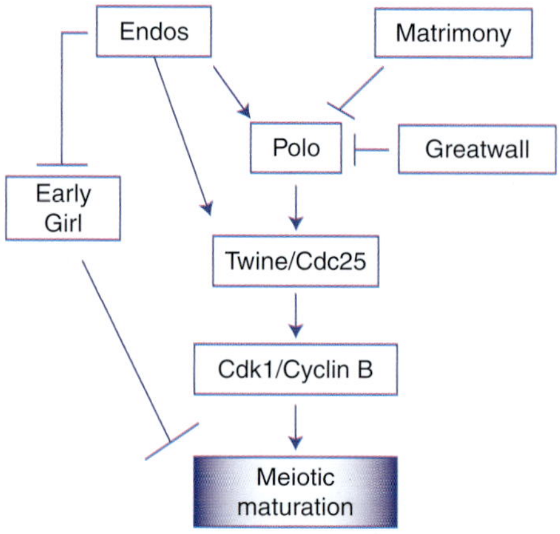

Figure 4. Regulatory genes affecting meiotic maturation in *Drosophila*. Evidence in *Drosophila* suggests that high levels of Cdk1/Cyclin B activity are required for meiotic maturation. The Polo kinase phosphorylates and activates the phosphatase Twine/Cdc25, which in turn phosphorylates and activates Cdk1. Before nuclear envelope breakdown (NEB), Matrimony sets the timing of meiotic maturation by inhibiting Polo activity. Endos positively regulates the timing of meiotic maturation by regulating the levels of Polo and Twine/Cdc25, which are required to promote Cdk1/Cyclin B activation. Independently, Endos inhibits the predicted E3 ubiquitin ligase Early Girl. The Greatwall kinase also inhibits Polo.

Twine/Cdc25 and Polo (Fig. 4) (Von Stetina et al. 2008). Although an in vitro reduction of Cdk1/Cyclin B kinase activity toward its substrate histone 1 (H1) was not observed in *endos* mutant oocytes, *endos* might affect the activity of Cdk1/Cyclin B toward other key substrates other than H1 in vivo (Von Stetina et al. 2008). Consistent with this, in *endos* mutants there is a marked reduction in the phosphorylation of mitotic protein monoclonal 2 (MPM2) conserved epitopes (Davis et al. 1983), normally phosphorylated by both Polo and Cdk1/Cyclin B in *Drosophila* (Logarinho and Sunkel 1998; Von Stetina et al. 2008). Finally, Endos physically interacts with a predicted E3 ubiquitin ligase encoded by *early girl* (*elgi*) (Von Stetina et al. 2008), which is also required for the timing of oocyte maturation. Contrary to *endos* mutants, some oocytes from mutants of *elgi* undergo premature nuclear envelope breakdown (Von Stetina et al. 2008). How Endos controls meiosis via Elgi is currently unknown.

Whereas *endos* is needed for normal levels of Polo in oocytes, premature activation of Polo prior to maturation is prevented by the protein encoded by *Matrimony* (*Mtrm*), which physically interacts with Polo (Fig. 4) (Xiang et al. 2007). *Mtrm*/ + heterozygous females undergo NEB precociously, and this phenotype can be fully suppressed by reducing the dose of *polo* (Xiang et al. 2007). The current model predicts that an excess of Mtrm inhibits Polo proteins prior to the onset of NEB (Xiang et al. 2007). On meiotic maturation at stage 13, Polo proteins exceed the available amounts of inhibitory Mtrm proteins and activate Twine/Cdc25, which in turn leads to the activation of Cdk1/Cyclin B and NEB. Independent results also show that the conserved Greatwall kinase (Gwl), discovered in flies (Yu et al. 2004), antagonizes Polo activity during meiosis (Archambault et al. 2007). Consistent with this role, homozygosity of the *Scott of the Antarctic* (*Scant*) mutation in the *gwl* gene, which results in kinase hyperactivity of Gwl, reduces Polo levels and suppresses the *Mtrm*/ + meiotic phenotypes (Archambault et al. 2007).

Currently, the extrinsic signal(s) that triggers oocyte meiotic maturation in *Drosophila* is unknown. Prostaglandin hormones or the steroid hormone ecdysone could serve as potential stimulatory signals. *Drosophila* cyclooxygenase (COX), one of the enzymes responsible for synthesis of prostaglandins, promotes early ovarian follicle maturation (Tootle and Spradling 2008). Ecdysone signaling is also required for progression of oogenesis and egg chamber maturation during mid-oogenesis (Buszczak et al. 1999). Because the follicles in mutants affecting either signaling pathways do not reach stage 13 (when meiotic maturation takes place), it is not known whether prostaglandins or ecdysone are required for meiotic maturation in flies.

Metaphase I Arrest and Release of Arrest

On meiotic maturation, the *Drosophila* oocyte arrests for a second time at metaphase I at stage 14 (Figs. 1, 2C) (King 1970). During this arrest

 Cite this article as *Cold Spring Harb Perspect Biol* doi: 10.1101/cshperspect.a005553

the exchange and nonexchange chromosomes cluster into a "lemon-shaped" DNA structure (Hughes et al. 2009). The metaphase I arrest is maintained by (1) the tension on the kinetochores resulting from the early exchange events between the homologs (chiasmata) (McKim et al. 1993; Jang et al. 1995), and (2) heterochromatin pairings between the nonexchange chromosomes (achiasmate) and their congression to the spindle midzone via heterochromatin threads (Hawley et al. 1992; Dernburg et al. 1996; Hughes et al. 2009).

Some evidence points to regulation of Cdk1/Cyclin B activity as being critical for maintenance of the secondary meiotic arrest in *Drosophila*. Degradation of Cyclin B is required for release of the metaphase I arrest. Oocytes that express nondegradable Cyclin B and thus cannot inactivate Cdk1/Cyclin B, do not resume meiosis (Swan and Schupbach 2007). There is an oocyte-specific form of the APC/C, activated by the Cortex Cdc20 family member, needed solely for the completion of meiosis in the oocyte (Pesin and Orr-Weaver 2007; Swan and Schupbach 2007). Both APC/C^{CORT} and APC/$C^{FZY(Cdc20)}$ forms are present in oocytes and appear redundant for the metaphase I-to-anaphase I transition, whereas APC/C^{CORT} is essential for the metaphase II-to-anaphase II transition (Pesin and Orr-Weaver 2007; Swan and Schupbach 2007).

Several genes involved in controlling different aspects of meiosis in the oocyte also fail to undergo metaphase I arrest when mutated. For example, in addition to the delay in the onset of meiotic maturation, *endos* and *twe* mutants fail to arrest at metaphase I (Xiang et al. 2007; Von Stetina et al. 2008). Oocytes with a meiotic-specific mutation in the *gwl* gene (gwl^{Sr18}), display scattered chromosomes and do not align properly on the metaphase plate because of premature loss of (or failure to establish) arm cohesion (Archambault et al. 2007). Mutants of *ald*, the *Drosophila* homolog of the conserved checkpoint component *mps1*, bypass the metaphase I arrest and enter anaphase I precociously (Gilliland et al. 2007). The defective spindle assembly checkpoint the (SAC) response in *ald* mutants is because of loss of sister-chromatin cohesion along the euchromatic arms and leads to high incidence of chromosome nondisjunction. Interestingly, the SAC is dispensable for cytostastic factor (CSF) arrest in mammals (Tsurumi et al. 2004).

Despite the strong requirement for the MAPK pathway for many meiotic events in vertebrates and *C. elegans*, MOS signaling is not essential in *Drosophila* (Ivanovska et al. 2004). *Drosophila* MOS (DMOS) has CSF activity when injected into two-cell *Xenopus* embryos and is required for MAPK phosphorylation in the *Drosophila* ovary. *dmos* mutant oocytes, nevertheless, have a normal metaphase I arrest. MAPK phosphorylation, however, is not completely eliminated in these mutants. Therefore, we cannot rule out a role of MOS in *Drosophila* meiosis or the presence of a redundant MAPKKK.

Egg Activation/Entry into Embryogenesis

Egg activation in *Drosophila* occurs independently of fertilization (Table 1) (Horner and Wolfner 2008). Instead, mechanical stimulation via passage through the oviduct and rehydration trigger egg activation and completion of meiosis (Mahowald et al. 1983). Recent studies point to a link between Ca^{2+}-dependent signaling and the resumption of meiosis. Oocytes carrying mutations in *Drosophila* Calcipressin, *sarah* (*sra*), an inhibitor of the Ca^{2+}/calmodulin-dependent phosphatase calcineurin, arrest at anaphase I and show high levels of Cyclin B (Horner et al. 2006; Takeo et al. 2006, 2010). These phenotypes suggest that inactivation of Cdk1/Cyclin B does not occur in *sra* mutants because the APC/C is not active (Horner and Wolfner 2008). No roles for CamkII in *Drosophila* egg activation have been established yet.

There are specialized regulators required for the onset of mitotic divisions in the *Drosophila* embryo following the completion of meiosis. The PAN GU (PNG) kinase complex, composed of the PNG kinase subunit and two activating subunits PLU and GNU, is necessary for the onset of mitotic activity following meiosis (Lee et al. 2003). This is first manifested in a requirement for PNG activity in the chromosome condensation that follows a transient interphase for the meiotic products in the egg after the completion of both meiotic divisions. In the absence of PNG function the meiotic products remain decondensed and inappropriately undergo DNA replication, and the male and female pronuclei do not undergo mitosis. This failure to restart mitosis following meiosis in *png* mutants is caused by reduced Cyclin B levels, as PNG promotes translation of Cyclin B at egg activation (Lee et al. 2001; Vardy and Orr-Weaver 2007a). PNG also is needed for the translation of Smaug (SMG), which later in embryogenesis facilitates the transition from maternal to zygotic control by promoting degradation of maternal mRNAs (Tadros et al. 2007). The nuclear protein YA is activated after egg activation and necessary for the first mitotic division (Sackton et al. 2009).

CONCLUDING REMARKS

Relationship between Reproductive Strategy and the Regulation of Oocyte Meiosis

Comparison of the control mechanisms for oocyte maturation, meiotic arrest, and egg activation between mammals, *C. elegans*, and *Drosophila* highlights evolutionary conservation in general strategies and molecular regulators while illustrating unique adaptations for specific reproductive goals (Table 1). Thus, distinct regulatory steps between these organisms provide informative insights into their reproductive success.

The self-fertilizing hermaphrodite *C. elegans* differs from most animals in having a limited number of sperm. Sperm are produced in the germline in the last larval stage and, during the adult lifespan, solely oocytes are produced. The critical role of sperm signaling via MSP to control oocyte maturation and egg activation puts the sperm in control. Making oocyte maturation dependent on the presence of sperm helps guarantee that each sperm will be able to fertilize an egg thereby ensuring optimal usage of the limiting gamete.

In contrast, in *Drosophila* egg activation and release of the secondary meiotic arrest are independent of sperm and fertilization. Solely movement of the oocyte into the uterus is sufficient for the completion of meiosis, although sperm proteins do stimulate egg laying in the female (Avila et al. 2010). Because some *Drosophila* species such as *D. mercatorum* are capable of parthenogenic development (Eisman and Kaufman 2007), we speculate that the independence of egg activation from sperm signals is a remnant of *Drosophila*'s capability to reproduce asexually. Although most *Drosophila* species have acquired dependence on fertilization for embryogenesis, they retain independence for egg activation. Given the investment in egg production in a *Drosophila* female, it is puzzling that eggs can be activated and laid in the absence of fertilization. *Drosophila* females may respond more to the nutritional state of their environment than to the presence of sperm. An important question will be to define the developmental cues regulating oocyte maturation in *Drosophila* and to delineate how these may be linked to environmental factors such as nutritional status and mating.

In mammals, a finite number of oocytes are produced during fetal development, differentiating to the primary arrest point. Although during each estrus cycle an oocyte undergoes maturation, no further oocytes are produced postnatally. Consequently, each oocyte is a precious commodity, explaining the precise regulation of oocyte maturation. Hormonal control links oocyte maturation with preparation of the uterus for potential implantation. The essential role of the somatic support cells in triggering oocyte maturation may serve as a control to ensure that the oocyte has grown and developed sufficiently prior to resumption of meiosis. This may also guarantee that the oocyte is ready to be delivered to the oviduct by the time of ovulation. Linking the completion of meiosis to fertilization provides a mechanism that permits development of the oocyte with a diploid genomic content while guaranteeing that haploid male and female pronuclei are present simultaneously to allow for syngamy and the onset of zygotic divisions. The dependency of completion of meiosis on fertilization also provides an additional mechanism to prevent parthenogenesis.

The Role of Cell–Cell Communication and Second Messengers

The recent developments in mouse and *C. elegans* define striking parallels in the regulation of oocyte maturation by somatic support cells. In mammals high cAMP in the oocyte maintains the prophase I arrest, but in both mammals and *C. elegans* elevation of cAMP in surrounding support cells in response to LH and MSP, respectively, triggers maturation. In *C. elegans* it will be informative to know whether the increase in Protein Kinase A resulting from cAMP acts in the sheath cell or the oocyte, as well as what molecules transit through the gap junctions to overcome the inhibition of maturation. Identification and regulation of the receptor in the sheath cell responding to MSP signaling also will be an important next step. A fascinating parallel between the mammalian and *C. elegans* systems is that maturation in mouse is associated with inactivation of gap junctions between the oocyte and the surrounding follicle cells, as in *C. elegans*. Given these shared mechanisms, it will be interesting to determine whether *Drosophila* follicle cells contribute to oocyte maturation and whether the innexin gap junctions between the oocyte and follicle cells inhibit maturation during the primary arrest. Although *Drosophila* mutations in protein kinase A are female sterile (Lane and Kalderon 1995), to date it has not been possible to address the function of protein kinase A in oocyte maturation because the mutants arrest too early in oogenesis.

Coordination between Developmental and Cell Cycle Control in the Oocyte

In metazoan oocytes, it is crucial that progression through meiosis be linked properly to differentiation, but recent discoveries show that meiotic progression can also influence development. Hormonal control of oocyte maturation in mammals and *Xenopus* illustrates a mechanism by which developmental input impinges on the meiotic cell cycle. In *Drosophila*, the onset of maturation coincides with a distinct stage of egg chamber differentiation, but the developmental cues accompanying this have yet to be unraveled.

The recent advances in *C. elegans* and *Drosophila* show that a reciprocal relationship exists between developmental and meiotic regulation, uncovered by examples in which cell cycle events and progression affect developmental control. In the worm, activation of Cyclin B/Cdk1 kinase at oocyte maturation leads to phosphorylation of MBK-2, making it active to phosphorylate its substrates once it is released from the cortex. Failure to repair double-strand breaks in *Drosophila* meiosis causes defects in dorsal-ventral patterning of the oocyte because of a defect in translation of Gurken. In addition, proper regulation of Polo levels and activity by Endos and Mtrm are necessary for the timing of meiosis I. It is interesting that in both *C. elegans* and *Drosophila* specific kinases act uniquely to control the oocyte to zygote transition: in *C. elegans*, MBK-2 alters the activity or level of maternal proteins and, in *Drosophila*, PNG promotes translation to start the embryonic mitotic divisions and later degrade maternal mRNAs.

Much remains to be uncovered to understand fully the regulatory mechanisms coordinating oocyte meiosis and development. The recent discoveries in mouse, *C. elegans*, and *Drosophila* have identified new proteins essential to link these processes as well as illustrated the roles played by conserved signal transduction pathways and second messengers. The parallels between these model organisms lead to powerful synergy in the analysis and discovery of these regulatory mechanisms.

ACKNOWLEDGMENTS

We would like to thank Tom DiCesare for preparing Figures 1–3, David Greenstein for assistance with Figure 3B, David Greenstein, Andreas Hochwagen, Cintia Hongay, and Laurinda Jaffe for many helpful comments on the manuscript, and Laurinda Jaffe and Takeo Kishimoto for providing manuscripts in press. T.L.O.-W. is an American Cancer Society Research Professor and support for the preparation of this article was provided also by NIH grant GM39341.

REFERENCES

Abdu U, Brodsky M, Schupbach T. 2002. Activation of a meiotic checkpoint during *Drosophila* oogenesis regulates the translation of Gurken through Chk2/Mnk. *Curr Biol* **12:** 1645–1651.

Abrieu A, Brassac T, Galas S, Fisher D, Labbe JC, Doree M. 1998. The Polo-like kinase Plx1 is a component of the MPF amplification loop at the G_2/M-phase transition of the cell cycle in *Xenopus* eggs. *J Cell Sci* **111:** 1751–1757.

Alphey L, Jimenez J, White-Cooper H, Dawson I, Nurse P, Glover DM. 1992. *twine*, a cdc25 homolog that functions in the male and female germline of *Drosophila*. *Cell* **69:** 977–988.

Archambault V, Zhao X, White-Cooper H, Carpenter AT, Glover DM. 2007. Mutations in *Drosophila Greatwall/Scant* reveal its roles in mitosis and meiosis and interdependence with Polo kinase. *PLoS Genet* **3:** e200.

Avila FW, Ravi Ram K, Bloch Qazi MC, Wolfner MF. 2010. Sex peptide is required for the efficient release of stored sperm in mated *Drosophila* females. *Genetics* **186:** 595–600.

Backs J, Stein P, Backs T, Duncan FE, Grueter CE, McAnally J, Qi X, Schultz RM, Olson EN. 2010. The γ isoform of CaM kinase II controls mouse egg activation by regulating cell cycle resumption. *Proc Natl Acad Sci* **107:** 81–86.

Bohrmann J, Zimmermann J. 2008. Gap junctions in the ovary of *Drosophila* melanogaster: Localization of innexins 1, 2, 3 and 4 and evidence for intercellular communication via innexin-2 containing channels. *BMC Dev Biol* **8:** 111.

Boxem M, Srinivasan DG, van den Heuvel S. 1999. The *Caenorhabditis elegans* gene *ncc-1* encodes a cdc2-related kinase required for M phase in meiotic and mitotic cell divisions, but not for S phase. *Development* **126:** 2227–2239.

Brisbin S, Liu J, Boudreau J, Peng J, Evangelista M, Chin-Sang I. 2009. A role for *C. elegans* Eph RTK signaling in PTEN regulation. *Dev Cell* **17:** 459–469.

Buszczak M, Freeman MR, Carlson JR, Bender M, Cooley L, Segraves WA. 1999. Ecdysone response genes govern egg chamber development during mid-oogenesis in *Drosophila*. *Development* **126:** 4581–4589.

Chase D, Serafinas C, Ashcroft N, Kosinski M, Longo D, Ferris DK, Golden A. 2000. The Polo-like kinase PLK-1 is required for nuclear envelope breakdown and the completion of meiosis in *Caenorhabditis elegans*. *Genesis* **26:** 26–41.

Cheng H, Govindan JA, Greenstein D. 2008. Regulated trafficking of the MSP/Eph receptor during oocyte meiotic maturation in *C. elegans*. *Curr Biol* **18:** 705–714.

Cheng KC, Klancer R, Singson A, Seydoux G. 2009. Regulation of MBK-2/DYRK by CDK-1 and the pseudophosphatases EGG-4 and EGG-5 during the oocyte-to-embryo transition. *Cell* **139:** 560–572.

Clary LM, Okkema PG. 2010. The EGR family gene *egrh-1* functions non-autonomously in the control of oocyte meiotic maturation and ovulation in *C. elegans*. *Development* **137:** 3129–3137.

Colledge WH, Carlton MB, Udy GB, Evans MJ. 1994. Disruption of c-mos causes parthenogenetic development of unfertilized mouse eggs. *Nature* **370:** 65–68.

Conti M, Andersen CB, Richard F, Mehats C, Chun SY, Horner K, Jin C, Tsafriri A. 2002. Role of cyclic nucleotide signaling in oocyte maturation. *Mol Cell Endocrinol* **187:** 153–159.

Conti M, Hsieh M, Park JY, Su YQ. 2006. Role of the epidermal growth factor network in ovarian follicles. *Mol Endocrinol* **20:** 715–723.

Courtot C, Fankhauser C, Simanis V, Lehner CF. 1992. The *Drosophila* cdc25 homolog *twine* is required for meiosis. *Development* **116:** 405–416.

Davis FM, Tsao TY, Fowler SK, Rao PN. 1983. Monoclonal antibodies to mitotic cells. *Proc Natl Acad Sci* **80:** 2926–2930.

Dernburg AF, Sedat JW, Hawley RS. 1996. Direct evidence of a role for heterochromatin in meiotic chromosome segregation. *Cell* **86:** 135–146.

Detwiler MR, Reuben M, Li X, Rogers E, Lin R. 2001. Two zinc finger proteins, OMA-1 and OMA-2, are redundantly required for oocyte maturation in *C. elegans*. *Dev Cell* **1:** 187–199.

DiLuigi A, Weitzman VN, Pace MC, Siano LJ, Maier D, Mehlmann LM. 2008. Meiotic arrest in human oocytes is maintained by a G_s signaling pathway. *Biol Reprod* **78:** 667–672.

Ducibella T, Fissore R. 2008. The roles of Ca^{2+}, downstream protein kinases, and oscillatory signaling in regulating fertilization and the activation of development. *Dev Biol* **315:** 257–279.

Edry I, Sela-Abramovich S, Dekel N. 2006. Meiotic arrest of oocytes depends on cell-to-cell communication in the ovarian follicle. *Mol Cell Endocrinol* **252:** 102–106.

Eisman R, Kaufman TC. 2007. Cytological investigation of the mechanism of parthenogenesis in *Drosophila mercatorum*. *Fly (Austin)* **1:** 317–329.

Freudzon L, Norris RP, Hand AR, Tanaka S, Saeki Y, Jones TL, Rasenick MM, Berlot CH, Mehlmann LM, Jaffe LA. 2005. Regulation of meiotic prophase arrest in mouse oocytes by GPR3, a constitutive activator of the G_s G protein. *J Cell Biol* **171:** 255–265.

Ghabrial A, Schupbach T. 1999. Activation of a meiotic checkpoint regulates translation of Gurken during *Drosophila* oogenesis. *Nat Cell Biol* **1:** 354–357.

Gilliland WD, Hughes SE, Cotitta JL, Takeo S, Xiang Y, Hawley RS. 2007. The multiple roles of Mps1 in *Drosophila* female meiosis. *PLoS Genet* **3:** e113.

Gonzalez-Reyes A, Elliott H, St Johnston D. 1997. Oocyte determination and the origin of polarity in *Drosophila*: The role of the spindle genes. *Development* **124:** 4927–4937.

Govindan JA, Cheng H, Harris JE, Greenstein D. 2006. $G\alpha_{o/i}$ and $G\alpha_s$ signaling function in parallel with the MSP/Eph receptor to control meiotic diapause in *C. elegans*. *Curr Biol* **16:** 1257–1268.

Govindan JA, Nadarajan S, Kim S, Starich TA, Greenstein D. 2009. Somatic cAMP signaling regulates MSP-dependent oocyte growth and meiotic maturation in *C. elegans*. *Development* **136:** 2211–2221.

Guven-Ozkan T, Nishi Y, Robertson SM, Lin R. 2008. Global transcriptional repression in *C. elegans* germline precursors by regulated sequestration of TAF–4. *Cell* **135:** 149–160.

Hambleton R, Krall J, Tikishvili E, Honeggar M, Ahmad F, Manganiello VC, Movsesian MA. 2005. Isoforms of cyclic nucleotide phosphodiesterase PDE3 and their contribution to cAMP hydrolytic activity in subcellular fractions of human myocardium. *J Biol Chem* **280:** 39168–39174.

Han SJ, Chen R, Paronetto MP, Conti M. 2005. Wee1B is an oocyte-specific kinase involved in the control of meiotic arrest in the mouse. *Curr Biol* **15:** 1670–1676.

Hansen DV, Tung JJ, Jackson PK. 2006. CaMKII and Polo-like kinase 1 sequentially phosphorylate the cytostatic factor Emi2/XErp1 to trigger its destruction and meiotic exit. *Proc Natl Acad Sci* **103:** 608–613.

Harris JE, Govindan JA, Yamamoto I, Schwartz J, Kaverina I, Greenstein D. 2006. Major sperm protein signaling promotes oocyte microtubule reorganization prior to fertilization in *Caenorhabditis elegans. Dev Biol* **299:** 105–121.

Hashimoto N, Watanabe N, Furuta Y, Tamemoto H, Sagata N, Yokoyama M, Okazaki K, Nagayoshi M, Takeda N, Ikawa Y, et al. 1994. Parthenogenetic activation of oocytes in c-mos-deficient mice. *Nature* **370:** 68–71.

Hawley RS, Irick H, Zitron AE, Haddox DA, Lohe A, New C, Whitley MD, Arbel T, Jang J, McKim K, et al. 1992. There are two mechanisms of achiasmate segregation in *Drosophila* females, one of which requires heterochromatic homology. *Dev Genet* **13:** 440–467.

Hinckley M, Vaccari S, Horner K, Chen R, Conti M. 2005. The G-protein-coupled receptors GPR3 and GPR12 are involved in cAMP signaling and maintenance of meiotic arrest in rodent oocytes. *Dev Biol* **287:** 249–261.

Hong A, Lee-Kong S, Iida T, Sugimura I, Lilly MA. 2003. The p27cip/kip ortholog dacapo maintains the *Drosophila* oocyte in prophase of meiosis I. *Development* **130:** 1235–1242.

Horner VL, Wolfner MF. 2008. Transitioning from egg to embryo: Triggers and mechanisms of egg activation. *Dev Dyn* **237:** 527–544.

Horner K, Livera G, Hinckley M, Trinh K, Storm D, Conti M. 2003. Rodent oocytes express an active adenylyl cyclase required for meiotic arrest. *Dev Biol* **258:** 385–396.

Horner VL, Czank A, Jang JK, Singh N, Williams BC, Puro J, Kubli E, Hanes SD, McKim KS, Wolfner MF, et al. 2006. The *Drosophila* calcipressin sarah is required for several aspects of egg activation. *Curr Biol* **16:** 1441–1446.

Hughes SE, Gilliland WD, Cotitta JL, Takeo S, Collins KA, Hawley RS. 2009. Heterochromatic threads connect oscillating chromosomes during prometaphase I in *Drosophila* oocytes. *PLoS Genet* **5:** e1000348.

Hunt PA, Hassold TJ. 2008. Human female meiosis: What makes a good egg go bad? *Trends Genet* **24:** 86–93.

Ivanovska I, Lee E, Kwan KM, Fenger DD, Orr-Weaver TL. 2004. The *Drosophila* MOS ortholog is not essential for meiosis. *Curr Biol* **14:** 75–80.

Jaffe LA, Norris RP. 2010. Initiation of the meiotic prophase-to-metaphase transition in mammalian oocytes. In *Oogenesis: The universal process* (ed. Verlhac M-H, Villenuve A), pp. 181–198. Wiley, New York.

Jang JK, Messina L, Erdman MB, Arbel T, Hawley RS. 1995. Induction of metaphase arrest in *Drosophila* oocytes by chiasma-based kinetochore tension. *Science* **268:** 1917–1919.

Jones KT. 2004. Turning it on and off: M-phase promoting factor during meiotic maturation and fertilization. *Mol Hum Reprod* **10:** 1–5.

Jones KT. 2005. Mammalian egg activation: From Ca^{2+} spiking to cell cycle progression. *Reproduction* **130:** 813–823.

Kalinowski RR, Berlot CH, Jones TL, Ross LF, Jaffe LA, Mehlmann LM. 2004. Maintenance of meiotic prophase arrest in vertebrate oocytes by a G_s protein-mediated pathway. *Dev Biol* **267:** 1–13.

Karaiskou A, Lepretre AC, Pahlavan G, Du Pasquier D, Ozon R, Jessus C. 2004. Polo-like kinase confers MPF autoamplification competence to growing *Xenopus* oocytes. *Development* **131:** 1543–1552.

King RC. 1970. The meiotic behavior of the *Drosophila* oocyte. *Int Rev Cytol* **28:** 125–168.

Kishimoto T. 2003. Cell-cycle control during meiotic maturation. *Curr Opin Cell Biol* **15:** 654–663.

Kosinski M, McDonald K, Schwartz J, Yamamoto I, Greenstein D. 2005. *C. elegans* sperm bud vesicles to deliver a meiotic maturation signal to distant oocytes. *Development* **132:** 3357–3369.

Kovo M, Kandli-Cohen M, Ben-Haim M, Galiani D, Carr DW, Dekel N. 2006. An active protein kinase A (PKA) is involved in meiotic arrest of rat growing oocytes. *Reproduction* **132:** 33–43.

Kubiak JZ, Ciemerych MA, Hupalowska A, Sikora-Polaczek M, Polanski Z. 2008. On the transition from the meiotic to mitotic cell cycle during early mouse development. *Int J Dev Biol* **52:** 201–217.

Kuwabara PE. 2003. The multifaceted *C. elegans* major sperm protein: An ephrin signaling antagonist in oocyte maturation. *Genes Dev* **17:** 155–161.

Lane ME, Kalderon D. 1995. Localization and functions of protein kinase A during *Drosophila* oogenesis. *Mech Dev* **49:** 191–200.

Lee LA, Elfring LK, Bosco G, Orr-Weaver TL. 2001. A genetic screen for suppressors and enhancers of the *Drosophila* PAN GU cell cycle kinase identifies cyclin B as a target. *Genetics* **158:** 1545–1556.

Lee LA, Van Hoewyk D, Orr-Weaver TL. 2003. The *Drosophila* cell cycle kinase PAN GU forms an active complex with PLUTONIUM and GNU to regulate embryonic divisions. *Genes Dev* **17:** 2979–2991.

Lee MH, Ohmachi M, Arur S, Nayak S, Francis R, Church D, Lambie E, Schedl T. 2007. Multiple functions and dynamic activation of MPK-1 extracellular signal-regulated kinase signaling in *Caenorhabditis elegans* germline development. *Genetics* **177:** 2039–2062.

Lincoln AJ, Wickramasinghe D, Stein P, Schultz RM, Palko ME, De Miguel MP, Tessarollo L, Donovan PJ. 2002. Cdc25b phosphatase is required for resumption of meiosis during oocyte maturation. *Nat Genet* **30:** 446–449.

Liu J, Maller JL. 2005. Calcium elevation at fertilization coordinates phosphorylation of XErp1/Emi2 by Plx1 and CaMK II to release metaphase arrest by cytostatic factor. *Curr Biol* **15:** 1458–1468.

Logarinho E, Sunkel CE. 1998. The *Drosophila* POLO kinase localises to multiple compartments of the mitotic apparatus and is required for the phosphorylation of MPM2 reactive epitopes. *J Cell Sci* **111:** 2897–2909.

Madgwick S, Jones KT. 2007. How eggs arrest at metaphase II: MPF stabilisation plus APC/C inhibition equals cytostatic factor. *Cell Div* **2:** 4.

Mahowald AP, Goralski TJ, Caulton JH. 1983. In vitro activation of *Drosophila* eggs. *Dev Biol* **98:** 437–445.

Marangos P, Verschuren EW, Chen R, Jackson PK, Carroll J. 2007. Prophase I arrest and progression to metaphase I in mouse oocytes are controlled by Emi1-dependent regulation of APC(Cdh1). *J Cell Biol* **176:** 65–75.

Markoulaki S, Matson S, Ducibella T. 2004. Fertilization stimulates long-lasting oscillations of CaMKII activity in mouse eggs. *Dev Biol* **272:** 15–25.

Masui Y, Markert CL. 1971. Cytoplasmic control of nuclear behavior during meiotic maturation of frog oocytes. *J Exp Zool* **177:** 129–145.

Matzuk MM, Lamb DJ. 2002. Genetic dissection of mammalian fertility pathways. *Nat Cell Biol* **4:** s41–49.

McKim KS, Jang JK, Theurkauf WE, Hawley RS. 1993. Mechanical basis of meiotic metaphase arrest. *Nature* **362:** 364–366.

Mehlmann LM. 2005a. Oocyte-specific expression of Gpr3 is required for the maintenance of meiotic arrest in mouse oocytes. *Dev Biol* **288:** 397–404.

Mehlmann LM. 2005b. Stops and starts in mammalian oocytes: Recent advances in understanding the regulation of meiotic arrest and oocyte maturation. *Reproduction* **130:** 791–799.

Mehlmann LM, Jones TL, Jaffe LA. 2002. Meiotic arrest in the mouse follicle maintained by a G_s protein in the oocyte. *Science* **297:** 1343–1345.

Mehlmann LM, Saeki Y, Tanaka S, Brennan TJ, Evsikov AV, Pendola FL, Knowles BB, Eppig JJ, Jaffe LA. 2004. The G_s-linked receptor GPR3 maintains meiotic arrest in mammalian oocytes. *Science* **306:** 1947–1950.

Miller MA, Nguyen VQ, Lee MH, Kosinski M, Schedl T, Caprioli RM, Greenstein D. 2001. A sperm cytoskeletal protein that signals oocyte meiotic maturation and ovulation. *Science* **291:** 2144–2147.

Miller MA, Ruest PJ, Kosinski M, Hanks SK, Greenstein D. 2003. An Eph receptor sperm-sensing control mechanism for oocyte meiotic maturation in *Caenorhabditis elegans*. *Genes Dev* **17:** 187–200.

Nishiyama TT, Kishimoto T. 2010. *Cytostatic arrest: Post-ovulation arrest until fertilization in metazoan oocytes.* Wiley, New York.

Norris RP, Freudzon M, Mehlmann LM, Cowan AE, Simon AM, Paul DL, Lampe PD, Jaffe LA. 2008. Luteinizing hormone causes MAP kinase-dependent phosphorylation and closure of connexin 43 gap junctions in mouse ovarian follicles: One of two paths to meiotic resumption. *Development* **135:** 3229–3238.

Norris RP, Ratzan WJ, Freudzon M, Mehlmann LM, Krall J, Movsesian MA, Wang H, Ke H, Nikolaev VO, Jaffe LA. 2009. Cyclic GMP from the surrounding somatic cells regulates cyclic AMP and meiosis in the mouse oocyte. *Development* **136:** 1869–1878.

Page SL, Hawley RS. 2001. *c(3)G* encodes a *Drosophila* synaptonemal complex protein. *Genes Dev* **15:** 3130–3143.

Panigone S, Hsieh M, Fu M, Persani L, Conti M. 2008. Luteinizing hormone signaling in preovulatory follicles involves early activation of the epidermal growth factor receptor pathway. *Mol Endocrinol* **22:** 924–936.

Parry JM, Velarde NV, Lefkovith AJ, Zegarek MH, Hang JS, Ohm J, Klancer R, Maruyama R, Druzhinina MK, Grant BD, et al. 2009. EGG-4 and EGG-5 link events of the oocyte-to-embryo transition with meiotic progression in *C. elegans*. *Curr Biol* **19:** 1752–1757.

Peng XR, Hsueh AJ, LaPolt PS, Bjersing L, Ny T. 1991. Localization of luteinizing hormone receptor messenger ribonucleic acid expression in ovarian cell types during follicle development and ovulation. *Endocrinology* **129:** 3200–3207.

Pesin JA, Orr-Weaver TL. 2007. Developmental role and regulation of cortex, a meiosis-specific anaphase-promoting complex/cyclosome activator. *PLoS Genet* **3:** e202.

Piontkewitz YaD, N. 1993. Heptanol, an alkanol that blocks gap junctions, induces oocyte maturation. *Endocrine* **1:** 365–372.

Pirino G, Wescott MP, Donovan PJ. 2009. Protein kinase A regulates resumption of meiosis by phosphorylation of Cdc25B in mammalian oocytes. *Cell Cycle* **8:** 665–670.

Racowsky C, Baldwin KV. 1989. In vitro and in vivo studies reveal that hamster oocyte meiotic arrest is maintained only transiently by follicular fluid, but persistently by membrana/cumulus granulosa cell contact. *Dev Biol* **134:** 297–306.

Radford HE, Meijer HA, de Moor CH. 2008. Translational control by cytoplasmic polyadenylation in *Xenopus* oocytes. *Biochim Biophys Acta* **1779:** 217–229.

Rauh NR, Schmidt A, Bormann J, Nigg EA, Mayer TU. 2005. Calcium triggers exit from meiosis II by targeting the APC/C inhibitor XErp1 for degradation. *Nature* **437:** 1048–1052.

Ray RP, Schupbach T. 1996. Intercellular signaling and the polarization of body axes during *Drosophila* oogenesis. *Genes Dev* **10:** 1711–1723.

Reis A, Chang HY, Levasseur M, Jones KT. 2006. APCcdh1 activity in mouse oocytes prevents entry into the first meiotic division. *Nat Cell Biol* **8:** 539–540.

Resnick TD, Dej KJ, Xiang Y, Hawley RS, Ahn C, Orr-Weaver TL. 2009. Mutations in the chromosomal passenger complex and the condensin complex differentially affect synaptonemal complex disassembly and metaphase I configuration in *Drosophila* female meiosis. *Genetics* **181:** 875–887.

Riechmann V, Ephrussi A. 2001. Axis formation during *Drosophila* oogenesis. *Curr Opin Genet Dev* **11:** 374–383.

Rose KL, Winfrey VP, Hoffman LH, Hall DH, Furuta T, Greenstein D. 1997. The POU gene ceh-18 promotes gonadal sheath cell differentiation and function required for meiotic maturation and ovulation in *Caenorhabditis elegans*. *Dev Biol* **192:** 59–77.

Roshak AK, Capper EA, Imburgia C, Fornwald J, Scott G, Marshall LA. 2000. The human *polo*-like kinase, PLK, regulates cdc2/cyclin B through phosphorylation and activation of the cdc25C phosphatase. *Cell Signal* **12:** 405–411.

Sackton KL, Lopez JM, Berman CL, Wolfner MF. 2009. YA is needed for proper nuclear organization to transition between meiosis and mitosis in *Drosophila*. *BMC Dev Biol* **9:** 43.

Sagata N. 1996. Meiotic metaphase arrest in animal oocytes: Its mechanisms and biological significance. *Trends Cell Biol* **6:** 22–28.

Schindler K, Schultz RM. 2009. CDC14B acts through FZR1 (CDH1) to prevent meiotic maturation of mouse oocytes. *Biol Reprod* **80:** 795–803.

Sela-Abramovich S, Chorev E, Galiani D, Dekel N. 2005. Mitogen-activated protein kinase mediates luteinizing hormone-induced breakdown of communication and oocyte maturation in rat ovarian follicles. *Endocrinology* **146:** 1236–1244.

Sela-Abramovich S, Edry I, Galiani D, Nevo N, Dekel N. 2006. Disruption of gap junctional communication within the ovarian follicle induces oocyte maturation. *Endocrinology* **147:** 2280–2286.

Sela-Abramovich S, Galiani D, Nevo N, Dekel N. 2008. Inhibition of rat oocyte maturation and ovulation by nitric oxide: Mechanism of action. *Biol Reprod* **78:** 1111–1118.

Stebbings LA, Todman MG, Phillips R, Greer CE, Tam J, Phelan P, Jacobs K, Bacon JP, Davies JA. 2002. Gap junctions in *Drosophila*: Developmental expression of the entire innexin gene family. *Mech Dev* **113:** 197–205.

Stitzel ML, Pellettieri J, Seydoux G. 2006. The *C. elegans* DYRK kinase MBK-2 marks oocyte proteins for degradation in response to meiotic maturation. *Curr Biol* **16:** 56–62.

Stitzel ML, Cheng KC, Seydoux G. 2007. Regulation of MBK-2/Dyrk kinase by dynamic cortical anchoring during the oocyte-to-zygote transition. *Curr Biol* **17:** 1545–1554.

Sun QY, Miao YL, Schatten H. 2009. Towards a new understanding on the regulation of mammalian oocyte meiosis resumption. *Cell Cycle* **8:** 2741–2747.

Swan A, Schupbach T. 2007. The Cdc20 (Fzy)/Cdh1-related protein, Cort, cooperates with Fzy in cyclin destruction and anaphase progression in meiosis I and II in *Drosophila*. *Development* **134:** 891–899.

Tadros W, Goldman AL, Babak T, Menzies F, Vardy L, Orr-Weaver T, Hughes TR, Westwood JT, Smibert CA, Lipshitz HD. 2007. SMAUG is a major regulator of maternal mRNA destabilization in *Drosophila* and its translation is activated by the PAN GU kinase. *Dev Cell* **12:** 143–155.

Takeo S, Tsuda M, Akahori S, Matsuo T, Aigaki T. 2006. The calcineurin regulator sra plays an essential role in female meiosis in *Drosophila*. *Curr Biol* **16:** 1435–1440.

Takeo S, Hawley RS, Aigaki T. 2010. Calcineurin and its regulation by Sra/RCAN is required for completion of meiosis in *Drosophila*. *Dev Biol* **344:** 957–967.

Tatone C, Delle Monache S, Iorio R, Caserta D, Di Cola M, Colonna R. 2002. Possible role for Ca^{2+} calmodulin-dependent protein kinase II as an effector of the fertilization Ca^{2+} signal in mouse oocyte activation. *Mol Hum Reprod* **8:** 750–757.

Tootle TL, Spradling AC. 2008. *Drosophila* Pxt: A cyclooxygenase-like facilitator of follicle maturation. *Development* **135:** 839–847.

Tsurumi C, Hoffmann S, Geley S, Graeser R, Polanski Z. 2004. The spindle assembly checkpoint is not essential for CSF arrest of mouse oocytes. *J Cell Biol* **167:** 1037–1050.

Tunquist BJ, Maller JL. 2003. Under arrest: Cytostatic factor (CSF)-mediated metaphase arrest in vertebrate eggs. *Genes Dev* **17:** 683–710.

Vaccari S, Horner K, Mehlmann LM, Conti M. 2008. Generation of mouse oocytes defective in cAMP synthesis and degradation: Endogenous cyclic AMP is essential for meiotic arrest. *Dev Biol* **316:** 124–134.

Vaccari S, Weeks JL, 2nd, Hsieh M, Menniti FS, Conti M. 2009. Cyclic GMP signaling is involved in the luteinizing hormone-dependent meiotic maturation of mouse oocytes. *Biol Reprod* **81:** 595–604.

Vardy L, Orr-Weaver TL. 2007a. The *Drosophila* PNG kinase complex regulates the translation of cyclin B. *Dev Cell* **12:** 157–166.

Vardy L, Orr-Weaver TL. 2007b. Regulating translation of maternal messages: multiple repression mechanisms. *Trends Cell Biol* **17:** 547–554.

Verlhac MH, Kubiak JZ, Weber M, Geraud G, Colledge WH, Evans MJ, Maro B. 1996. Mos is required for MAP kinase activation and is involved in microtubule organization during meiotic maturation in the mouse. *Development* **122:** 815–822.

Von Stetina JR, Tranguch S, Dey SK, Lee LA, Cha B, Drummond-Barbosa D. 2008. α-Endosulfine is a conserved protein required for oocyte meiotic maturation in *Drosophila*. *Development* **135:** 3697–3706.

Whitaker M. 1996. Control of meiotic arrest. *Rev Reprod* **1:** 127–135.

White-Cooper H, Alphey L, Glover DM. 1993. The cdc25 homologue *twine* is required for only some aspects of the entry into meiosis in *Drosophila*. *J Cell Sci* **106:** 1035–1044.

Whitten SJ, Miller MA. 2007. The role of gap junctions in *Caenorhabditis elegans* oocyte maturation and fertilization. *Dev Biol* **301:** 432–446.

Xiang Y, Takeo S, Florens L, Hughes SE, Huo LJ, Gilliland WD, Swanson SK, Teeter K, Schwartz JW, Washburn MP, et al. 2007. The inhibition of polo kinase by matrimony maintains G_2 arrest in the meiotic cell cycle. *PLoS Biol* **5:** e323.

Yamamoto I, Kosinski ME, Greenstein D. 2006. Start me up: Cell signaling and the journey from oocyte to embryo in *C. elegans*. *Dev Dyn* **235:** 571–585.

Yu J, Fleming SL, Williams B, Williams EV, Li Z, Somma P, Rieder CL, Goldberg ML. 2004. Greatwall kinase: A nuclear protein required for proper chromosome condensation and mitotic progression in *Drosophila*. *J Cell Biol* **164:** 487–492.

Zhang M, Su YQ, Sugiura K, Xia G, Eppig JJ. 2010. Granulosa cell ligand NPPC and its receptor NPR2 maintain meiotic arrest in mouse oocytes. *Science* **330:** 366–369.

Mammalian Genomic Imprinting

Marisa S. Bartolomei[1] and Anne C. Ferguson-Smith[2]

[1]Department of Cell and Developmental Biology, University of Pennsylvania School of Medicine, Philadelphia, Pennsylvania 19063

[2]Department of Physiology, Development, and Neuroscience, University of Cambridge, Cambridge, United Kingdom

Correspondence: bartolom@mail.med.upenn.edu

Normal mammalian development requires a maternal and paternal contribution, which is attributed to imprinted genes, or genes that are expressed from a single parental allele. Approximately 100 imprinted genes have been reported in mammals thus far. Imprinted genes are controlled by *cis*-acting regulatory elements, termed imprinting control regions (ICRs), which have parental-specific epigenetic modifications, including DNA methylation. ICRs are methylated by de novo DNA methyltransferases during germline development; these parental-specific modifications must be maintained following fertilization when the genome is extensively reprogrammed. Many imprinted genes reside in ~1-megabase clusters, with two major mechanisms of imprinting regulation currently recognized, CTCF-dependent insulators and long noncoding RNAs. Unclustered imprinted genes are generally regulated by germline-derived differential promoter methylation. Here, we describe the identification and functions of imprinted genes, *cis*-acting control sequences, *trans*-acting factors, and imprinting mechanisms in clusters. Finally, we define questions that require more extensive research.

In mammals, a small number of genes are marked with their parental origin with the result that only a single parental allele is expressed. These genes, which are termed "imprinted," are dependent on the epigenetic machinery for their initial designation of parental identity as well as establishment and maintenance of their parent-of-origin-specific gene expression (Fig. 1). Although organisms other than mammals harbor imprinted genes, such as *Arabidopsis* (Kinoshita et al. 2008), we will confine our discussion to what is known about the identification and regulation of imprinted genes in mammals. Moreover, it is important to note that chromosomal imprinting in mammals was first described from studies on the X chromosome, where paternal-specific inactivation of one of the X chromosomes in all cells of female marsupials and in the extraembryonic tissues of the mouse occur (Cooper et al. 1971).

FIRST SUGGESTIONS OF IMPRINTING IN MAMMALS

The elegant pronuclear transplantation experiments performed by the Solter and Surani laboratories in the 1980s were among the first to suggest that the mammalian (specifically the

Cite this article as *Cold Spring Harb Perspect Biol* doi: 10.1101/cshperspect.a002592

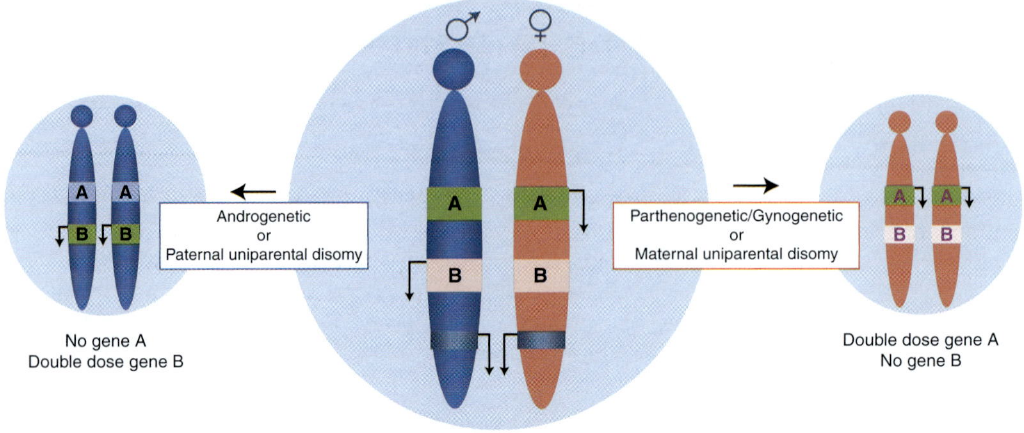

Figure 1. Imprinted genes are expressed in a parental-origin-specific manner. In the center, a cell is depicted with a chromosome inherited maternally (red) or paternally (blue), and two imprinted genes. Gene A is transcribed from the maternal allele (green box and arrow indicating transcription) and repressed on the paternal allele (light blue box). In contrast, gene B is transcribed from the paternal allele (green box and arrow) and repressed on the maternal allele (light pink box). Uniparental embryos and embryos that are uniparental for a given chromosome have been used to show that imprinted genes are critical for development as well as to identify imprinted genes (see text).

mouse) genome possessed imprinted genes (McGrath and Solter 1983, 1984; Surani and Barton 1983; Surani et al. 1984). These experiments showed that mammalian development required both maternal and paternal contributions to proceed normally. Maternal uniparental embryos (gynogenotes or parthogenotes) developed into tissues predominantly of embryonic origin, with a failure of the extraembryonic lineages, whereas paternal uniparental embryos (androgenotes) developed into conceptuses derived of extraembryonic lineages. The investigators reasoned that absence or overexpression of imprinted genes exclusively expressed from either the maternal or paternal genome caused the developmental failure. Experiments by Kono and colleagues have largely shown these ideas to be correct by generating viable mice that are exclusively maternal in origin (Kono et al. 2004; Kawahara et al. 2007). Briefly, these investigators engineered offspring by combining nuclei from nongrowing and fully-grown oocytes that also had mutations at two different imprinted loci. The resulting bimaternal offspring, which were produced at a slightly lower frequency than similarly manipulated wild-type embryos, had normalized the imprinted gene dosage to what is observed in wild-type offspring, suggesting that imprinted gene expression is the main barrier to parthenogenetic development in mammals (Kawahara et al. 2007).

Genetic experiments by Cattanach, Searle, Beechey, and colleagues pointed to specific regions of the genome that harbored imprinted genes (Searle and Beechey 1978, 1990; Cattanach 1982, 1986; Cattanach and Kirk 1985). Offspring with uniparental disomies for specific chromosomal regions were generated by mating mice that were heterozygous for Robertsonian or balanced translocations. That is, rather than having an entire uniparental genome, these experimental manipulations produced offspring that had uniparental origin of only part of the genome—usually both homologs of an individual, or fraction of a, chromosome. If defects resulted, they surmised a requirement for both parental chromosomal regions. This allowed the dissection of the genome approximately one chromosome at a time and the cytogenetic mapping of regions that were imprinted. In some cases, uniparental disomies caused obvious parent-of-origin-specific defects such as growth abnormalities or death, demonstrating a major role of imprinted genes for normal development. Behavioral abnormalities were also detected (Cattanach and Kirk 1985). It is important to note, however, that imprinted genes whose altered dosage exert more subtle or tissue-specific effects would not necessarily be identified in these tests. Nonetheless, more that 90% of the imprinted genes that have been identified in the mouse to date, map to the regions originally identified through these genetic studies.

At approximately the same time as the mouse genetic experiments were performed, genetic disorders in humans were suggested to arise from parental-specific gene expression. One early set of observations came from patients with Prader–Willi syndrome (PWS). PWS, which is characterized by hypotonia in infancy and subsequent hyperphagia, hypogonadism, etc. (MIM176270), was shown to be associated with chromosomal deletions that were paternally derived (Knoll et al. 1989; Nicholls and Knepper 2001). Subsequent experiments identified a number of paternally expressed genes that are likely involved in the diverse phenotypes of PWS, including *SNPRN*, *NECDIN*, and *HBII-85* snoRNAs (Horsthemke and Wagstaff 2008).

IDENTIFICATION OF IMPRINTED GENES

The first endogenous imprinted genes were identified less than a decade after the original nuclear transplantation and genetic experiments. Diverse strategies were employed to identify these imprinted loci. Some relied on molecularly characterizing regions that likely contained imprinted genes (Barlow et al. 1991; Ferguson-Smith et al. 1991), whereas others incorporated genome-wide studies (Kaneko-Ishino et al. 1995; Wang et al. 2008), and still others were identified as a consequence of gene targeting experiments (DeChiara et al. 1991). *Igf2r* (insulin-like growth factor type 2 receptor gene) was the first of three imprinted genes that were reported in 1991. The identification of *Igf2r* was based on a targeted strategy to determine which gene was responsible for the lethality of Hairpin-tail deletion mice that inherited a large deletion of chromosome 17 specifically from their mothers (Johnson 1974; Barlow et al. 1991). Here, positional cloning discovered genes in the deleted region and mice inheriting the deletion from one parent were used to identify the gene(s) showing expression exclusively from the maternal allele. A different strategy was employed to show imprinting of the mouse *H19* gene, which encodes a noncoding RNA (ncRNA). In this case, maternal-specific expression of *H19* was revealed using single nucleotide polymorphisms (SNPs) that reside in RNA (Bartolomei et al. 1991). By generating F1 hybrid mice between divergent species of *Mus*, the mouse strain, and therefore the parental allele, from which the RNA product is derived can be tracked. In fact, now that the sequence of multiple mouse strains is known and high densities of strain-specific SNPs are available, whole transcriptome analysis is being used to assess which RNAs are monoallelically expressed (Babak et al. 2008; Wang et al. 2008). With time, this will allow the identification of all imprinted genes. Finally, a few imprinted genes have been identified that result from gene targeting experiments initially designed to test the function of the gene product. In the case of *Igf2* (Insulin-like growth factor 2), gene targeting showed that mice inheriting the deletion from the fathers were phenotypically identical to homozygous null mice for the *Igf2* gene, whereas mice inheriting the deletion from their mothers were phenotypically wild-type (DeChiara et al. 1991). The conclusion from these breeding studies was that the *Igf2* gene is imprinted, with the expressed copy being paternal in origin. In a different approach, using uniparental disomy conceptuses derived from the genetic experiments discussed above and analysis of candidate genes influencing growth, expression of *Igf2* was shown to be absent in embryos with maternal uniparental disomy of distal chromosome 7 indicating expression from the paternally inherited chromosome (Ferguson-Smith et al. 1991).

To date, approximately 100 imprinted genes have been described in mammals (for a complete list, see http://www.har.mrc.ac.uk/research/genomic_imprinting/ or http://igc.otago.ac.nz/home.html). Although singletons have been identified, many imprinted genes reside in 1 Mb clusters throughout the genome. These clusters typically contain at least one ncRNA and both maternally and paternally expressed imprinted genes. The imprinting of the multiple genes in the cluster is usually under the control of a discrete DNA element, termed an imprinting control region (ICR; see below for more details). ICRs show parent-of-origin-specific epigenetic modifications that are set up in the germline, such as DNA methylation and posttranslational histone modifications. Where it has been tested, deletion of the ICR results in loss of imprinting of multiple genes in the cluster, emphasizing the importance of this element and its regulatory influence over a multigenic region (Wutz et al. 1997; Thorvaldsen et al. 1998; Yang et al. 1998; Fitzpatrick et al. 2002; Lin et al. 2003; Williamson et al. 2006).

FUNCTIONS OF IMPRINTED GENES

Given their monoallelic expression status and the complexities of their regulation, it is perhaps not surprising that many imprinted genes appear to be dosage sensitive with functional consequences associated with changes in their expression levels. Over the years, phenotypic analyses of humans and mice with altered dosages of individual or multiple imprinted genes have provided insights into their roles. Recurrent themes have emerged indicating that imprinted genes are important for prenatal growth control, the development of particular lineages, for normal brain function and in postnatal energy homeostasis. Some understanding of imprinted gene function in human has come from genotype-phenotype studies in patients with imprinting disorders (Table 1). These include the behavioral and neurodevelopmental disorders Prader–Willi and Angelman syndromes (Horsthemke and Buiting 2006), the imprinted growth disorders Beckwith–Wiedemann and Silver–Russell syndromes (Weksberg et al. 2005; Abu-Amero et al. 2008), the maternal and paternal uniparental disomy 14 syndromes (Ogata et al. 2008), and transient neonatal diabetes (Temple 2007). In general, the features described in patients are consistent with those observed in the corresponding mouse mutants.

In mouse, the dosage of imprinted genes has been manipulated by targeted mutagenesis of the active parental allele, through transgenic overexpression studies and by mutation targeted at the ICR resulting in either activation of the repressed allele or inappropriate repression of active alleles. Phenotypic analysis of mouse mutants combined with descriptive studies localizing imprinted gene products in developing and adult cell types in vivo has shown many imprinted genes to be expressed in a range of developing organ systems where defects are manifest in the mutants (for example, see da Rocha et al. 2009). Imprinted genes, in general, are expressed widely and highly during prenatal stages, however, are predominantly down-regulated after birth. The placenta and the brain are sites of expression of many imprinted transcripts consistent with the growth and neurodevelopmental effects seen in human imprinting disorders (Coan et al. 2005; Wilkinson et al. 2007). In particular defective placental development and physiology is a characteristic of perturbed imprinting. For example, absence of expression of a placenta-specific isoform of *Igf2* causes impaired nutrient transport to the growing fetus (Sibley et al. 2004). *Ascl2* (achaete-scute homolog complex 2) expressed from the maternally inherited chromosome is required for the normal differentiation of spongiotrophoblast cells early within the developing murine placenta (Guillemot et al. 1995). Interestingly, *Peg10* and *Rtl1* are two imprinted genes required for normal placental function that have evolved from retrotransposons (Ono et al. 2006; Sekita et al. 2008). Although imprinting in the placenta provides a critical role in the control of resources at the interface between mother and fetus, perturbed imprinting in the embryo can influence its growth independent of an influence of the placenta. For example, a double dose of *Dlk1* (*Delta-like homolog 1*) equivalent to expression from both parental chromosomes instead of just the paternally inherited one, results in a growth enhanced fetus even when the placenta expresses normal levels of the gene (da Rocha et al. 2009). Despite a clear role in

 Cite this article as *Cold Spring Harb Perspect Biol* doi: 10.1101/cshperspect.a002592

Table 1. Human imprinted disorders

Syndrome	Clinical features	Etiology	Mouse chromosome
Angelman syndrome (AS)	Mental retardation, speech impairment, ataxia, seizure, microcephaly	15q11.2-q13 deletion (70%) PatUPD15 (7%), *UBE3A* mutation (11%), methylation defects (3%), epimutation	7C
Prader–Willi syndrome (PWS)	Neonatal hypotonia, childhood obesity, cognitive impairment, behavioral characteristics, hypogonadism	De novo paternal deletion in 15q11-q13 (70%), MatUPD15 (29%), imprinting defects (1%)	7C
Beckwith–Wiedemann syndrome (BWS)	Pre/postnatal overgrowth, neonatal hypoglycemia, exompholos, macroglossia, hemihypertrophy, increased embryonal tumors	Epimutation of *IGF2/H19* DMR1, epimutation of *KCNQ1/CDKN1C* DMR2 both on 11p15, hypomethylation of DMR2 (50%), hypermethylation DMR1 (2%–7%), PatUPD11, *CDKN1C* mutation	7F5
Silver–Russell syndrome (SRS)	Intrauterine/postnatal growth retardation, variable features (inc. 5th finger clinodactyl, learning disabilities)	Paternal DMR1 hypomethylation at 11p15 (>50%), MatUPD7 (5%) Matdup11p15, unknown (30%)	7F5
Maternal UPD14 (and UPD14 mat-like) syndrome	Low birth weight, short stature, characteristic facies, premature puberty, hypotonia	MatUPD14, paternal microdeletions at 14q32.2, hypomethylated DMRs at *DLK1/GTL2*	12F1
Paternal UPD14 (and UPD14 pat-like) syndrome	Bell-shaped thoracic cage, mental retardation, placentomegaly, polyhydramnios	PatUPD 14, maternal microdeletions at 14q32.2, hypermethylation at DMRs at *DLK1/GTL2*	12F1
Pseudo-hypoparathyroidism 1b	Resistance to parathyroid hormone, hypocalcaemia, hyperphophatemia	Microdeletion upstream of *GNAS* at 20q, maternal hypomethylation, PatUPD20	2H4
Transient neonatal diabetes mellitus	Growth retardation, hyperglycemia with low/undetectable insulin resolved by 6 months old, 40% Type2 diabetes later in life	Paternal UPD6, paternal duplication 6q22-q23, maternal hypomethylation at *ZAC1/PLAGL1* DMR	10A2

prenatal growth, the experimental challenges of functionally separating imprinted genes expressed in embryonic and extraembryonic compartments have made it hard to determine the relative contributions of placental and embryonic expression of imprinted genes to the in utero acquisition of resources and growth control of the fetus.

In the brain, imprinted genes have been implicated in processes modulating metabolic axes, behavior, learning, and maternal care. For example, it was recently shown that the imprinted gene, $G_s\alpha$, expressed from the maternally inherited chromosome in the hypothalamus, controls melanocortin-mediated energy expenditure (Chen et al. 2009). $G_s\alpha$ is a G protein regulating receptor-mediated cAMP production. Mice with brain-specific deletion of the maternally inherited copy of $G_s\alpha$, have glucose intolerance, insulin resistance, and obesity. In other examples, *Peg1* and *Peg3* are two paternally expressed imprinted genes strongly transcribed in the brain and which, when mutated, result in reduced maternal care by mutant mothers (Lefebvre et al. 1998; Li et al. 1999). These findings implicate imprinted genes in the brain in the postnatal control of nutritional resources either directly through the central control of energy homeostasis or indirectly through maternal–offspring interactions.

PWS and Angelman syndrome (AS) are two phenotypically distinct disorders mapping to the same imprinted domain on human chromosome 15q11-q13, the location of a cluster of genes expressed and imprinted in the brain (Horsthemke and Wagstaff 2008). Patients with PWS show hypotonia and early neonatal failure to thrive. Later they have mild to moderate mental retardation, behavioral problems, and hyperphagia in early childhood leading to obesity. Paternally expressed transcripts from the PWS/AS locus including the protein-coding SNRPN gene and its associated noncoding C/D snoRNAs (small nucleolar RNAs), are not expressed in PWS and are implicated in the syndrome though their precise contribution to the etiology of the disorder is not known. Angelman Syndrome, in contrast, is associated with absence of the *UBE3A* transcript, which is expressed from the maternally inherited chromosome and located at the 3′ end of the imprinted cluster on 15q11-q13. Individuals with AS have ataxia, severe mental retardation, seizures, and absence of speech. Although much has been learnt about the mechanisms of imprinting through the study of the *SNRPN/UBE3A* imprinted cluster of genes, little is known about how the absence or altered dosage of these genes influences neuronal function and causes the phenotypes that characterize the syndromes.

Despite the emphasis on placenta and brain functions for imprinted genes and an apparently related function in the control of nutrient acquisition, it is clear that imprinted genes also influence the development of other organ systems. For example, the imprinted cyclin-dependent kinase inhibitor, *Cdkn1c* is strongly expressed in several somatic tissues during development such as muscle, lung, kidney, and the eye. Normal expression of *Cdkn1c* is associated with cessation of the cell cycle and several studies have shown defects in organ size when *Cdkn1c* levels are perturbed (Jia et al. 2007b). The paternally expressed *Dlk1* gene encodes a protein that is related to ligands functioning in Notch signaling, one of the key signaling pathways regulating the development of multiple lineages. *Dlk1* mutants have growth retardation, skeletal abnormalities, adiposity defects and abnormalities of hematopoietic lineages (Moon et al. 2002; Raghunandan et al. 2008).

Findings that defective imprinting of multiple genes results in related phenotypic anomalies mostly involving pre- and postnatal growth control, suggests that perhaps imprinted genes might function in common pathways. Some evidence for this exists. For example, genes whose transcription was perturbed in mice mutant for the paternally expressed transcription factor *Plagl1/Zac1* gene, included a large number of other imprinted genes. The work suggested that *Plagl1/Zac1* might be a *trans*-acting regulator of a network of imprinted genes involved in prenatal growth control including *Igf2, H19, Cdkn1c,* and *Dlk1* (Varrault et al. 2006). More detailed analysis of the in vivo targets of *Plagl/Zac1* and the functional interactions that might exist between imprinted genes in the proposed network will determine whether this hypothesis is true.

PROPERTIES OF THE IMPRINTING MECHANISM

How does the transcription machinery of the nucleus distinguish between maternally and paternally inherited chromosome homologs and express only one of the two alleles of an imprinted gene—and furthermore, always the same one? How are the two homologs "marked" such that they know their parental origin? The process regulating genomic imprinting has four important properties: (1) The "mark" must be able to influence transcription; (2) it must be heritable in somatic lineages such that a memory of parental origin is faithfully propagated into daughter cells during cell division; (3) the "mark" is likely to be placed on the paternally and maternally inherited chromosomes at a time when they are not in the same nucleus (i.e., during gametogenesis or perhaps immediately after fertilization); (4) there must be a mechanism of erasure of the "mark" in order that paternally inherited chromosomes in the female germline can establish a new "mark" indicative of their maternal origin and vice versa; that maternally inherited chromosomes contributing to spermatogenesis in the developing male would lose their maternal identity and become "marked" as paternal in origin.

DNA methylation is the epigenetic modification that fulfills these criteria. In mammalian cells, it is well established that 5-methylcytosine at CpG dinucleotides can influence transcription. CpG dinucleotides are not evenly distributed throughout the genome but are often found as CG-rich sequences called CpG-islands. CpG-islands, which are usually unmethylated, are predominantly associated with the promoters of housekeeping genes (i.e., genes that are ubiquitously expressed and seldom repressed [Bird 2002]). CpG-islands are also found at the promoters of imprinted genes and most, though not all, CG-island promoters of imprinted genes are differentially methylated, where the repressed allele is methylated and the active allele unmethylated. These differentially methylated regions are called DMRs.

DNA methylation fulfills the second criterion of somatic heritability because DNA methyltransferase 1 (DNMT1) has a maintenance mode of action that recognizes newly replicated DNA comprising an old methylated strand and a new unmethylated strand (hemimethylated DNA) and places methylation on the newly replicated CpG; hence the critically important heritability of DNA methylation is maintained (Goll and Bestor 2005). In terms of establishment, it is now well-known that DNA is methylated in different places on the maternally and paternally inherited chromosomes during gametogenesis by the de novo methyltransferase DNMT3A (see below). The establishment of these germline DMRs is essential for imprinting after fertilization and initiates the heritable memory that forms the identity "mark" of the two parental chromosomes. Interestingly, in the female germline, methylation marks always act at the promoters of imprinted genes resulting in their heritable repression from the maternally inherited chromosome. In contrast, methylation in the paternal germline is not placed at promoters but rather is deposited within intergenic regions (Fig. 2). Targeted deletion studies of maternally methylated promoter germline DMRs and paternally methylated intergenic germline DMRs in mouse have indicated that these are the imprinting control regions that can regulate multiple imprinted genes at some distance from the ICR (Wutz et al. 1997; Thorvaldsen et al. 1998; Yang et al. 1998; Fitzpatrick et al. 2002; Lin et al. 2003; Williamson et al. 2006). ICRs can therefore be long-range *cis*-acting control sequences whose function is modulated by their epigenetic state. Finally, DNA methylation imprints are erased in the male and female germlines during a process of germ cell-specific genome-wide reprogramming that occurs in the developing embryo as the cells destined to become germ cells are finishing their migration along the genital ridges toward the developing gonads. This is discussed in more detail below.

How do paternally methylated ICRs located in an intergenic regions influence the monoallelic activity and repression of multiple imprinted genes? The best-studied example is that of the *Igf2-H19* locus. The paternally expressed *Igf2* gene is located 90 kb away from the maternally expressed *H19* gene (Bartolomei et al. 1991; DeChiara et al. 1991). The two genes, in general, share common sites of expression and share enhancer sequences located downstream of *H19*. The ICR resides 2–4 kb upstream from the transcriptional start site of *H19* (Tremblay et al. 1997). This ICR contains multiple binding sites for the zinc-finger

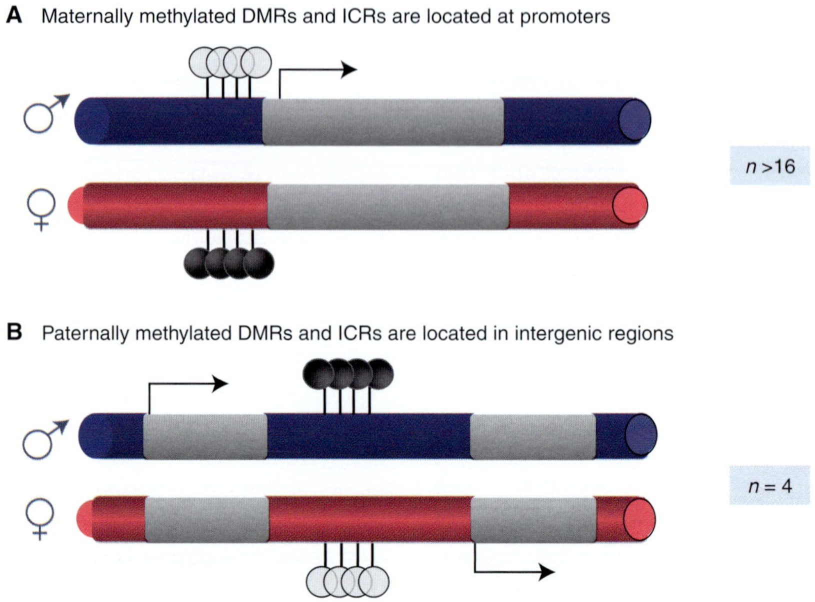

Figure 2. DNA methylation is essential for establishment and maintenance of imprinting. Shown are the two types of DNA methylation present at imprinted loci. (*A*) Greater than 16 imprinted loci are associated with maternal-specific methylation. For these loci, DNA methylation (filled lollipops) is associated with the promoter of a repressed gene, whereas hypomethylation (gray lollipops) is associated with the promoter of the expressed gene (arrow). (*B*) Four imprinted loci are associated with DNA methylation in intergenic regions. In one case (*H19* and *Igf2*), the DNA methylation regulates a CTCF-dependent insulator element (see text).

insulator protein, CTCF, which binds exclusively to the unmethylated maternal chromosome (Bell and Felsenfeld 2000; Hark et al. 2000). On the maternal chromosome, the presence of CTCF blocks the enhancers from interacting with *Igf2* promoters rendering the gene inactive. The enhancers on the maternal chromosome drive activity from *H19* instead. On the paternally inherited chromosome where the ICR is methylated, CTCF cannot bind, and the unblocked enhancers are able to drive activity from the *Igf2* promoters. Such methylation-sensitive parental-chromosome-specific promoter–enhancer interaction has been validated by chromatin conformation studies where the expected loops of enhancer–promoter interactions have been identified, modulated by differential DNA methylation and differential CTCF binding on the two parental chromosomes (Murrell et al. 2004; Kurukuti et al. 2006; Engel et al. 2008). However, one should not assume that this CTCF-mediated control of ICRs, which regulates imprinting at the *Igf2-H19* locus, acts at all intergenic ICRs. Indeed, there is considerable evidence that this mechanism is not operating at the paternally methylated ICR regulating the *Dlk1-Dio3* imprinted domain (da Rocha et al. 2008).

Maternally methylated ICRs, which are located at gene promoters, exert a different effect to mediate imprinting over distances compared to that described for *H19* and *Igf2*. The most extensively studied example of this mode of imprinting control comes from the analysis of the *Igf2r* cluster of imprinted genes (Koerner et al. 2009). Here, the ICR is located at a CpG island within an intron of the *Igf2r* gene (Stöger et al. 1993). The *Igf2r* gene is expressed from the maternally inherited chromosome (Barlow et al. 1991), which has intronic methylation. In contrast, on the paternally inherited chromosome, the CpG-island in the intron is unmethylated and can act as a promoter for a large nonprotein-coding RNA (ncRNA), which is designated *Airn*, and is transcribed in an antisense direction to *Igf2r* (Wutz et al. 1997; Lyle et al. 2000). This antisense transcription crosses the *Igf2r* promoter and is required for its repression on the paternal chromosome (Sleutels et al. 2002). Interestingly, two genes located downstream from *Igf2r* are also expressed on the maternal chromosome and repressed on the paternally inherited one, even though the direction of antisense transcription is away from those two genes. One hypothesis for which there is some supporting evidence, proposes that the *Airn* ncRNA interacts in *cis* with the promoter of at least one of these genes and recruits repressive histone modifications that inactivate the promoter on the paternal chromosome (Nagano et al. 2008). A mechanism in which transcription of a long ncRNA regulated by maternal germline DNA methylation can control multiple genes in an imprinted cluster has also been shown for the *Kcnq1* cluster (Fitzpatrick et al. 2002), the *Snrpn* cluster (Horsthemke and Wagstaff 2008), and for the *Gnas* cluster (Williamson et al. 2006; Chotalia et al. 2009). Importantly, studies of the mechanisms whereby large ncRNAs act in *cis* to control the activity and repression of imprinted genes have revealed a useful paradigm for understanding the functional significance of genome-wide expression of ncRNAs in mammalian cells.

GERMLINE ESTABLISHMENT OF IMPRINTS

As mentioned above, an important question in imprinting is how parental alleles are marked with their parental origin. If it is assumed that ICRs are key for allelic identity of imprinted genes in *cis* and that differential epigenetic modifications to the ICRs mediate parental identity, then the question becomes when and how is this ICR marking achieved. It has long been hypothesized that parental-specific marks are assigned in the germline as this is the time when the genomes are in distinct compartments and can be epigenetically modified according to the sex of the transmitting gametes (Fig. 3). Although multiple epigenetic modifications could designate parental origin, DNA methylation has been most widely investigated because of its (1) differential presence on the ICRs as assayed in mature gametes, (2) ease of study on the few cells present in the germline, and (3) proven heritability by the maintenance DNA methyltransferase, DNMT1.

Analysis of methylation patterns in primordial germ cells (PCGs) shows that during embryonic migration imprinted genes have the methylation patterns that are characteristic of somatic cells (Hajkova et al. 2002). During their final migration into the genital ridges, the level of DNA methylation in PGCs is reduced. The mechanisms that control DNA demethylation in PGCs and erasure of imprints are unclear, although a recent study has indicates that these processes are linked to changes in chromatin structure and histone modifications (Hajkova et al. 2008). Following demethylation and differentiation of the PGCs, methylation is imposed on the ICRs in a sex-specific manner (Fig. 3). That is, male and female germ cells show DNA remethylation at different times of development, with ICRs in male germ cells initially methylated in prospermatogonia during the period between mitotic arrest and birth (Davis et al. 1999; Li et al. 2004b) and maternal-specific methylation established after birth during the oocyte growth phase prior to ovulation (Lucifero et al. 2002).

Intriguingly, most ICRs are methylated on the maternal allele, with only four paternally methylated ICRs (*H19*/*Igf2*, *Dlk1*/*Dio3*, *Rasgrf1*, *Zdbf2*) identified thus far. All of the ICRs tested to date, with the exception of *Rasgrf1*, use the de novo DNA methyltransferase DNMT3A and its stimulatory protein DNMT3L to confer DNA methylation on the ICRs in the respective germ cells (Bourc'his et al. 2001; Hata et al. 2002; Bourc'his and Bestor 2004; Kaneda et al. 2004). Nevertheless, it is still unclear how the epigenetic machinery recognizes ICRs in the presence of all of the sequences in the mammalian genome. One clue comes from the X-ray crystallography studies of the complexed carboxy-terminal domains of DNMT3A and DNMT3L (Jia et al. 2007a). A tetrameric complex consisting of these two enzymes preferentially methylates a pair

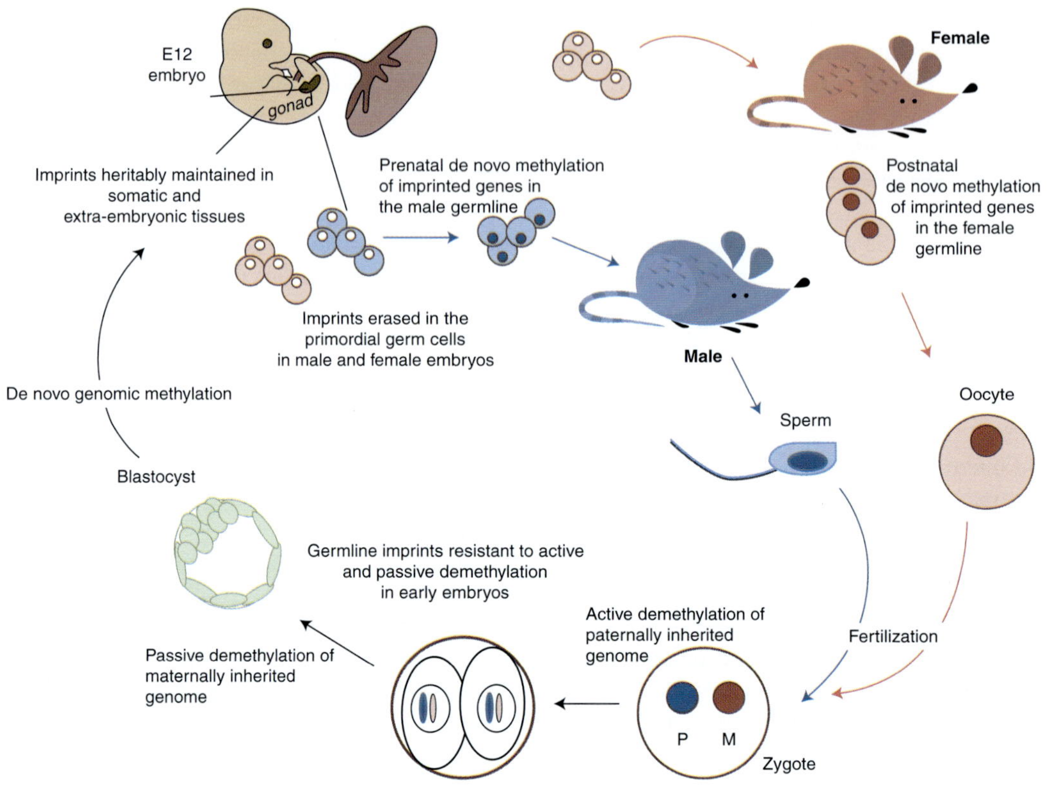

Figure 3. Cycle of Imprinting. Imprints are acquired in a sex-specific manner in the germline: Maternally and paternally DNA methylated ICRs gain DNA methylation in oocytes and sperm, respectively. Imprints are maintained despite reprogramming and global changes in DNA methylation after fertilization. Paternal imprints are resistant to the active demethylation of the paternal genome in the zygote, and maternal methylation of ICRs is retained despite passive maternal demethylation in the preimplantation embryo. De novo DNA methylation of the genome begins at the end of the preimplantation stage of development. Imprints are maintained in somatic cells throughout the lifetime of the organism. In the germline, DNA methylation is erased during migration of PGCs into the genital ridge. Imprints are then reestablished appropriately during gametogenesis for transmission to the next generation.

of CpGs that were 8–10 base pairs apart. Such spacing is found in maternally methylated, but not paternally methylated, imprinted loci, although this CpG spacing is widespread in the genome (Ferguson-Smith and Greally 2007). Additional specificity may arise from the observation that DNMT3L interacts with the amino terminus of histone H3 and this interaction is inhibited by H3 lysine 4 methylation (Ooi et al. 2007).

Recently transcription has been suggested to be important for acquisition of maternal imprints in oocytes. Chotalia and colleagues described transcription, some being oocyte-specific, across differentially methylated regions that they propose are required for the establishment of DNA methylation imprints in the oocyte (Chotalia et al. 2009). In contrast to the well-established role for transcription of ncRNAs in regulating the imprinting of adjacent genes described above, only protein-coding transcripts traversing the germline ICRs appear to be involved in methylation establishment. Although an apparent paradox that transcription is attracting the DNA methylation machinery, the authors suggest that transcription across ICRs is required to establish or maintain open chromatin domains that are permissive for establishment of DNA methylation. Whereas temporal relationships require additional investigation, CpG spacing, posttranslational histone modifications, and transcription in oocytes provide a compelling starting point for the acquisition of maternal-specific DNA methylation imprints.

MAINTENANCE OF GENOMIC IMPRINTS

After imprints are set in the germline, it is imperative that they are maintained so that appropriate expression patterns can be achieved in the developing organism. This is especially difficult right after fertilization in the zygote as this is the time when mammals undergo rapid and extensive reprogramming of the gamete genome into a pluripotent genome (Morgan et al. 2005). As part of this reprogramming, DNA methylation and chromatin modifications are erased and subsequently reset (Fig. 3). Nevertheless, genomic imprints must somehow survive these extensive epigenetic changes. Thus, not only must the imprints survive erasure, they must also be propagated during cleavage divisions. In this case, DNMT1, which has well described role in maintaining DNA methylation at imprinted loci (Li et al. 1993), is present at very low levels in preimplantation embryos. Although the precise mechanism is not known and is, in fact, somewhat controversial, it is probable that DNA methylation is maintained in the preimplantation embryos through a combination of the oocyte-specific form of DNMT1 (DNMT1o) and low levels of the somatic form of DNMT1 (Howell et al. 2001; Cirio et al. 2008; Hirasawa et al. 2008).

Maintenance of parental identity most likely involves the unique combination of *cis*-acting sequences and *trans*-acting factors. The *cis*-acting sequences are presumably ICRs but additional sequences cannot be ruled out. Moreover, with respect to *trans*-acting factors, it is doubtful that the diverse array of sequences present at imprinted loci would attract a single protein. Consistent with this idea, a number of proteins have been identified in the past few years that contribute to maintenance of imprints after fertilization. Some of these proteins impact DNA methylation maintenance at multiple imprinted loci, whereas others appear to have a more restricted role.

One recently defined factor that is critical for DNA methylation maintenance is ZFP57, a KRAB zinc finger protein. This class of transcription factors represses transcription by recruiting KAP-1/TIF1β corepressor complexes (Friedman et al. 1996; Abrink et al. 2001). Mutations in *Zfp57* in the mouse cause both maternal- and zygotic-effect lethality, and result in aberrant DNA methylation and expression patterns of imprinted genes (Li et al. 2008). Lack of both maternal and zygotic ZFP57 resulted in embryonic lethality and complete loss of methylation at numerous imprinted loci, including *Snrpn*, *Peg1*, *Peg3*, *Peg5*, and *Dlk1* DMRs, whereas zygotic disruption alone resulted in partial loss of methylation at these sites and partial lethality. Li and colleagues determined that maternal ZFP57 was additionally required for the establishment of DNA methylation at the *Snrpn* ICR in oocytes, but surprisingly, zygotic *Zfp57* expression compensated for this loss and *Snrpn* methylation was reestablished after E3.5 (Li et al. 2008). Such a result indicates that despite the lack of DNA methylation at the *Snrpn* ICR in the oocyte, *Snprn* harbored a residual imprint that was able to direct de novo methylation; this residual mark was either outside the assayed region or derived from an epigenetic modification other than DNA methylation at the ICR, the latter being more difficult to ascertain in preimplantation embryos. The role of ZFP57 in the maintenance of ICR methylation was independently discovered in cases of transient neonatal diabetes (TND), which is caused by hypomethylation of the promoter of the imprinted gene *PLAGL1* (Temple and Shield 2002; Mackay et al. 2008). Autosomal recessive *ZFP57* mutations were identified in multiple affected pedigrees; patients with these mutations typically had hypomethylation of other ICRs and DMRs, as well as clinical features not normally associated with TND. These results show conservation in the role of ZFP57 in maintenance of DNA methylation between mice and humans.

Additional proteins have been identified that are involved in the stability of imprints, some of which have restricted roles and others of which have a more global role in the maintenance of DNA methylation. In the former category, RBBP1/ARID4A and RBBP1L1/ARID4B are involved in maintenance of imprinting at the *Snprn* locus (Wu et al. 2006), whereas RNAi-mediated depletion of MBD3 in the oocyte and preimplantation embryo resulted in biallelic expression of *H19* in blastocysts, suggesting a role for components of the NuRD complex in maintenance of *H19* imprinting (Reese et al. 2007). A general role in imprinting maintenance is indicated for PGC7/STELLA (Nakamura et al. 2007); *Stella* −/− eggs have normal ICR methylation but embryos derived from these eggs are hypomethylated at multiple loci with maternally or paternally methylated ICRs. Zygotes also show premature global loss of DNA methylation on the maternal pronucleus, indicating a more widespread role for STELLA. Thus, these results suggest that the maternal genome must be protected from the active demethylation occurring on the paternal genome immediately after fertilization.

Two highly related members of the CATERPILLER protein family have also been suggested to be involved in early imprinting regulation in humans, although it is not clear whether these proteins function specifically in imprinting establishment in the germline or early imprinting maintenance, or both. *NLRP7* (*NALP7*), which has no mouse ortholog, was identified as the causative gene for familial recurrent hydatidiform mole, a rare maternal-effect autosomal recessive disorder in which affected women have recurrent molar pregnancies of diploid biparental origin (Murdoch et al. 2006). These women generate eggs that are capable of fertilization, but the resulting conceptus has extensive hypomethylation at the ICRs of imprinted genes. Mutation in a second family member, *NLRP2* (*NALP2*), has recently been associated with a more restricted loss of ICR methylation (Meyer et al. 2009). A family that has two siblings with Beckwith–Wiedemann syndrome was reported in which the mother was homozygous for an *NLRP2* mutation. The siblings showed hypomethylation in the ICR that regulates the imprinting of the *KCNQ1* cluster, although one child also had a partial loss of methylation of the *PEG1* DMR. Together these experiments suggest that NLRP family members are involved in the regulation of imprinting. Consistent with this, these genes are highly expressed early in mammalian development. Nevertheless, their mechanism of action is uncertain, as some family members are involved in the inflammasome, which detects and reacts to extracellular pathogens.

IMPRINTING IN VARIOUS MAMMALIAN SPECIES

Gene imprinting is found in flowering plants and in mammals and it is likely that imprinting in the two organisms evolved convergently. Interestingly, in plants, the altered dosage of imprinted genes influences the development of the endosperm, a vegetative tissue that can be considered extraembryonic and, like in mammals, supports the development of the embryo. Most imprinted genes identified to date in plants are endosperm genes. Similarly, in plants, imprinting depends on DNA methylation for its maintenance, but as endosperm is a vegetative tissue, there is no requirement for germline erasure like in mammals (Feil and Berger 2007). The dependence on imprinting in extraembryonic tissues to control nutritional resources and influence the growth of the embryo perhaps indicates common evolutionary pressures acting to establish and maintain this process of dosage control in both plants and mammals.

Interestingly, not all genes are imprinted in all mammals. For example, some genes that are imprinted in the mouse are not imprinted in the human. This is true for the placenta-specific imprinted genes of the *Kcnq1* cluster (Monk et al. 2006). In contrast, the L3MBTL gene is imprinted in human but not imprinted in mouse (Li et al. 2004a, 2005). To add further complexity, the *Igf2r* gene, which is imprinted in the mouse, has been shown to exhibit polymorphic imprinting in human—it is imprinted in some individuals but not in others (Xu et al. 1993). These findings suggest that imprinting of particular

 Cite this article as *Cold Spring Harb Perspect Biol* doi: 10.1101/cshperspect.a002592

genes varies between different mammalian species and that imprinting may be adaptable and flexible both within and between species.

It is not known how or why mammalian imprinting evolved although its emergence appears to be associated with the evolution of a placenta (Constancia et al. 2004; Kaneko-Ishino et al. 2006), suggesting that even distantly related placental mammals such as metatherians (marsupials) will have imprinting, whereas oviparous mammals, the prototherians (monotremes), will not. Recently, this idea has been put to the test and data investigating the imprinting status of a few individual mammalian imprinted genes in marsupials and monotremes is consistent with this idea. These studies have shown that some eutherian imprinted genes such as *Igf2-H19* are imprinted in marsupials. In the tammar wallaby, like in mouse and human, a DMR located upstream of *H19* containing CTCF binding sites suggests conservation of the imprinting control mechanism as well (Smits et al. 2008). *Peg10*, essential for placentation in mouse, is also imprinted in marsupials and has a differentially methylated promoter (Suzuki et al. 2007). However, other eutherian imprinted genes such as those of the *Dlk1* imprinted cluster are either absent or not imprinted in the marsupial (Edwards et al. 2008). This suggests that imprinting has evolved independently at different individual loci in some mammalian clades as the need has arisen. It will be interesting to determine why selective pressures have acted to control gene dosage by imprinting at some loci and not at others in metatherian mammals. No evidence exists for imprinting in oviparous mammals such as the platypus or the echidna (Suzuki et al. 2007; Edwards et al. 2008).

CONCLUDING REMARKS

Whereas much has been learned about the numbers and types of imprinted genes, imprinting mechanisms and epigenetic machinery that impacts imprinted gene expression, much remains to be determined. For example, how many more imprinted genes remain to be identified and to what functional pathways do they contribute? Questions remain about the imprinting mechanisms and in particular, what are the underlying germline features that confer maternal- and paternal-specific imprints differentially in egg and sperm? What are the chromatin-associated complexes that recognize, establish, and maintain genomic imprints? What are the mechanistic relationships between DNA methylation and epigenetic modifications to core histones at imprinted domains, and how do these modifications interact with large ncRNAs to influence gene regulation in *cis* over large genomic regions containing multiple genes? Mouse models and rare human patients continue to be a useful paradigm for exploring this and results can continue to be applied more generally to other nonimprinted regions contributing to a wider understanding of genome function. The advent of technologies in which epigenetic modifications in germ cells and early embryos can be characterized in vivo during key stages of development, applying emerging molecular technologies to very small populations of cells, will help determine the controlling factors and temporal order of events that define the developmental epigenetic program.

The dynamic and variable nature of imprinted gene activity and repression within and between mammalian species raises questions about the extent to which imprints can be adaptable to normal and abnormal environmental influences. Experiments addressing the functional importance of gene dosage control at these loci have the potential to shed light on this. Such functional and mechanistic studies will contribute to our understanding of the complex developmental pathways that have evolved to make a mammal, will provide deeper insights into the relationships between DNA sequence, chromatin structure, and genome function, and along the way, might help us know how and why this remarkable process evolved in the first place.

ACKNOWLEDGMENTS

We acknowledge our many colleagues whose hard work and numerous ideas have led to the numerous breakthroughs discussed in this article. Work in the Bartolomei laboratory was supported by U.S. Public Health Service grants GM51279 and HD42026 and in the Ferguson-Smith lab by grants from the MRC, BBSRC, and Wellcome Trust.

REFERENCES

Abrink M, Ortiz JA, Mark C, Sanchez C, Looman C, Hellman L, Chambon P, Losson R. 2001. Conserved interaction between distinct Krüppel-associated box domains and the transcriptional intermediary factor 1 β. *Proc Natl Acad Sci* **98:** 1422–1426.

Abu-Amero S, Monk D, Frost J, Preece M, Stanier P, Moore GE. 2008. The genetic aetiology of Silver–Russell syndrome. *J Med Genet* **45:** 193–199.

Babak T, Deveale B, Armour C, Raymond C, Cleary MA, van der Kooy D, Johnson JM, Lim LP. 2008. Global survey of genomic imprinting by transcriptome sequencing. *Curr Biol* **18:** 1735–1741.

Barlow DP, Stoger R, Herrmann BG, Saito K, Schweifer N. 1991. The mouse insulin-like growth factor type-2 receptor is imprinted and closely linked to the *Tme* locus. *Nature* **349:** 84–87.

Bartolomei MS, Zemel S, Tilghman SM. 1991. Parental imprinting of the mouse *H19* gene. *Nature* **351:** 153–155.

Bell AC, Felsenfeld G. 2000. Methylation of a CTCF-dependent boundary controls imprinted expression of the *Igf2* gene. *Nature* **405:** 482–485.

Bird A. 2002. DNA methylation patterns and epigenetic memory. *Genes Dev* **16:** 6–21.

Bourc'his D, Bestor TH. 2004. Meiotic catastrophe and retrotransposon reactivation in male germ cells lacking Dnmt3L. *Nature* **431:** 96–99.

Bourc'his D, Xu GL, Lin CS, Bollman B, Bestor TH. 2001. Dnmt3L and the establishment of maternal genomic imprints. *Science* **294:** 2536–2539.

Cattanach BM. 1982. Non-disjunction tests with Robertsonian translocations. *Mouse News Lett* **66:** 62–63.

Cattanach BM. 1986. Parental origin effects in mice. *J Embrol Exp Morph Suppl* **97:** 137–150.

Cattanach BM, Kirk M. 1985. Differential activity of maternally and paternally derived chromosome reigons in mice. *Nature* **315:** 496–498.

Chen M, Wang J, Dickerson KE, Kelleher J, Xie T, Gupta D, Lai EW, Pacak K, Gavrilova O, Weinstein LS. 2009. Central nervous system imprinting of the G protein G(s)α and its role in metabolic regulation. *Cell Metab* **9:** 548–555.

Chotalia M, Smallwood SA, Ruf N, Dawson C, Lucifero D, Frontera M, James K, Dean W, Kelsey G. 2009. Transcription is required for establishment of germline methylation marks at imprinted genes. *Genes Dev* **23:** 105–117.

Cirio MC, Martel J, Mann M, Toppings M, Bartolomei M, Trasler J, Chaillet JR. 2008. DNA methyltransferase 1o functions during preimplantation development to preclude a profound level of epigenetic variation. *Dev Biol* **324:** 139–150.

Coan PM, Burton GJ, Ferguson-Smith AC. 2005. Imprinted genes in the placenta—A review. *Placenta* **26 Suppl A:** S10–S20.

Constancia M, Kelsey G, Reik W. 2004. Resourceful imprinting. *Nature* **432:** 53–57.

Cooper DW, VandeBerg JL, Sharman GB, Poole WE. 1971. Phosphoglycerate kinase polymorphism in kangaroos provides further evidence for paternal X inactivation. *Nature New Biology* **230:** 155–157.

da Rocha ST, Edwards CA, Ito M, Ogata T, Ferguson-Smith AC. 2008. Genomic imprinting at the mammalian *Dlk1-Dio3* domain. *Trends Genet* **24:** 306–316.

da Rocha S, Charalambous M, Lin SP, Gutteridge I, Ito Y, Gray D, Dean W, Ferguson-Smith AC. 2009. Gene dosage effects of the imprinted Delta-like homologue 1 (*Dlk1*/*Pref1*) in development: Implications for the evolution of imprinting. *PLoS Genet* **5:** e1000392.

Davis TL, Trasler JM, Moss SB, Yang GJ, Bartolomei MS. 1999. Acquisition of the *H19* methylation imprint occurs differentially on the parental alleles during spermatogenesis. *Genomics* **58:** 18–28.

DeChiara TM, Robertson EJ, Efstratiadis A. 1991. Parental imprinting of the mouse insulin-like growth factor II gene. *Cell* **64:** 849–859.

Edwards CA, Mungall AJ, Matthews L, Ryder E, Gray DJ, Pask AJ, Shaw G, Graves JA, Rogers J, Dunham I, et al. 2008. The evolution of the DLK1-DIO3 imprinted domain in mammals. *PLoS Biol* **6:** e135.

Engel N, Raval AK, Thorvaldsen JL, Bartolomei MS. 2008. Three-dimensional conformation at the H19/Igf2 locus supports a model of enhancer tracking. *Hum Mol Genet* **17:** 3021–3029.

Feil R, Berger F. 2007. Convergent evolution of genomic imprinting in plants and mammals. *Trends Genet* **23:** 192–199.

Ferguson-Smith AC, Greally JM. 2007. Epigenetics: Perceptive enzymes. *Nature* **449:** 148–149.

Ferguson-Smith AC, Cattanach BM, Barton SC, Beechey CV, Surani MA. 1991. Embryological and molecular investigations of parental imprinting on mouse chromosome 7. *Nature* **351:** 667–670.

Fitzpatrick GV, Soloway PD, Higgins MJ. 2002. Regional loss of imprinting and growth deficiency in mice with a targeted deleion of *KvDMR1*. *Nat Genet* **32:** 426–431.

Friedman JR, Fredericks WJ, Jensen DE, Speicher DW, Huang XP, Neilson EG, Rauscher FJ III. 1996. KAP-1, a novel corepressor for the highly conserved KRAB repression domain. *Genes Dev* **10:** 2067–2078.

Goll MG, Bestor TH. 2005. Eukaryotic cytosine methyltransferases. *Annu Rev Biochem* **74:** 481–514.

Guillemot F, Caspary T, Tilghman SM, Copeland NG, Gilbery DJ, Jenkins NA, Anderson DJ, Joyner AL, Rossant J, Nagy A. 1995. Genomic imprinting of Mash-2, a mouse gene required for trophoblast development. *Nature Genet* **9:** 235–241.

Hajkova P, Ancelin K, Waldmann T, Lacoste N, Lange UC, Cesari F, Lee C, Almouzni G, Schneider R, Surani MA. 2008. Chromatin dynamics during epigenetic reprogramming in the mouse germ line. *Nature* **452:** 877–881.

Hajkova P, Erhardt S, Lane N, Haaf T, El-Maarri O, Reik W, Walter J, Surani MA. 2002. Epigenetic reprogramming in mouse primordial germ cells. *Mech Dev* **117:** 15–23.

Hark AT, Schoenherr CJ, Katz DJ, Ingram RS, Levorse JM, Tilghman SM. 2000. CTCF mediates methylation-sensitive enhancer-blocking activity at the *H19/Igf2* locus. *Nature* **405:** 486–489.

Hata K, Okano M, Lei H, Li E. 2002. Dnmt3L cooperates with the Dnmt3 family of de novo DNA methyltransferases to establish maternal imprints in mice. *Development* **129:** 1983–1993.

Hirasawa R, Chiba H, Kaneda M, Tajima S, Li E, Jaenisch R, Sasaki H. 2008. Maternal and zygotic Dnmt1 are necessary and sufficient for the maintenance of DNA methylation imprints during preimplantation development. *Genes Dev* **22:** 1607–1616.

Horsthemke B, Buiting K. 2006. Imprinting defects on human chromosome 15. *Cytogenet Genome Res* **113:** 292–299.

Horsthemke B, Wagstaff J. 2008. Mechanisms of imprinting of the Prader–Willi/Angelman region. *Am J Med Genet A* **146A:** 2041–2052.

Howell CY, Bestor TH, Ding F, Latham KE, Mertineit C, Trasler JM, Chaillet JR. 2001. Genomic imprinting disrupted by a maternal effect mutation in the *Dnmt1* gene. *Cell* **104:** 829–838.

Jia D, Jurkowska RZ, Zhang X, Jeltsch A, Cheng X. 2007a. Structure of Dnmt3a bound to Dnmt3L suggests a model for de novo DNA methylation. *Nature* **449:** 248–251.

Jia J, Lin M, Zhang L, York JP, Zhang P. 2007b. The Notch signaling pathway controls the size of the ocular lens by directly suppressing p57Kip2 expression. *Mol Cell Biol* **27:** 7236–7247.

Johnson DR. 1974. Hairpin-tail: A case of post-reductional gene action in the mouse egg. *Genetics* **76:** 795–805.

Kaneda M, Okano M, Hata K, Sado T, Tsujimoto N, Li E, Sasaki H. 2004. Essential role for de novo DNA methyltransferase Dnmt3a in paternal and maternal imprinting. *Nature* **429:** 900–903.

Kaneko-Ishino T, Kohda T, Ono R, Ishino F. 2006. Complementation hypothesis: The necessity of a monoallelic gene expression mechanism in mammalian development. *Cytogenet Genome Res* **113:** 24–30.

Kaneko-Ishino T, Kuroiwa Y, Miyoshi N, Kohda T, Suzuki R, Yokoyama M, Viville S, Barton SC, Ishino F, Surani MA. 1995. *Peg1/Mest* imprinted gene on chromosome 6 identified by cDNA subtraction hybridization. *Nat Genet* **11:** 52–59.

Kawahara M, Wu Q, Takahashi N, Morita S, Yamada K, Ito M, Ferguson-Smith AC, Kono T. 2007. High-frequency generation of viable mice from engineered bi-maternal embryos. *Nat Biotechnol* **25:** 1045–1050.

Kinoshita T, Ikeda Y, Ishikawa R. 2008. Genomic imprinting: A balance between antagonistic roles of parental chromosomes. *Semin Cell Dev Biol* **19:** 574–579.

Knoll JHM, Nicholls RD, Magenis RE, Graham JMJ, Lalande M, Latt SA. 1989. Angelman and Prader–Willi syndromes share a common chromosome 15 deletion but differ in parental origin of the deletion. *Am J Med Genet* **32:** 285–290.

Koerner MV, Pauler FM, Huang R, Barlow DP. 2009. The function of non-coding RNAs in genomic imprinting. *Development* **136:** 1771–1783.

Kono T, Obata Y, Wu Q, Niwa K, Ono Y, Yamamoto Y, Park ES, Seo JS, Ogawa H. 2004. Birth of parthenogenetic mice that can develop to adulthood. *Nature* **428:** 860–864.

Kurukuti S, Tiwari VK, Tavoosidana G, Pugacheva E, Murrell A, Zhao Z, Lobanenkov V, Reik W, Ohlsson R. 2006. CTCF binding at the H19 imprinting control region mediates maternally inherited higher-order chromatin conformation to restrict enhancer access to Igf2. *Proc Natl Acad Sci* **103:** 10684–10689.

Lefebvre L, Viville S, Barton SC, Ishino F, Keverne EB, Surani MA. 1998. Abnormal maternal behaviour and growth retardation associated with loss of the imprinted gene Mest. *Nat Genet* **20:** 163–169.

Li E, Beard C, Jaenisch R. 1993. Role for DNA methylation in genomic imprinting. *Nature* **366:** 362–365.

Li J, Bench AJ, Piltz S, Vassiliou G, Baxter EJ, Ferguson-Smith AC, Green AR. 2005. L3mbtl, the mouse orthologue of the imprinted L3MBTL, displays a complex pattern of alternative splicing and escapes genomic imprinting. *Genomics* **86:** 489–494.

Li J, Bench AJ, Vassiliou GS, Fourouclas N, Ferguson-Smith AC, Green AR. 2004a. Imprinting of the human L3MBTL gene, a polycomb family member located in a region of chromosome 20 deleted in human myeloid malignancies. *Proc Natl Acad Sci* **101:** 7341–7346.

Li JY, Lees-Murdock DJ, Xu GL, Walsh CP. 2004b. Timing of establishment of paternal methylation imprints in the mouse. *Genomics* **84:** 952–960.

Li X, Ito M, Zhou F, Youngson N, Zuo X, Leder P, Ferguson-Smith AC. 2008. A maternal-zygotic effect gene, *Zfp57*, maintains both maternal and paternal imprints. *Dev Cell* **15:** 547–557.

Li L, Keverne EB, Aparicio SA, Ishino F, Barton SC, Surani MA. 1999. Regulation of maternal behavior and offspring growth by paternally expressed Peg3. *Science* **284:** 330–333.

Lin SP, Youngson N, Takada S, Seitz H, Reik W, Paulsen M, Cavaille J, Ferguson-Smith AC. 2003. Asymmetric regulation of imprinting on the maternal and paternal chromosomes at the Dlk1-Gtl2 imprinted cluster on mouse chromosome 12. *Nat Genet* **35:** 97–102.

Lucifero D, Mertineit C, Clarke HJ, Bestor TH, Trasler JM. 2002. Methylation dynamics of imprinted genes in mouse germ cells. *Genomics* **79:** 530–538.

Lyle R, Watanabe D, te Vruchte D, Lerchner W, Smrzka OW, Wutz A, Schageman J, Hahner L, Davies C, Barlow DP. 2000. The imprinted antisense RNA at the Igf2r locus overlaps but does not imprint Mas1. *Nat Genet* **25:** 19–21.

Mackay DJ, Callaway JL, Marks SM, White HE, Acerini CL, Boonen SE, Dayanikli P, Firth HV, Goodship JA, Haemers AP, et al. 2008. Hypomethylation of multiple imprinted loci in individuals with transient neonatal diabetes is associated with mutations in ZFP57. *Nat Genet* **40:** 949–951.

McGrath J, Solter D. 1983. Nuclear transplantation in mouse embryos. *J Exp Zool* **228:** 355–362.

McGrath J, Solter D. 1984. Completion of mouse embryogenesis requires both the maternal and paternal genomes. *Cell* **37:** 179–183.

Meyer E, Lim D, Pasha S, Tee LJ, Rahman F, Yates JR, Woods CG, Reik W, Maher ER. 2009. Germline mutation in NLRP2 (NALP2) in a familial imprinting disorder (Beckwith–Wiedemann syndrome). *PLoS Genet* **5:** e1000423.

Monk D, Arnaud P, Apostolidou S, Hills FA, Kelsey G, Stanier P, Feil R, Moore GE. 2006. Limited evolutionary conservation of imprinting in the human placenta. *Proc Natl Acad Sci* **103:** 6623–6628.

Moon YS, Smas CM, Lee K, Villena JA, Kim KH, Yun EJ, Sul HS. 2002. Mice lacking paternally expressed Pref-1/Dlk1 display growth retardation and accelerated adiposity. *Mol Cell Biol* **22:** 5585–5592.

Morgan HD, Santos F, Green K, Dean W, Reik W. 2005. Epigenetic reprogramming in mammals. *Hum Mol Genet* **14 Spec No 1:** R47–R58.

Murdoch S, Djuric U, Mazhar B, Seoud M, Khan R, Kuick R, Bagga R, Kircheisen R, Ao A, Ratti B, et al. 2006. Mutations in NALP7 cause recurrent hydatidiform moles and reproductive wastage in humans. *Nat Genet* **38:** 300–302.

Murrell A, Heeson S, Reik W. 2004. Interaction between differentially methylated regions partitions the imprinted genes *Igf2* and *H19* into parent-specific chromatin loops. *Nat Genet* **36:** 889–893.

Nagano T, Mitchell JA, Sanz LA, Pauler FM, Ferguson-Smith AC, Feil R, Fraser P. 2008. The Air noncoding RNA epigenetically silences transcription by targeting G9a to chromatin. *Science* **322:** 1717–1720.

Nakamura T, Arai Y, Umehara H, Masuhara M, Kimura T, Taniguchi H, Sekimoto T, Ikawa M, Yoneda Y, Okabe M, et al. 2007. PGC7/Stella protects against DNA demethylation in early embryogenesis. *Nat Cell Biol* **9:** 64–71.

Nicholls RD, Knepper JL. 2001. Genome organization, function, and imprinting in Prader–Willi and Angelman syndromes. *Annu Rev Genomics Hum Genet* **2:** 153–175.

Ogata T, Kagami M, Ferguson-Smith AC. 2008. Molecular mechanisms regulating phenotypic outcome in paternal and maternal uniparental disomy for chromosome 14. *Epigenetics* **3:** 181–187.

Ono R, Nakamura K, Inoue K, Naruse M, Usami T, Wakisaka-Saito N, Hino T, Suzuki-Migishima R, Ogonuki N, Miki H, et al. 2006. Deletion of *Peg10*, an imprinted gene acquired from a retrotransposon, causes early embryonic lethality. *Nat Genet* **38:** 101–106.

Ooi SK, Qiu C, Bernstein E, Li K, Jia D, Yang Z, Erdjument-Bromage H, Tempst P, Lin SP, Allis CD, et al. 2007. DNMT3L connects unmethylated lysine 4 of histone H3 to de novo methylation of DNA. *Nature* **448:** 714–717.

Raghunandan R, Ruiz-Hidalgo M, Jia Y, Ettinger R, Rudikoff E, Riggins P, Farnsworth R, Tesfaye A, Laborda J, Bauer SR. 2008. Dlk1 influences differentiation and function of B lymphocytes. *Stem Cells Dev* **17:** 495–507.

Reese KJ, Lin S, Verona RI, Schultz RM, Bartolomei MS. 2007. Maintenance of paternal methylation and repression of the imprinted *H19* gene requires MBD_3. *PLoS Genetics* **3:** e137.

Searle AG, Beechey CV. 1978. Complementation studies with mouse translocations. *Cytogenet Cell Genet* **20:** 282–303.

Searle AG, Beechey CV. 1990. Genome imprinting phenomena on mouse chromosome 7. *Genet Res* **56:** 237–244.

Sekita Y, Wagatsuma H, Nakamura K, Ono R, Kagami M, Wakisaka N, Hino T, Suzuki-Migishima R, Kohda T, Ogura A, et al. 2008. Role of retrotransposon-derived imprinted gene, *Rtl1*, in the feto–maternal interface of mouse placenta. *Nat Genet* **40:** 243–248.

Sibley CP, Coan PM, Ferguson-Smith AC, Dean W, Hughes J, Smith P, Reik W, Burton GJ, Fowden AL, Constancia M. 2004. Placental-specific insulin-like growth factor 2 (Igf2) regulates the diffusional exchange characteristics of the mouse placenta. *Proc Natl Acad Sci* **101:** 8204–8208.

Sleutels F, Zwart R, Barlow DP. 2002. The non-coding Air RNA is required for silencing autosomal imprinted genes. *Nature* **415:** 810–813.

Smits G, Mungall AJ, Griffiths-Jones S, Smith P, Beury D, Matthews L, Rogers J, Pask AJ, Shaw G, Vandeberg JL, et al. 2008. Conservation of the H19 noncoding RNA and H19-IGF2 imprinting mechanism in therians. *Nat Genet* **40:** 971–976.

Stöger R, Kubicka P, Liu C-G, Kafri T, Razin A, Cedar H, Barlow DP. 1993. Maternal-specific methylation of the imprinted mouse *Igf2r* locus identifies the expressed locus as carrying the imprinting signal. *Cell* **73:** 61–71.

Surani MAH, Barton SC. 1983. Development of gynogenetic eggs in the mouse: implications for parthenogenetic embryos. *Science* **222:** 1034–1036.

Surani MAH, Barton SC, Norris ML. 1984. Development of reconstituted mouse eggs suggest imprinting of the genome during gametogenesis. *Nature* **308:** 548–550.

Suzuki S, Ono R, Narita T, Pask AJ, Shaw G, Wang C, Kohda T, Alsop AE, Marshall Graves JA, Kohara Y, et al. 2007. Retrotransposon silencing by DNA methylation can drive mammalian genomic imprinting. *PLoS Genet* **3:** e55.

Temple IK. 2007. Imprinting in human disease with special reference to transient neonatal diabetes and Beckwith–Wiedemann syndrome. *Endocr Dev* **12:** 113–123.

Temple IK, Shield JP. 2002. Transient neonatal diabetes, a disorder of imprinting. *J Med Genet* **39:** 872–875.

Thorvaldsen JL, Duran KL, Bartolomei MS. 1998. Deletion of the *H19* differentially methylated domain results in loss of imprinted expression of *H19* and *Igf2*. *Genes Dev* **12:** 3693–3702.

Tremblay KD, Duran KL, Bartolomei MS. 1997. A 5′ 2-kilobase-pair region of the imprinted mouse *H19* gene exhibits exclusive paternal methylation throughout development. *Mol Cell Biol* **17:** 4322–4329.

Varrault A, Gueydan C, Delalbre A, Bellmann A, Houssami S, Aknin C, Severac D, Chotard L, Kahli M, Le Digarcher A, et al. 2006. Zac1 regulates an imprinted gene network critically involved in the control of embryonic growth. *Dev Cell* **11:** 711–722.

Wang X, Sun Q, McGrath SD, Mardis ER, Soloway PD, Clark AG. 2008. Transcriptome-wide identification of novel imprinted genes in neonatal mouse brain. *PLoS ONE* **3:** e3839.

Weksberg R, Shuman C, Smith AC. 2005. Beckwith–Wiedemann syndrome. *Am J Med Genet C Semin Med Genet* **137:** 12–23.

Wilkinson LS, Davies W, Isles AR. 2007. Genomic imprinting effects on brain development and function. *Nat Rev Neurosci* **8:** 832–843.

Williamson CM, Turner MD, Ball ST, Nottingham WT, Glenister P, Fray M, Tymowska-Lalanne Z, Plagge A, Powles-Glover N, Kelsey G, et al. 2006. Identification of an imprinting control region affecting the expression of all transcripts in the *Gnas* cluster. *Nat Genet* **38:** 350–355.

Wu MY, Tsai TF, Beaudet AL. 2006. Deficiency of *Rbbp1/Arid4a* and *Rbbp1l1/Arid4b* alters epigenetic modifications and suppresses an imprinting defect in the PWS/AS domain. *Genes Dev* **20:** 2859–2870.

Wutz A, Smrzka OW, Schweifer N, Schellander K, Wagner EF, Barlow DP. 1997. Imprinted expression of the *Igf2r* gene depends on an intronic CpG island. *Nature* **389:** 745–749.

Xu Y, Goodyer CG, Deal C, Polychronakos C. 1993. Functional polymorphism in the parental imprinting of the human *IGF2R* gene. *Biochem Biophys Res Commun* **197:** 747–754.

Yang T, Adamson TE, Resnick JL, Leff S, Wevrick R, Francke U, Jenkins NA, Copeland NG, Brannan CI. 1998. A mouse model for Prader–Willi syndrome imprinting-centre mutations. *Nature Genet* **19:** 25–31.

Nuclear Transfer to Eggs and Oocytes

J.B. Gurdon[1] and Ian Wilmut[2]

[1]Wellcome Trust Cancer Research UK Gurdon Institute, Cambridge CB2 1QN, United Kingdom, and Department of Zoology, University of Cambridge, Cambridge CB2 1QN, United Kingdom

[2]MRC Centre for Regenerative Medicine, Edinburgh EH16 4SB, United Kingdom

Correspondence: j.gurdon@gurdon.cam.ac.uk

We review experiments in which somatic cell nuclei are transplanted singly to enucleated eggs (metaphase II) in amphibia and mammals and as multiple nuclei to the germinal vesicle of amphibian oocytes (prophase I). These experiments have shown the totipotency of some somatic cell nuclei, as well as switches in cell type and changes in gene expression. Abnormalities of nuclear transplant embryo development increase greatly as nuclei are taken from progressively more differentiated donor cells. The molecular changes that accompany the reprogramming of transplanted nuclei help to indicate the mechanisms used by eggs and oocytes to reprogram gene expression. We discuss the importance of chromosomal protein exchange, of transcription factor supply, and of chromatin access in reprogramming.

Cells of the female germline are unique because they form eggs that are the direct ancestors of all cells of the body including the future eggs and sperm. After normal fertilization, 100% of eggs can form complete fertile individuals. Even when not fertilized, activated eggs of most species, especially when made diploid, can develop into remarkably complete organisms with nearly all somatic cells though usually not the germline. In accord with this totipotency, eggs of many species can also develop entirely normally when provided with the nucleus of a somatic cell in place of egg chromosomes or a sperm nucleus. Somatic cells do not have the ability to generate a complete organism and the nucleus of a somatic cell must be reprogrammed if it is to participate in normal development with an enucleated egg. We review the extent to which a transplanted somatic nucleus can, in combination with an enucleated egg, generate a normal individual. We first describe the extent to which normal development results from somatic cell nuclear transfer. We then discuss the extent to which this does not happen, especially when nuclei from differentiated somatic cells are used. Finally, we discuss possible mechanisms by which the reprogramming of the somatic nucleus is induced after transfer to eggs or oocytes.

The original reason for wishing to carry out nuclear transfer to eggs was to determine whether the genome of somatic cells is complete in the sense of containing copies of all genes in the genome. Up until the 1950s, it was thought possible that genes could become lost or permanently inactivated in those cells that follow different lineages in which certain genes would never normally be required. Over the last half

Cite this article as *Cold Spring Harb Perspect Biol* doi: 10.1101/cshperspect.a002659

century, nuclear transfer and some other procedures have established the general principle that the genome is conserved during development, so that almost all somatic cells contain a complete copy of the original zygote genome (Gurdon and Byrne 2003). In more recent time this situation has been used as a basis of procedures for cell replacement. It has become possible to derive all kinds of cells of the body from a somatic cell already committed to a particular lineage (Takahashi and Yamanaka 2006). This ability has opened up the possibility of providing replacement cells of many different kinds starting from a specialized somatic cell. In this way it is, in principle, possible to provide an individual with replacement cells of their own genetic constitution, thereby avoiding the need for immunosuppression in any cell replacement therapy.

We have not attempted to give a detailed review of the nuclear transfer literature, but refer to several other reviews for different aspects of this problem (Kikyo and Wolffe 2000; Cibelli et al. 2002; Morgan et al. 2005; Collas and Taranger 2006; Meissner and Jaenisch 2006; Yang et al. 2007).

EXPERIMENTAL SYSTEMS

The basic procedure, by which a living cell nucleus is transplanted to an egg or oocyte, was established by Briggs and King (1952). They used *Rana pipiens* and sucked a blastula cell into a micropipette so that the cell wall was broken but the nucleus remained intact and covered by cytoplasm. The whole cell was injected into an unfertilized egg in second meiotic metaphase (M2). The egg was enucleated manually by removing the metaphase spindle with its chromosomes from the surface of the egg. The same procedure is used for eggs of *Xenopus*, except that the egg chromosomes are destroyed by ultraviolet irradiation; the large egg size of amphibia and the low penetrance of ultraviolet irradiation ensures that no significant damage is done to the rest of the egg. A key characteristic of early *Xenopus* egg nuclear transfer experiments was the ability to make use of the one-nucleolated mutant as a genetic marker to prove that development resulted from the implanted nucleus and not from a failure of enucleation (Elsdale et al. 1960). Amphibian eggs do not need activation when they have been penetrated with a nuclear transfer pipette. In nearly all mammalian nuclear transfer experiments, unfertilized eggs in second meiotic metaphase are used as recipients. Nuclear injection in mammals is usually achieved manually or by cell fusion with a micropipette as in amphibia, but in some species piezo-electric needle penetration is used. Enucleation is achieved by removal of the female pronucleus in second meiotic metaphase with a microinjection pipette. Activation is by procedures that differ between the species, and in the mouse is usually performed by the addition of $SrCl_2$ to the medium. Activation is normally performed within a few hours of nuclear injection and enucleation.

An altogether different nuclear transfer procedure is to use growing eggs in first meiotic prophase (M1); in amphibia, these cells (usually called oocytes) are present in huge abundance in the ovary. Somatic nuclei can be injected into the specialized and enlarged nucleus (germinal vesicle, GV) of the ovary (Byrne et al. 2003) the contents of the GV are essential for normal development (Gao et al. 2002). Injected nuclei remain in the germinal vesicle for many days; there is no cell division and no DNA replication but intensely active transcription. The advantage of this experimental system is that it is useful for analyzing the changes undergone by transplanted nuclei as they become transcriptionally reprogrammed. Twenty-five thousand oocytes per frog, and up to five hundred injected nuclei per oocyte, make molecular analysis of transcriptional reprogramming practicable. In contrast to oocytes, eggs are primarily active in DNA replication, and transcription does not start until the four thousand-cell stage of the amphibian blastula (5 h) or until the two-cell stage in mouse embryos (20 h). The immediate responses of somatic nuclei to injected eggs (M2) are therefore more likely to be associated with the induction of DNA synthesis than with transcriptional reprogramming.

BACKGROUND

The background to the field of somatic cell nuclear transfer has been outlined by Gurdon (2006). Following the initial success of Briggs and King (1957), subsequent work with *Rana pipiens* found that nuclei from postgastrula cells could no longer support normal development (Briggs and King 1957). *Xenopus* experiments on the other hand continued to give entirely normal development in some nuclear transplant embryos even with nuclei from differentiated intestinal epithelial cells of tadpoles. The sequence of events in mammalian nuclear transfer has been summarized by Meissner and Jaenisch (2006) and others. The finding of McGrath and Solter (1984) that even nuclei from eight cell mouse embryos did not support normal development was later attributed to the use of fertilized eggs as recipients rather than of unfertilized eggs as used in earlier amphibian experiments and in most subsequent mammalian work. Initial success was obtained with embryo nuclei transplanted to eggs of sheep (Willasden et al. 1986) and to those of cows and pigs (Prather et al. 1989). A major advance took place when adult sheep were obtained from nuclei of a cultured line of cells grown from sheep embryo cells (Campbell et al. 1996) and then from nuclei of an adult sheep cell line (Wilmut et al. 1997). Since then a wide range of mammalian species has been cloned (production of fertile adults). Nuclear transfer, and hence evidence for nuclear programming, has been successful in fish, chickens, and *Drosophila* within the invertebrates. However, nuclei from fully differentiated or adult cells have not yielded normal adults in these latter species.

The reversal of the differentiated state is seen in experiments other than nuclear transfer to eggs. For example, switches in gene expression follow fusion of somatic cells (Pereira et al. 2008; Zhou et al. 2008; Cowan et al. 2005), the overexpression of transcription factors (Weintraub et al. 1989; Graf 2002; Takahashi and Yamanaka 2006) and at least temporarily after the resealing of permeabilized cells incubated in cell extracts (Collas and Taranger 2006). Induced pluripotency by transcription factor overexpression has the great practical benefit of deriving proliferating ES cells, and hence all other cells types, from accessible adult cells without the need to use eggs which would be hard to obtain in humans. Nuclear transplantation into eggs or oocytes is of interest because it uses the natural reprogramming activity of eggs on the sperm after normal fertilization without the need for induced gene overexpression, and because it gives a relatively high efficiency of transcriptional activation. Eventually, it may be possible to identify the molecules and mechanisms of nuclear reprogramming used by eggs, and hence to use these molecules to enhance the efficiency of reprogramming by other procedures.

NORMAL DEVELOPMENT

Totipotency

The ultimate test of reprogramming of a transplanted somatic cell nucleus is to ask if it can yield a normal fertile sexually mature adult when transplanted to an enucleated egg. This was performed in *Xenopus* when a considerable number of fertile males and females were grown from enucleated eggs injected with nuclei from intestinal epithelial cells of feeding tadpoles (Gurdon and Uehlinger 1966). The demonstration that the nucleus of a cell in an adult animal can yield, after nuclear transfer, a complete sexually mature adult was achieved with sheep nuclear transfers (Wilmut et al. 1997). Dolly, the sheep, lived for 6 years, a life span within the normal range of sheep reared indoors. Dolly was as fertile as a normal sheep raised indoors. Since that time, a somatic cell nuclear transfer has been successful in a wide range of mammal species (Lanza et al. 2001b; Wakayama et al. 1998). In some cases, technical maneuvers such as the timing and type of activating stimulus are needed to give the best results. In a few mammalian species tested, a small proportion of nuclei from differentiated or adult tissues have given normal adult nuclear transfer development (terminal lymphocytes [Hochedlinger and Jaenisch 2002]; postmitotic neurons [Eggan et al. 2004; Li et al. 2004]). In these cases tetraploid rescue was used, but Inoue et al. (2004)

obtained adults from postmitotic neurons without tetraploid rescue. In conclusion, at least 1%–3% of both differentiated and adult cells contain totipotent nuclei. Embryonic stem cell lines have been derived in the mouse, cow, and monkey (Munsie et al. 2000; Wakayama et al. 2001; Byrne et al. 2007). A claim of human ES cells derived from nuclear transfers was retracted (Kennedy 2006).

There is one exception to the results summarized above, and this is the particular case of antibody-forming cells, in which gene rearrangement accounts for the huge diversity of antibodies that cells can make. Nuclei transplanted from T-cells form normal mice except that they make only one kind of antibody (Hochedlinger and Jaenisch 2002).

We can conclude that, with the special exception of cells in which genes are known to undergo rearrangement, the genome is conserved during cell differentiation.

Switches in Cell Type

From most points of view including the prospect of cell replacement therapy, we are much more interested in the extent to which one particular type of cell can be derived from cells of another unrelated cell-type, rather than in totipotency. For example, can brain or heart cells be derived from skin or blood? The efficiency of reprogramming at this more limited level is much higher than when assessed by the production of fertile adults. In *Xenopus* nuclear transfer experiments from intestinal epithelium, we can ask with what frequency functional muscle and nerve cells are obtained, as judged by the formation of embryos that make muscular movements when stimulated with a needle. We can take into account those embryos that result from serial nuclear transfer after taking donor nuclei from the normal-appearing cells of an otherwise defective nuclear transplant embryo, as well as muscle and nerve cells resulting from the grafting of such abnormal embryo-derived cells to fertilized egg-derived recipients. GFP-marked donor nuclei in the combination of these experiments give an overall efficiency of about 30%; that is to say, nearly one third of all intestinal epithelial cell nuclear transfers produce functional muscle and nerve cells. In other *Xenopus* work, nuclei from a range of adult tissues including foot-web skin, lung, heart, etc., yielded normal feeding tadpoles (Laskey and Gurdon 1970; Gurdon et al. 1975). Although the proportion of total nuclear transfers that developed to feeding tadpoles was very small (2%–3%), these results established the principle that the nuclei of differentiated and adult cells can be reprogrammed to generate a wide range of unrelated cell-types. The same conclusion is probably true of mammals, because ES cells, and hence a range of somatic cell types, can be derived from many adult mammalian tissues.

Early Gene Expression

It was found long ago that embryos derived from transplanted muscle nuclei discontinue muscle gene transcription soon after nuclear transfer, but recommence muscle gene expression in muscle but not in other cell-types (Gurdon et al. 1984). Microarray analyses give an approximate measure of transcription of many genes; these conclude that over 95% of several thousand genes assayed are correctly transcribed in early mouse nuclear transplant embryos (Humpherys et al. 2002; Beyhan et al. 2007; Vassena et al. 2007). We can conclude that nuclear transfer to eggs causes a major reprogramming of the great majority of genes. The genes that are not reprogrammed correctly may account for some of the abnormal embryo development always seen in nuclear transfer experiments.

Cancer Nuclei

Nuclei transplanted from the Herpes virus related frog Lucké adenocarcinoma cells yielded tadpoles with a wide range of normal cell types (McKinnell et al. 1969). Blelloch et al (2004) transplanted nuclei from mouse embryonal carcinoma cells to mouse eggs; the resulting blastocysts gave rise to embryonic stem cells that, when transferred to hosts, displayed the same range of abnormalities as the donor embryonal carcinoma cells. There is therefore no evidence that nuclear transfer changes the genetic state of donor cells. We can conclude that a cancerous state is compatible with normal early development and with some kinds of somatic cell differentiation.

Cross-Species Nuclear Transfer

An appealing idea is to generate human ES cells from a human cell nucleus transplanted to enucleated eggs of cows, pigs, rabbits, etc. The earliest nuclear transfer experiments across related species within the genera *Rana* or *Xenopus* always resulted in embryo death (Moore 1960; Gurdon 1962b). In mammals, most such experiments have not been published, but it is generally agreed that nuclear transfers between mammalian species (defined as ones in which cross-fertilization gives no survival) die at close to the stage when zygotic transcription starts; this is at the two-cell stage in mice, at the 8–16 cell stage in cows, pigs, and humans. In spite of this, very early postnuclear transfer events may succeed in cross-species combination, such as sperm demethylation (Beaujean et al. 2004b). A report of proliferating ES cells from a human-rabbit nuclear transplant combination (Chen et al. 2003) has yet to be confirmed.

ABNORMAL DEVELOPMENT

Increase with Progressive Cell Differentiation

It has been known since the earliest nuclear transfer experiments that the normality of nuclear transplant embryo development decreases dramatically as nuclei are taken from increasingly differentiated cells (Briggs and King 1957; Gurdon 1962a). The same is true of mammalian experiments (Hochedlinger and Jaenisch 2002).

A question of great interest is whether the type of nuclear transplant abnormalities is related to the type of donor cells. In amphibia, nuclear transplant embryos die progressively throughout development and there is no morphological indication of any particular type of developmental abnormality in relation to donor cell-type. In mammals, the most obvious defects all relate to the placenta and it appears that reprogramming is much less successful in the trophectoderm than in the inner cell mass. Commonly observed abnormalities include shorter life span, enlarged placenta, obesity, respiratory problems, and large offspring syndrome (Yang et al. 2007). It was suggested that developmental abnormalities observed in amphibian nuclear transplant embryos might be inversely related to the cell-type from which nuclei were taken (Briggs and King 1960). However, the numbers in this experiment were too small to be significant, and subsequent experiments with *Xenopus* neural cell nuclei failed to show any morphological correlation (Simnett 1964). When assessing the developmental abnormalities of nuclear transplant embryos it is important to appreciate that such embryos have had their zona layers removed and their development should therefore be compared to that of ICSI embryos (intracytoplasmic sperm injection), because the culture medium of zona-free mouse embryos can adversely affect the normality of their development.

Epigenetic Memory

An entirely unexpected result was encountered when a transcriptional analysis was made of *Xenopus* nuclear transplant embryo development (Ng and Gurdon 2008). In this case, nuclei were transplanted from a defined cell type such as somite muscle, neural cells, or from the endoderm (future intestine). When the resulting embryos reached the blastula stage, they were divided into future neurectoderm, mesoderm (including) muscle, and endoderm regions and the embryo parts were cultured for another day until they reached a stage of lineage-specific gene expression, when they were analyzed for marker gene expression. The surprise was that, in about half of all embryos analyzed, an excessive abundance of transcripts characteristic of the donor cell-type was observed in an inappropriate lineage. For example, the neurectoderm and endoderm parts of embryos resulting from the transfer of muscle nuclei had expressed muscle genes to an excessive extent, even though the same regions of control embryos grown from fertilized eggs showed no such aberrant transcription. This memory of a former state of gene expression was especially remarkable because *Xenopus* embryos, whether from fertilized eggs or from transplanted nuclei, are inactive in transcription for their first 12-cell cycles. This example of epigenetic memory is therefore able to persist through multiple cell divisions in the absence of transcription. This effect is seen in genes that mark lineage determination, such as MyoD, and not in those that represent terminal differentiation. A somewhat related phenomenon has been described in mouse muscle nuclear transfers when the culture of these in myoblast culture medium promotes a continued expression of muscle cell markers (Gao et al. 2002). We can conclude that, judged by gene expression, reprogramming by eggs is about 50% efficient, at least for nuclei of cells that are undergoing, but have not yet completed, differentiation. Incomplete reprogramming may account for some of the nuclear transplantation abnormalities observed.

Early Gene Expression

Most attempts to estimate the normality or defects in early gene expression of nuclear transplant embryos have used a microarray technology. The general finding is that the vast majority of genes, which are active in early development but not in adult cells, show clear evidence of reprogramming. Humpherys et al. (2002) saw abnormalities in 4% of the genes tested. Smith et al. (2005) report that 99% of genes tested appeared to be normally expressed and therefore potentially reprogrammed. The methods used are not very quantitative and an incomplete reprogramming of a gene would not always be seen. For genes in which the amount of product may be critical for early development (e.g., *Oct4*) incomplete reprogramming may be damaging (Boiani 2002; Bortvin et al. 2003).

MOLECULAR CHANGES

Somatic Mutation

Assuming a gene mutation rate of 10^{-6} per gene per cell, a mouse embryo containing 10^9 cells might have accumulated 10^3 mutations in its cells. Most of these are unlikely to be in genes needed for early development, and most mutations are also unlikely to inactivate a gene. Furthermore, even nuclei transplanted from embryo cells, with a theoretically much lower load of accumulated mutations, result in a substantial number of early developmental abnormalities. We conclude that naturally accumulated mutations in donor cells are very unlikely to account for the majority of nuclear transplant embryo abnormalities.

DNA Methylation

It is well established that the control regions of genes that are wholly inactive in most somatic cells undergo methylation, and that this contributes to the silencing of those genes. A good example is the promoter of *Oct4*, a gene that must be expressed soon after nuclear transfer to eggs if development is to proceed (Niwa et al. 2000). DNA demethylation hardly ever takes place in differentiated somatic cells, but takes place to some extent, though not efficiently, in nuclear transplant embryos (Beaujean et al. 2004a; review by Morgan et al. 2005). DNA demethylation is known to be imperfect in some embryos as judged by antibody staining of transplanted nuclei (Bortvin 2003; Boiani 2002). Because the level of *Oct4* expression is critical for early mouse development, it is likely that incomplete *Oct4* DNA demethylation may account for abnormalities of development in nuclear transplant embryos. Because demethylation of the *Oct4* promoter is seen in nuclei transplanted to nondividing *Xenopus* oocytes (Simonsson and Gurdon 2004), the process of demethylation cannot depend on loss by nonreplacement in dividing cells as it is supposed to do for the mammalian female pronucleus. A candidate facilitator for DNA demethylation that has been proposed is Gadd4a (Barreto et al. 2007). Other candidates include components of the DNA repair process (Niehrs 2009).

DNA methylation is believed to be involved in mammalian female X chromosome inactivation, and in mammalian imprinting. Although some degree of X chromosome reactivation has been reported in transplanted nuclei (Nolen et al. 2005) the efficiency of this process is not clear and an incomplete reactivation of Xi probably contributes to the abnormal development of nuclear transplant embryos.

Telomere Replacement

The low expression of telomerase in most somatic cells compared to the high level in embryo and embryo stem cells leads us to expect that normal nuclear transplant embryos would contain an increased content of telomeres on their chromosomes compared to the somatic cells from which they are derived. The terminal restriction fragment of DNA (that contains telomeres) was smaller in all three nuclear transfer sheep analyzed than in age-related controls (Wilmut et al. 1997). On the other hand, six calves had a higher level of an age-related gene transcript than controls of similar age (Lanza et al. 2000). The fact that the serial transfer of nuclei in mice is successful for seven generations argues against an irreversible shortening of telomeres (Wakayama 1998). In general, telomere length seems to be restored in cloned fetuses. However, it is not possible to determine the telomere content of a single cell that is used to donate a nucleus for successful nuclear transfer; the telomere characteristic of that donor nucleus must be assumed to be like the average of the donor cell population. We can therefore conclude that some telomere replacement probably occurs after nuclear transfer.

Epigenetic Modification

Modifications to histones in transplanted nuclei have been studied in both mammals and frogs (review by Morgan et al. 2005). The most commonly used procedure is to stain transplanted nuclei with antibodies specific for particular histone modifications. This type of assay has revealed variable changes in nuclei transplanted to eggs (Wang et al. 2007). For example, H3K9 is deacetylated but later reacetylated but other acetylations show no change. H3K9 is demethylated in some cases, but not in others.

A general observation is that these changes are seen in nuclei transplanted from differentiated cells. These then come to resemble, in this respect, transplanted ES nuclei, which do not themselves undergo these changes. It has to be assumed that these changes are made on histones that are components of chromatin rather than in the nucleoplasm of transplanted nuclei. These changes are seen in nuclei transplanted to both eggs and oocytes, and are not therefore correlated with transcription rather than DNA replication. In future it will be important to know whether these are global changes to the whole chromatin or are localized to those particular genes that undergo reprogramming. There is some evidence that trichostatin A treatment of nuclear transplant embryos can improve their survival (Kishigami et al. 2006).

It is interesting that some of the chromatin features of embryo nuclei are already seen in sperm (Hammoud et al. 2009). Because transplanted somatic nuclei are not exposed to conditions that prevail in sperm, it may be hard for these chromatin characteristics to be correctly imposed by eggs.

Chromatin Decondensation and Protein Exchange

The most obvious first response of transplanted nuclei to eggs and oocytes is a huge increase in volume. In frogs, this can be as much as 30 fold in 1 h for eggs and similarly for oocytes on a slower time scale. The same effect is seen in mammalian eggs, but more slowly. In *Xenopus* oocytes this enlargement includes the dispersion of chromatin and is correlated with transcriptional activation (Gurdon et al. 1976). A massive nuclear volume increase is also seen in sperm soon after fertilization in mammals and frogs. Nucleoplasmin is present at an enormous concentration of 7 mg/mL in *Xenopus* eggs (Leno et al. 1996), and is in part, but not wholly, responsible for chromatin dispersal (Tamada et al. 2006). The major function of nucleoplasmin is to act as a chaperone for histone H2a and b. N1 and N2 serve a similar function for histone H3 and H4. The oocyte-specific linker histones are also present in a very high concentration in eggs and oocytes. The loosening or decompaction of chromatin is a primary event in nuclear reprogramming, and seems to set the stage for ensuring transcription or DNA replication.

It has long been known that somatic nuclei transplanted to eggs and oocytes of *Xenopus* soon exchange proteins between nucleus and cytoplasm; about 80% of the proteins brought in with a transplanted nucleus are lost within a few hours, and there is an obvious immigration of cytoplasmic proteins into transplanted nuclei (Gurdon et al. 1976). This may reflect the normal rapid exchange of proteins between nucleus and cytoplasm in most kinds of cells; the very large volume of eggs and oocytes will result in loss being much more evident than uptake. The abundant proteins of oocytes are among those taken up by transplanted nuclei; labeled histones present in oocyte cytoplasm, whether by protein or mRNA injection, become intensely concentrated in the germinal vesicle and in transplanted nuclei. The rapid replacement of the mammalian linker histone H1foo may be related to chromatin decondensation (Teranishi et al. 2004). There is, at present, no clear evidence that the loss or gain of a particular protein is causally connected with induced transcription or replication of transplanted nuclei.

Cell Extracts

A preferred route by which to identify functional components of cells is to use extracts of cells that have a desired activity. Depletion of an extract by antibodies or by other means can identify components of the extract that make a necessary contribution to the activity being analyzed. This experimental design has been particularly successful in identifying components of *Xenopus* eggs that are required for DNA replication (Gonzalez et al. 2005). However, a substantial problem is that, so far, no one has succeeded in making an extract of cells that can carry out transcriptional reprogramming in vitro. Three types of experiment go some way toward this objective.

One is to use a cell-free extract derived from eggs or oocytes to describe changes that take place in nuclei or permeabilized cells, even though these changes do not include new transcription. Important work within this category is that of Kikyo and Wolffe (2000) who described the ISWI-induced loss of TBP (TATA box binding protein) from *Xenopus* cultured cell nuclei incubated in *Xenopus* egg extracts. Subsequent work found that FRGY2 in egg extracts causes a reversible disassembly of nucleoli in somatic nuclei (Gonda et al. 2003), and that nucleoplasmin in eggs is required for chromatin dispersal (Tamada et al. 2006). These and other changes are listed in Table 1.

A second route based on cell-free extracts involves the addition of extracts to lightly permeabilized cells, which are then resealed by Ca^{++} and cultured until transcription starts (Teranger et al. 2005; Collas and Taranger 2006). With this experimental design, Hansis et al. (2004) found that Brg-1 is needed to permit reactivation of Oct4. All these experiments analyze extracts of eggs, and may therefore identify components needed for DNA replication as opposed to transcription (see above). Extracts of porcine oocytes show the entry of linker histone B4 and lamin III into

Table 1. In vitro culture

Donor nuclei	Extract of eggs or oocytes	Duration	Responses	Reference
Xenopus XTC cells	*Xenopus* S phase	2 h	Tata protein release by ISWI	Kikyo and Wolffe 2000
Bovine fetal fibroblast	*Xenopus* egg or oocyte	3 h	Lamin A/C removed; PolI+II transcription maintained	Alberio et al. 2005
Xenopus embryo fibroblasts	*Xenopus* S phase egg	2 h	FRGY 2a+b; nucleolus disassembly	Gonda et al. 2003
Xenopus embryo fibroblasts	*Xenopus* eggs	2 h	Chromatin decondensation; H3 histone phosphorylation + acetylation; nucleoplasmin	Tamada et al. 2006

 Cite this article as *Cold Spring Harb Perspect Biol* doi: 10.1101/cshperspect.a002659

permeabilized fibroblasts, which when resealed and cultured, can activate pluripotency genes (Miyamoto et al. 2007, 2008, 2009). An extract of axolotl oocytes is able to demethylate DNA to reduce H3K9Me3 and to increase H3K9Ac so that nuclei of differentiated cells become like those of ES cells (Bian et al. 2009). These latter changes are seen in extracts of oocytes, and are therefore likely to be relevant to transcriptional reprogramming. See Table 2 for a list of these results.

MECHANISMS

The aim of this section is to discuss possible mechanisms by which eggs or oocytes can successfully reprogram somatic cell nuclei to behave like embryo nuclei. In view of the slow or nil division rate and specialized gene expression of most somatic cells, it is remarkable that normal development ever results from nuclear transfer. However, there are two respects in which reprogramming does not succeed well. Compared to fertilized eggs, almost all of which develop normally, only about 30% of even blastula or blastocyst nuclei elicit normal development, and with the nuclei of differentiated cells only about 2% of total transfers become adults. The incomplete functioning of any of the mechanisms we discuss could help to explain developmental defects in nuclear transplant embryos.

The Initiation of DNA Replication

In amphibia, the first cell division takes place 1.5 h after fertilization or nuclear transfer, and DNA replication of transplanted nuclei in GI needs to precede this. Eggs have a strong DNA synthesis inducing activity (Graham et al. 1966). This is brought about by a great decrease in the spacing of origins of replication (Lemaitre et al. 2005). Even nondividing adult brain nuclei respond to egg cytoplasm by initiating DNA synthesis, but chromosome replication is sometimes incomplete by the time of the first cell division, and partially replicated chromosomes are pulled apart leading to damaged cleavage nuclei and defective cleavage (Di Berardino and King 1965). In mammals, the long (20 h) first cell cycle gives no reason to attribute the 50% of nuclear transplants, even from cumulus nuclei that fail to reach the two-cell stage, to failures of DNA replication.

Chromosomal Protein Exchange at Mitosis

An interesting idea that has surfaced over many years is that the exchange of chromosomal proteins that takes place in chromosomes during mitosis may facilitate changes in cell differentiation, and these could include nuclear reprogramming. There is indeed a general release of chromosomal proteins at mitosis. In the course of normal development, it is supposed that cells remaining in the same lineage

Table 2. In vitro culture followed by resealing of cells

Extract source	Nuclei	Duration	Responses	Reference
Xenopus eggs or oocytes	Human 293 T and leucocytes	30 min, then 7 d	Oct4 and GCAP; Brg required	Hansis et al. 2004
Xenopus egg	Porcine fibroblast	2 h, then 10 d	B4 histone and nuclear lamin uptake; Oct4 & Sox2 activation	Miyamoto et al. 2007
Porcine oocytes (GV)	Porcine fibroblast	2 h, then 10 d	Nanog activation & demethylation; histone acetylation	Miyamoto et al. 2009
Axolotl oocytes (GV)	Mouse embryo fibroblast	5 h, then 3 d	H3K9Me3, HP1α & DNA methylation reduced; Oct4 and Nanog activation	Bian et al. 2009
Mouse oocytes (GV)	Cumulus cells	45 min, then 4 wk	H3K9Me reduced; Oct4 activation	Bui et al. 2008

pathway will lose regulatory chromosomal proteins as they enter mitosis, but will regain the same kinds of proteins as they progress to interphase with no change of gene control. In the case of nuclear transfer to eggs, the proteins lost at the first mitosis will be massively diluted in the huge volume of an egg, and will be replaced by proteins that represent the maternal content of the egg, and a change in gene control would result. In amphibia, the egg is 100,000 times larger than a somatic cell, and a mouse egg is 350 times larger. Because there is no transcription for the first cell cycle in the mouse, nor during the first 12-cell cycles in the frog, chromosomal proteins of a transplanted somatic nucleus should be diluted out. This concept has been most recently presented by Eggan (2004). We point out that mitotic exchange of this kind cannot be the only mechanism of nuclear reprogramming, because the latter takes place efficiently in nuclear transfer to oocytes (no DNA synthesis of mitosis) and in heterokaryon cell fusion experiments (no cell division), as explained above. If chromosomal protein exchange in eggs worked perfectly, nuclear transplant embryos should all develop normally and the phenomenon of epigenetic memory (above) would not exist. It is likely, nevertheless, that mitotic chromosomal protein exchange is a contributory factor in reprogramming when this takes place over multiple cell cycles and in iPS and other transcription factor overexpression experiments.

Transcription Factor Supply

An obvious explanation for a dramatic switch in gene expression is the provision of new transcription factors. As noted above, the overexpression of transcription factors has long been known to be able to activate genes in inappropriate cell types. The remarkable results of Yamanaka (Takahashi et al. 2006) and Melton (Zhou et al. 2008) illustrate this effect very clearly. This would provide a simple explanation for the transcriptional activation of pluripotency genes in nuclear transfers to eggs or oocytes; for example *Xenopus* oocytes that can directly activate transcription of Sox2, Oct4, and Nanog might contain a maternal content of transcription factors required for these genes. Because these factors are not yet identified for *Xenopus*, this proposition cannot be tested directly. It has, however, been possible to test this idea in respect of the very well studied myogenic gene MyoD. This gene is, surprisingly, strongly expressed in *Xenopus* oocytes, and furthermore it is strongly induced in nonmuscle nuclei transplanted to oocytes (Biddle et al. 2009). MyoD and other myogenic genes are activated in normal development by MyoD protein itself, of which there is a maternal content in *Xenopus* oocytes. It is possible to eliminate the function of MyoD in *Xenopus* oocytes by overexpressing the inhibitory protein Id. When this is performed, it has no effect at all on the transcriptional activation of MyoD or other myogenic genes in nuclei transplanted to oocytes. This suggests that gene activation in nuclei transplanted to oocytes may not be achieved by transcription factor overexpression.

In contrast to this result, the overexpression of TPT-1, an oncogenic factor, does enhance *Oct4* activation in somatic nuclei in oocytes (Koziol et al. 2007). However, the genes immediately downstream from TPT-1 are not known, and the TPT-1 effect may be indirect.

We conclude that the maternal provision of gene-specific transcription factors may not be a general reprogramming mechanism in oocytes and eggs. At present, it seems more likely that oocytes (and perhaps eggs) cause a global derepression of genes, so that most genes become available for transcription by a nonspecific transcriptional apparatus. This would be consistent with the intensely active transcription of most of the genome during the lampbrush phase of amphibian oogenesis (Davidson 1986; Alberts et al. 2008).

Chromatin Access

It is generally believed that, as cells differentiate, an increasing proportion of genes become transcriptionally inactive, compared to embryonic cells. In concert with this, chromatin becomes increasingly condensed and compacted. An especially obvious effect of transplanting somatic nuclei to eggs or oocytes is that the nuclei undergo a huge volume increase that is accompanied by an increased dispersion of chromatin in the transplanted nuclei (see above). This could be interpreted as a general loosening of chromatin structure that might give the transcriptional apparatus of the cell more ready access to repressed genes.

FRAP provides a valuable technique for determining the dynamics of nuclear protein mobility (Hager et al. 2009). It has not yet been possible to apply this method to single copy proteins bound to individual genes; multiple fluorescent proteins need to be present at one point on a chromosome for fluorescence to be strong enough to detect loss and replacement of a protein. This technique has therefore been used when multiple copies of a protein, such as histones, HP1, etc., are associated with a gene in chromatin. In all such cases, the residence time of proteins has turned out to be remarkably short, in the order of seconds or minutes. If this is also true of repressor complexes such as components of the polycomb series, we would expect such complexes to be rapidly replaced in transplanted nuclei, which would then be rapidly derepressed. As explained above, nuclei from differentiated cells are remarkably resistant to reprogramming compared to nuclei of embryonic cells.

One explanation for this, in terms of chromosomal protein mobility is the following (Gurdon and Melton 2008). We suggest that the numerous proteins that comprise a repressor complex like PC1 are only loosely connected to chromatin in embryo nuclei. Each component of the complex would have a short dwell time, and so, in a limited time, all such components would have dissociated from chromatin; being only loosely associated with each other, the gene region that they occupy would soon become vacated, allowing access of the cell's transcriptional apparatus. In contrast, the components of a repressor complex in a differentiated cell are supposed to be extensively bound to each other in a compact structure. Each component of the complex would partially dissociate, but would not be fully released because of its association, at the same time to another member of the complex. As a result, a complete vacation of the site would require a far longer time before the transcriptional apparatus would have access to this site. It would therefore be understandable that a gene in a very specialized cell with highly condensed chromatin would need a very long time, compared to the same gene in an embryonic cell, before it could be transcribed. This concept seems to be in agreement with the results of somatic cell nuclear transfer and with iPS experiments.

ACKNOWLEDGMENTS

J.B.G. thanks the Wellcome Trust and Medical Research Council of the UK for having supported much of his work. I.W. thanks BBSRC, DEFRA, and EU-FP4 for supporting his work, and Jane Taylor and Keith Campbell for collaboration.

REFERENCES

Alberio R, Johnson AD, Stick R, Campbell KHS. 2005. Differential nuclear remodelling of mammalian somatic cells by *Xenopus laevis* oocyte and egg cytoplasm. *Exp Cell Res* **307:** 131–141.

Alberts B, Johnson A, Lewis J, Raff M, Roberts K, Walter P. 2008. *Molecular biology of the cell*, 5th ed, pp. 234–235. Garland Science, New York.

Barreto G, Schäfer A, Marhold J, Stach D, Swaminathan SK, Handa V, Döderlein G, Maltry N, Wu W, Lyko F, et al. 2007. Gadd45a promotes epigenetic gene activation by repair-mediated DNA demethylation. *Nature* **445:** 671–675.

Beaujean N, Taylor J, Gardner J, Wilmut I, Meehan R. Young L. 2004a. Effect of limited DNA methylation reprogramming in the normal sheep embryo on somatic cell nuclear transfer. *Biol Reprod* **71:** 185–193.

Beaujean N, Taylor JE, McGarry M, Gardner JO, Wilmut I, Loi P, Ptak G, Galli C, Lazzari G, Bird A, et al. 2004b. The effect of interspecific oocytes on demethylation of sperm DNA. *Proc Natl Acad Sci* **101:** 7636–7640.

Beyhan Z, Ross PJ, Iager AE, Kocabas AM, Cunniff K, Rosa GJ, Cibelli JB. 2007. Transcriptional reprogramming of somatic cell nuclei during preimplantation development of cloned bovine embryos. *Dev Biol* **305:** 637–649.

Bian Y, Alberio R, Allegrucci C, Campbell KH, Johnson AD. 2009. Epigenetic marks in somatic chromatin are remodelled to resemble pluripotent nuclei by amphibian oocyte extracts. *Epigenetics* **4:** 194–202.

Biddle A, Simeoni I, Gurdon JB. 2009. *Xenopus* oocytes reactivate muscle gene transcription in transplanted somatic nuclei independently of myogenic factors. *Development* **136:** 2695–2703.

Blelloch RH, Hochedlinger K, Yamada Y, Brennan C, Kim M, Mintz B, Chin L, Jaenisch R. 2004. Nuclear cloning of embryonal carcinoma cells. *Proc Natl Acad Sci* **101:** 13985–13990.

Bortvin A, Eggan K, Skaletsky H, Akutsu H, Berry DL, Yanagimachi R, Page DC, Jaenisch R. 2003. Incomplete reactivation of *Oct4*-related genes in mouse embryos cloned from somatic nuclei. *Development* **130:** 1673–1680.

Boiani M, Eckardt S, Scholer HR, McLaughlin KJ. 2002. Oct4 distribution and level in mouse clones: Consequences for pluripotency. *Genes Dev* **16:** 1209–1219.

Briggs R, King TJ. 1952. Transplantation of living nuclei from blastula cells into enucleated frogs' eggs. *Proc Nat Acad Sci* **38:** 455–463.

Briggs R, King TJ. 1957. Changes in the nuclei of differentiating endoderm cells as revealed by nuclear transplantation. *J Morph* **100:** 269–312.

Briggs R, King TJ. 1960. Nuclear transplantation studies on the early gastrula (*Rana pipiens*). I. Nuclei of presumptive endoderm. *Dev Biol* **2:** 252–270.

Byrne JA, Simonsson S, Western PS, Gurdon JB. 2003. Nuclei of adult mammalian somatic cells are directly reprogrammed to *oct-4* stem cell gene expression by amphibian oocytes. *Curr Biol* **13:** 1206–1213.

Bui H-T, Wakayama S, Kishigami S, Kim J-H, Van Thuan N, Wakayama T. 2008. The cytoplasm of mouse germinal vesicle stage oocytes can enhance somatic cell nuclear reprogramming. *Development* **135:** 3935–3945.

Byrne JA, Pedersen DA, Clepper LL, Nelson M, Sanger WG, Gokhale S, Wolf DP, Mitalipov SM. 2007. Producing primate embryonic stem cells by somatic cell nuclear transfer. *Nature* **450:** 497–502.

Campbell KH, McWhir J, Ritchie WA, Wilmut I. 1996. Sheep cloned by nuclear transfer from a cultured cell line. *Nature* **380:** 64–66.

Chen Y, He ZX, Liu A, Wang K, Mao WW, et al. 2003. Embryonic stem cells generated by nuclear transfer of human somatic nuclei into rabbit oocytes. *Cell Research* **13:** 251–263.

Cibelli J, Lanza RP, Campbell KHS, West MD, eds. 2002. *Principles of cloning*. Academic Press, New York.

Collas P, Taranger CK. 2006. Epigenetic reprogramming of nuclei using cell extracts. *Stem Cell Rev* **2:** 309–317.

Cowan CA, Atienza J, Melton DA, Eggan K. 2005. Nuclear reprogramming of somatic cells after fusion with human embryonic stem cells. *Science* **309:** 1369–1373.

Davidson EH. 1986. *Gene activity in early development*, 3rd ed, pp. 327–376. Academic Press, New York.

Di Berardino MA, King TJ. 1965. Transplantation of nuclei from the frog renal adenocarcinoma. II Chromosomal and histological analysis of tumor nuclear-transplant embryos. *Dev Biol* **11:** 217–242.

Eggan K, Baldwin K, Tackett M, Osborne J, Gogos J, Chess A, Axel R, Jaenisch R. 2004. Mice cloned from olfactory sensory neurons. *Nature* **428:** 44–49.

Elsdale TR, Gurdon JB, Fischberg M. 1960. A description of the technique for nuclear transplantation in *Xenopus laevis*. *J Embryol Exp Morph* **8:** 437–444.

Gao S, Gasparrini B, McGarry M, Ferrier T, Fletcher J, Harkness L, De Sousa P, Wilmut I. 2002. Germinal vesicle material is essential for nucleus remodeling after nuclear transfer. *Biol Reprod* **67:** 928–934.

Gonda K, Fowler J, Katoku-Kikyo N, Haroldson J, Wudel J, Kikyo N. 2003. Reversible disassembly of somatic nucleoli by the germ cell proteins FRGY2a and FRGY2b. *Nat Cell Biol* **5:** 205–210.

Gonzalez MA, Tachibana KE, Laskey RA, Coleman N. 2005. Control of DNA replication and its potential clinical exploitation. *Nat Rev Cancer* **5:** 135–141.

Graf T. 2002. Differentiation plasticity of hematopoietic cells. *Blood* **99:** 3089–3101.

Graham CF, Arms K, Gurdon JB. 1966. The induction of DNA synthesis by frog egg cytoplasm. *Dev Biol* **14:** 349–381.

Gurdon JB. 1962a. The transplantation of nuclei between two species of *Xenopus*. *Dev Biol* **5:** 68–83.

Gurdon JB. 1962b. The developmental capacity of nuclei taken from intestinal epithelium cells of feeding tadpoles. *J Embryol Exp Morph* **10:** 622–640.

Gurdon JB. 2006. From nuclear transfer to nuclear reprogramming: The reversal of cell differentiation. *Ann Rev Cell Dev Biol* **22:** 1–22.

Gurdon JB, Brennan S, Fairman S, Mohun TJ. 1984. Transcription of muscle-specific actin genes in early *Xenopus* development: Nuclear transplantation and cell dissociation. *Cell* **38:** 691–700.

Gurdon JB, Byrne JA. 2003. The first half-century of nuclear transplantation. *Proc Natl Acad Sci* **100:** 8048–8052.

Gurdon JB, Melton DA. 2008. Nuclear reprogramming in cells. *Science* **322:** 1811–1815.

Gurdon JB, Uehlinger V. 1966. "Fertile" intestine nuclei. *Nature* **210:** 1240–1241.

Gurdon JB, Laskey RA, Reeves OR. 1975. The developmental capacity of nuclei transplanted from keratinized skin cells of adult frogs. *J Embryol Exp Morph* **34:** 93–112.

Gurdon JB, Partington GA, De Robertis EM. 1976. Injected nuclei in frog oocytes: RNA synthesis and protein exchange. *J Embryol Exp Morph* **36:** 541–553.

Hager GL, McNally JG, Mistelli T. 2009. Transcriptional dynamics. *Mol Cell* **35:** 741–753.

Hammoud SS, Dix DA, Zhang H, Purwar J, Carrell DT, Cairns BR. 2009. Distinctive chromatin in human sperm packages genes for embryo development. *Nature* **460:** 473–478.

Hansis C, Barreto G, Maltry N, Niehrs C. 2004. Nuclear reprogramming of human somatic cells by *Xenopus* egg extract requires BRG1. *Curr Biol* **16:** 1475–1480.

Hochedlinger K, Jaenisch R. 2002. Monoclonal mice generated by nuclear transfer from mature B and T donor cells. *Nature* **415:** 1035–1038.

Hochedlinger K, Jaenisch R. 2002. Nuclear transplantation: Lessons from frogs and mice. *Curr Opin Cell Biol* **14:** 741.

Humpherys D, Eggan K, Akutsu H, Friedman A, Hochedlinger K, Yanagimachi R, Lander ES, Golub TR, Jaenisch R. 2002. Abnormal gene expression in cloned mice derived from embryonic stem cell and cumulus cell nuclei. *Proc Natl Acad Sci* **99:** 12889–12894.

Inoue K, Wakao H, Ogonuki N, Miki H, Seino K, Nambu-Wakao R, Noda S, Miyoshi H, Koseki H, Taniguchi M, et al. 2005. Generation of cloned mice by direct nuclear transfer from natural killer T cells. *Curr Biol* **15:** 1114–1118.

Kennedy D. 2006. Editorial retraction. *Science* **311:** 335.

Kikyo N, Wolffe AP. 2000. Reprogramming nuclei: Insights from cloning, nuclear transfer and heterokaryons. *J Cell Sci* **113:** 11–20.

Kishigami S, Mizutani E, Ohta H, Hikichi T, van Thuan N, Wakayama S, Bui H-T, Wakayama T. 2006. Significant improvement of mouse cloning technique by treatment with trichostatin A after somatic nuclear transfer. *Biochem Biophys Res Com* **340:** 183–189.

Koziol MJ, Garrett N, Gurdon JB. 2007. Tpt1 activates transcription of *oct4* and *nanog* in transplanted somatic nuclei. *Curr Biol* **17:** 801–807.

Laskey RA, Gurdon JB. 1970. Genetic content of adult somatic cells tested by nuclear transplantation from cultured cells. *Nature* **228:** 1332–1334.

Lanza RP, Cibelli JB, Blackwell C, Cristofalo VJ, Francis MK, Baerlocher GM, Mak J, Schertzer M, Chavez EA, Sawyer N, et al. 2000. Extension of cell lifespan and telomere length in animals cloned from senescent somatic cells. *Science* **288:** 665–669.

Lanza RP, Cibelli JB, Faber D, Sweeney RW, Henderson B, Nevala W, West MD, Wettstein PJ. 2001. Cloned cattle can be healthy and normal. *Science* **294:** 1893–1894.

Lemaitre JM, Danis E, Pasero P, Vassetzky Y, Mechali M. 2005. Mitotic remodelling of the replicon and chromosome structure. *Cell* **123:** 787–801.

Leno GH, Mills AD, Philpott A, Laskey RA. 1996. Hyperphosphorylation of nucleoplasmin facilitates *Xenopus* sperm decondensation at fertilization. *J Biol Chem* **271:** 7253–7256.

Li J, Ishii T, Feinstein P, Mombaerts P. 2004. Odorant receptor gene choice is reset by nuclear transfer from mouse olfactory sensory neurons. *Nature* **428:** 393–399.

Meissner A, Jaenisch R. 2006. Mammalian nuclear transfer. *Dev Dyn* **235:** 2460–2469.

McGrath J, Solter D. 1984. Inability of mouse blastomere nuclei transferred to enucleated zygotes to support development in vitro. *Science* **226:** 1317–1319.

McKinnell RG, Deggins BA, Labat DD.1969. Transplantation of pluripotential nuclei from triploid frog tumors. *Science* **165:** 394–396.

Miyamoto K, Furusawa T, Ohnuki M, Goel S, Tokunaga T, Minami N, Yamada M, Ohsumi K, Imai H. 2007. Reprogramming events of mammalian somatic cells induced by *Xenopus laevis* egg extracts. *Mol Reprod Dev* **74:** 1268–1277.

Miyamoto K, Tsukiyama T, Yang Y, Li N, Minami N, Yamada M, Imai H. 2009. Cell-free extracts from mammalian ocytes partially induce nuclear reprogramming in somatic cells. *Biol Reprod* **80:** 935–943.

Miyamoto K, Yamashita T, Tsukiyama T, Kitamura N, Minami N, Yamada M, Imai H. 2008. Reversible membrane permeabilization of mammalian cells treated with digitonin and its use for inducing nuclear reprogramming by *Xenopus* egg extracts. *Cloning and Stem Cells* **10:** 535–542.

Moore JA. 1960. Serial back-transfers of nuclei in experiments involving two species of frogs. *Dev Biol* **2:** 535–550.

Morgan HD, Santos F, Green K, Dean W, Reik W. 2005. Epigenetic reprogramming in mammals. *Human Mol Genet* **14:** R47–R58.

Munsie MJ, Michalska AE, O'Brien CM, Trounson AO, Pera MF, Mountford PS. 2000. Isolation of pluripotent embryonic stem cells from reprogrammed adult mouse somatic cell nuclei. *Curr Biol* **10:** 989–992.

Ng RK, Gurdon JB. 2008. Epigenetic memory of an active gene state depends on histone H3.3 incorporation into chromatin in the absence of transcription. *Nature Cell Biol* **10:** 102–109.

Niehrs C. 2009. Active DNA demethylation and DNA repair. *Differentiation* **77:** 1–11.

Niwa H, Miyazaki J, Smith AG. 2000. Quantitative expression of Oct-3/4 defines differentiation, dedifferentiation or self-renewal of ES cells. *Nat Genet* **24:** 372–376.

Nolen LD, Gao S, Han Z, Mann MR, Gie Chung Y, Otte AP, Bartolomei MS, Latham KE. 2005. X chromosome reactivation and regulation in cloned embryos. *Dev Biol* **279:** 525–540.

Pereira CF, Terranova R, Ryan NK, Santos J, Morris KJ, Cui W, Merkenschlager M, Fisher AG. 2008. Heterokaryon-based reprogramming of human B lymphocytes for pluripotency requires Oct4 but not Sox2. *PLoS Genet* **4:** e1000170.

Prather RS, Sims MM, First NL. 1989. Nuclear transplantation in early pig embryos. *Biol Reprod* **41:** 414–418.

Simnett JD. 1964. The development of embryos derived from the transplantation of neural ectoderm cell nuclei in *Xenopus laevis*. *Dev Biol* **10:** 467–486.

Simonsson S, Gurdon JB. 2004. DNA demethylation is necessary for the epigenetic reprogramming of somatic cell nuclei. *Nature Cell Biol* **6:** 984–990.

Smith SL, Everts RE, Tian XC, Du F, Sung LY, Rodriguez-Zas SL, Jeong BS, Renard JP, Lewin HA, Yang X. 2005. Global gene expression profiles reveal significant nuclear reprogramming by the blastocyst stage after cloning. *Proc Natl Acad Sci* **102:** 17582–17587.

Takahashi K, Yamanaka S. 2006. Induction of pluripotent stem cells from mouse embryonic and adult fibroblast cultures by defined factors. *Cell* **126:** 663–676.

Tamada H, Van Thuan N, Reed P, Nelson D, Katoku-Kikyo N, Wudel J, Wakayama T, Kikyo N. 2006. Chromatin decondensation and nuclear reprogramming by nucleoplasmin. *Mol Cell Biol* **26:** 1259–1271.

Taranger CK, Noer A, Sørensen AL, Håkelien AM, Boquest AC, Collas P. 2005. Induction of dedifferentiation, genomewide transcriptional programming, and epigenetic reprogramming by extracts of carcinoma and embryonic stem cells. *Mol Biol Cell* **16:** 5719–5735.

Teranishi T, Tanaka M, Kimoto S, Ono Y, Miyakoshi K, Kono T, Yoshimura Y. 2004. Rapid replacement of somatic linker histones with the oocyte-specific linker histone H1foo in nuclear transfer. *Dev Biol* **266:** 76–86.

Vassena R, Han Z, Gao S, Baldwin D, Schultz RM, Latham KE. 2007. Tough beginnings: Alterations in the transcriptome of cloned embryos during the first two cell cycles. *Dev Biol* **304:** 75–89.

Wakayama T, Perry AC, Zuccotti M, Johnson KR, Yanagimachi R. 1998. Full-term development of mice from enucleated oocytes injected with cumulus cell nuclei. *Nature* **394:** 369–374.

Wakayama T, Tabar V, Rodriguez I, Perry AC, Studer L, Mombaerts P. 2001. Differentiation of embryonic stem cell lines generated from adult somatic cells by nuclear transfer. *Science* **292:** 740–743.

Wang F, Kou Z, Zhang Y, Gao S. 2007. Dynamic reprogramming of histone acetylation and methylation in the first cell cycle of cloned mouse embryos. *Biol Reprod* **77:** 1007–1016.

Weintraub H, Tapscott SJ, Davis RL, Thayer MJ, Adam MA, Lassar AB, Miller AD. 1989. Activation of muscle-specific genes in pigment, nerve, fat, liver, and fibroblast cell lines by forced expression of MyoD. *Proc Natl Acad Sci* **86:** 5434–5438.

Willasden SM. 1986. Nuclear transplantation in sheep embryos. *Nature* **320:** 63–65.

Wilmut I, Beaujean N, de Sousa PA, Dinnyes A, King TJ, Paterson LA, Wells DN, Young LE. 2002. Somatic cell nuclear transfer. *Nature* **419:** 583–586.

Wilmut I, Schnieke AE, McWhir J, Kind AJ, Campbell KH. 1997. Viable offspring derived from fetal and adult mammalian cells. *Nature* **385:** 810–813.

Yang X, Smith SL, Tian XC, Lewin HA, Renard JP, Wakayama T. 2007. Nuclear reprogramming of cloned embryos and its implications for therapeutic cloning. *Nature Genetics* **39:** 295–302.

Zhou Q, Brown J, Kanarek A, Rajagopal J, Melton DA. 2008. In vivo reprogramming of adult pancreatic exocrine cells to β-cells. *Nature* **455:** 627–632.

Selection in the Rapid Evolution of Gamete Recognition Proteins in Marine Invertebrates

Victor D. Vacquier[1] and Willie J. Swanson[2]

[1]Marine Biology Research Division, Scripps Institution of Oceanography, University of California San Diego, La Jolla, California 92093-0202

[2]Department of Genome Sciences, University of Washington, Seattle, Washington 98195-7730

Correspondence: vvacquier@ucsd.edu

Animal fertilization is governed by the interaction (binding) of proteins on the surfaces of sperm and egg. In many examples presented herein, fertilization proteins evolve rapidly and show the signature of positive selection (adaptive evolution). This review describes the molecular evolution of fertilization proteins in sea urchins, abalone, and oysters, animals with external fertilization that broadcast their gametes into seawater. Theories regarding the selective forces responsible for the rapid evolution driven by positive selection seen in many fertilization proteins are discussed. This strong selection acting on divergence of interacting fertilization proteins might lead to prezygotic reproductive isolation and be a significant factor in the speciation process.

> Since only a fraction of all eggs are fertilized and only an infinitesimal fraction of male gametes succeed in fertilizing an egg, gametes are obviously a category of entities subjected to intense selection. It is curious that this is never mentioned in the literature dealing with selection, perhaps because we know so little about fitness differences among gametes.
>
> (Ernst Mayr, 1997)

Animal fertilization remains one of the least understood fundamental biological processes. Through fertilization, generations are bridged by fusion of two haploid cells. Fusion restores the diploid genome and sets the zygote on the path of embryogenesis. There are five events in sperm-egg interaction; not all can be shown in every species. These are chemotaxis of sperm to egg, attachment of sperm to the egg envelope, induction of the sperm acrosome reaction, penetration of sperm through the egg envelope, and fusion of gamete plasma membranes (Vacquier 1998). Forty years ago, researchers began to identify and isolate proteins mediating fertilization. The thinking then was that fertilization was such an important process that "fertilization proteins" would be conserved as a distinct family of genes common among animals. Today, the converse is the reality; proteins involved in all phases of sexual reproduction generally evolve rapidly and show positive Darwinian selection (adaptive evolution), meaning there is adaptive value in altering their sequences. By all phases of

Cite this article as *Cold Spring Harb Perspect Biol* doi: 10.1101/cshperspect.a002931

sexual reproduction we mean genes encoding protein transcription factors acting on reproductive genes (Wu and Su 2008), female and male genital secretions (Swanson et al. 2001), pregnancy (Xie et al. 1997; Chuong et al. 2010), and proteins on sperm and egg surfaces (Swanson and Vacquier 2002; Dorus et al. 2010).

The evolution of fertilization proteins has an extensive literature (reviewed in Swanson and Vacquier 2002; Clark et al. 2006; Turner and Hoekstra 2008; Turner et al. 2008; Palumbi 2009). We narrow this review to sea urchin, abalone, and oyster proteins that mediate sperm-egg interaction. We present data supporting theories attempting to explain selective forces responsible for driving rapid evolution and positive selection of fertilization proteins.

MARINE INVERTEBRATES AS MODEL ORGANISMS

The most used invertebrates for fertilization research are sea urchins, starfish, clams, oysters, and worms. In most experiments, in which gametes of two congeneric species are mixed in cross-species fertilizations, either no or only a low percentage of eggs are fertilized. In most cases, the barrier to cross-species fertilization is not absolute; if eggs are flooded with hetero-specific sperm, some cross-species fertilization usually occurs. However, when quantitative experiments are performed in which gametes of two species are cross-mixed, the percentage of hybrid zygotes formed can range from 0%–100% depending on the species combinations. Thus, in general, at equivalent gamete concentrations, homo-specific mixtures of sperm and eggs yield more zygotes than hetero-specific mixtures. This indicates that gamete interaction is mediated by sperm and egg proteins that bind each other before plasma membrane fusion.

GAMETE RECOGNITION AND SPECIATION

Species-specific fertilization in externally fertilizing marine invertebrates led to the idea that divergence of gamete recognition systems could be involved in speciation (Coyne and Orr 2004). Could the divergence of fertilization proteins lead the way in the speciation process, or would such molecular differentiation occur only after the genomes of two populations had diverged to the point of incompatibility? Here, we discuss some possible causes for the rapid evolution of gamete recognition proteins. One or all of these mechanisms could be acting simultaneously on the evolution of a gamete recognition system. Other, yet unknown selective forces could be causing the rapid evolution of fertilization proteins and the differentiation of populations into new species. In addition to habitat separation, there are no doubt many different ways that speciation can occur.

TYPES OF SELECTION THAT MAY DRIVE THE RAPID EVOLUTION OF FERTILIZATION PROTEINS

Reinforcement

One population could become two species by geographic separation (allopatry) and in time, their two genomes become incompatible, yet their two fertilization systems might remain compatible. If these two sister species make contact again in sympatry (same habitat), the gamete recognition systems might allow fertilization, leading to hybrid embryo death. To prevent cross-species fertilizations, natural selection would "reinforce" differences in the gamete recognition systems to prevent hetero-specific fertilization (Dobzhanski 1940). This could result in the evolution of species-specific fertilization in closely related marine invertebrate species (Minor et al. 1991; Palumbi and Metz 1991; Coyne and Orr 2004). Reinforcement could select for nonoverlapping breeding seasons, making inter-species hybrids impossible. For example, the two Mexican sand dollar species, *Encope grandis* and *Encope californicus*, which occur intermingled in equal numbers, have gametes at exactly opposite times of the year (V Vacquier, unpubl.).

Specialization Following Gene Duplication

New proteins may arise by gene duplication and selective pressure imposed to specialize the functions of each paralog. Gene duplication is thought to be a reason for positive selection in ungulate pregnancy associated glycoproteins (Xie et al. 1997). It may also be involved in the rapid divergence of the two abalone sperm acrosomal paralogs discussed below (Kresge et al. 2001a).

Sexual Conflict: The Balance between Sperm Competition and Egg Polyspermy Avoidance

Polyspermy is the fusion of more than one sperm with the egg, resulting in embryo death. Monospermy requires eggs to evolve mechanisms to insure fusion with only one sperm. Some species block polyspermy by rapidly reversing the electrical potential of the egg membrane after the first sperm fuses, which blocks fusion with a second sperm by an unknown mechanism (Gould and Stephano 2003). In sea urchins, the electrical block is followed by elevation of the fertilization envelope, a physical barrier to sperm penetration (Wong and Wessel 2008). Sexual conflict (Rice and Holland 1997) comes into play because sperm competition is always selecting for the fastest sperm to fuse with the egg. Eggs must evolve ways to slow sperm fusion to avoid polyspermy, and one of the ways they might do this is to alter their sperm surface receptors. Altered receptors, having a "poorer fit" with their cognate sperm proteins, slow sperm fusion. In theory, eggs survive to become adults by slowing sperm fusion by altering their egg surface sperm receptors. However, as a counter force, sperm competition continuously selects for sperm with the "best fit" to the altered receptor. To summarize, the egg surface protein changes first and its cognate sperm ligand adapts to the ever-changing female target. This concept has been termed "the coevolutionary chase" of female and male GRPs that could lead to the splitting of one population into two new species (Gavrilets 2000; Gavrilets and Waxman 2002; Gavrilets and Hayashi 2005).

Sexual Selection

This occurs solely if it is a benefit to reproduction. It can drive the rapid evolution of reproductive traits (Andersson 1994). Because of differences in sperm surface alleles, eggs might be fertilized more readily with certain alleles of the cognate sperm surface protein. Such a molecular recognition system could be termed "cryptic female choice" (Palumbi 1999).

RAPID EVOLUTION AND POSITIVE SELECTION IN MARINE INVERTEBRATE FERTILIZATION PROTEINS: ECHINODERM SPERM BINDIN

Sea urchin (su) bindin was the first gamete recognition protein isolated (Vacquier and Moy 1977). After the acrosome reaction, su-bindin coats the acrosomal process that attaches the sperm to the egg vitelline layer (Moy and Vacquier 1979; Vacquier et al. 1995). Su-bindin binds species-specifically with EBR1 (egg bindin receptor-1) (Kamei and Glabe 2003), a large, repetitive glycoprotein. Su-bindin is expressed in spermatocytes (Cameron et al. 1990) as a preproprotein of ~51-kDa (Gao et al. 1986). Mature su-bindin is packaged into the acrosomal vesicle. Su-bindin (mature su-bindins are 193–418 residues) has a central domain of ~60 residues that is conserved among sea urchin species. Inside this region are 29 residues that have been completely conserved for ~250 million years (Zigler and Lessios 2003a; Ziegler 2008). Within this sequence is an 18-residue peptide (B18) that is highly fusagenic (mediates the fusion of lipid bilayers) to biomembranes (Rocha et al. 2008). Thus, su-bindin has two functions, first to species-specifically attach the sperm to the egg, and second, to fuse the gamete membranes. The conserved domain is flanked by species-specific sequences, which are usually repeats of 7–9 residues (Minor et al. 1991; Biermann 1998). Correlating the percent interspecies zygotes formed with su-bindin sequences, shows that 10 differences out of ~240 residues are enough to produce complete reproductive isolation (Zigler et al. 2005).

Sea urchins and starfish (sf) diverged ~500 million years ago. The sf-bindin sequence is known for one starfish species (Patiño et al. 2009). The conserved features between su- and sf-bindins include: placement of cysteine residues in a long prepro region, a furin cleavage site between the prepro and mature sequences, repeating sequence elements, and the B18 fusagenic peptide that differs from sea urchin by one residue. Unique features of sf-bindin include: two collagen-like domains, longer more numerous repeats that are unrelated to other sequences, a larger mature mass of ~88-kDa and placement of the fusagenic B18 peptide at the carboxyl terminus. The structural similarities of these two echinoderm bindins over such long evolutionary time suggests that the B18 peptide will be found in bindins of the three other extant classes of the echinoderms.

EBR1, THE SEA URCHIN EGG RECEPTOR FOR SU-BINDIN

Su-bindin particles adhere species-specifically to unfertilized eggs (Glabe and Vacquier 1977; Glabe and Lennarz 1979; Metz et al. 1994), indicating the presence of species-specific egg receptors. cDNAs for EBR1 from two spe cies, *Strongylocentrotus purpuratus* (Spurp) and *Strongylocentrotus franciscanus* (Sfran) were sequenced (Kamei and Glabe 2003). These glycoproteins are homologous in their amino-terminal halves, yet strikingly different in their carboxy-terminal thirds. Sf-EBR1 is 4595 and Sp-EBR1 3710 residues (Fig. 1). From the amino terminus, both have signal sequences, propeptides, reprolysin, and ADAM domains. Next, both have nine tandem TSP domains (thrombospondin type-1), each being 53–59 residues. Progressing toward the carboxyl terminus, both have eight tandem repeats of CUB domains (C1s/Clr, uEGF, bone morphogenic protein) of 104–123 residues. The combination of one TSP domain and one CUB domain is an "EB repeat," averaging 171 residues. Spurp-EBR1 has 8.5 EBR repeats, whereas Sfran has 19. To this point, the amino-terminal halves of these two proteins are >80% identical. In Sfran, the terminal 10 EB repeats of the carboxyl terminus are different from the upstream EB repeats and are species-specific. In Spurp, after the 8.5 EB repeats, come 11 tandem hyalin repeats each of 81 residues. Hyalin repeats are not present in Sfran-EBR1. Thus, the carboxy-terminal thirds of these two EBR1s show no evolutionary resemblance to each other, except that they contain tandem repeats of roughly the same size and number (Kamei and Glabe 2003). The hyalin repeats are polymorphic in two Spurp individuals, one *Echinometra mathaei* and one *Echinometra oblonga* individual (Palumbi 2009). In the two Spurp individuals, 13 and 14 polymorphic hyalin repeats were found in 26 and 27 clones. This polymorphism in Spurp hyalin repeats could provide the raw material for selection to act on to insure fertilization success under a variety of sperm concentrations (Palumbi 2009).

Unfortunately, sequences of EBR1 from many individuals of one species, from the same and different populations, and from many different species, have not yet been determined. The EB repeats in Sfran-EBR1 are all homologous. The possession of terminal hyalin repeats

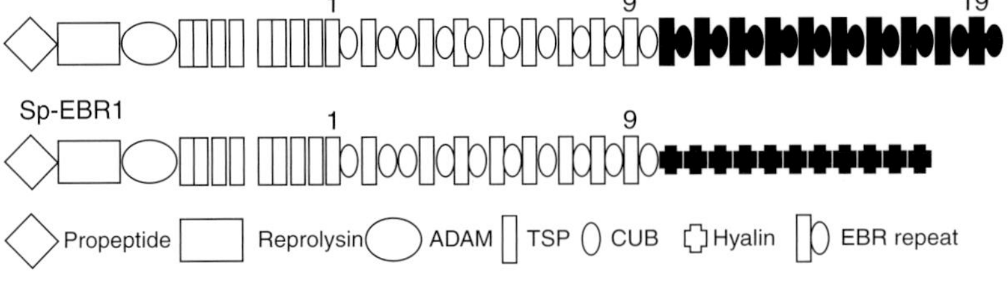

Figure 1. The protein domains of the sea urchin egg bindin receptor, EBR1, from *Strongylocentrotus franciscanus* and *S. purpuratus*. After the ninth EB repeat, the Sf TSP/CUB domains are species-specific in sequence (black). However, after the ninth EB repeat in Sp, there are 11 hyalin repeats of 81 residues each, which have no relationship to TSP/CUP sequences. (Figure redrawn from Kamei and Glabe [2003] and reprinted, with permission, from Cold Spring Harbor Press © 2003.)

in Spurp-EBR1 suggests that exon shuffling occurred to produce a chimeric EBR1, which was advantageous to fertilization success.

RAPID BINDIN EVOLUTION, SEQUENCE POLYMORPHISM, AND POSITIVE SELECTION IN SYMPATRIC SPECIES OF SEA URCHINS

The first su-bindin sequences of congeneric, sympatric sea urchin species were those of *Strongylocentrotus purpuratus* and *S. franciscanus.* The sequences were different in the regions flanking su-bindin's conserved central domain (Minor et al. 1991; Biermann 1998). These two are the most divergent species in this genus (Biermann et al. 2003; Lee 2003). Subsequent sequencing work on Hawaiian sea urchins, *Echinometra oblonga* and *E. mathaei*, living intermingled on the same reefs, showed that these two species diverged 0.5–2.0 million years ago. Cross-species mixtures of their gametes showed a strong block to hybrid fertilization (Palumbi and Metz 1991), caused by sperm su-bindin failing to attach the sperm to the egg (Metz et al. 1994).

These two Hawaiian *Echinometra* species and four unnamed incipient *Echinometra* species from the western Pacific and Panama (Metz and Palumbi 1996; McCartney and Lessios 2002, 2004; Zigler et al. 2005), and species of the genus *Strongylocentrotus* (Biermann 1998) all show that positive selection has acted on the rapid evolution of su-bindin in a region ~40 amino acids upstream of su-bindin's conserved central core. The greatest su-bindin rate of evolution and highest su-bindin sequence polymorphism correlate positively with the greatest number of sympatric species, suggesting that reinforcement might be the selective force operating on the evolution of su-bindins (Palumbi and Lessios 2005; Lessios 2007).

In the sea urchin, *Echinometra oblonga*, individuals from Guam and Okinawa have dramatically different sperm sizes and cross-fertilizations between these two locations are less successful than within the same location. These differences have arisen after an estimated divergence time of ~250,000 years (Landry et al. 2003). Likewise, within the circumpolar species, *Strongylocentrotus droebachiensis*, sperm dimensions are different between the Atlantic and Pacific oceans and also between various islands (Manier and Palumbi 2008; Marks et al. 2008). Thus, sperm morphometrics, su-bindin sequences and reproductive isolation can evolve rapidly depending on the sea urchin populations in question. It would be interesting to have EBR1 repeat sequences from these sea urchins.

Paracentrotus gaimardi is one of the two sea urchin species of this genus, which occurs in five distinct sympatric color morphs on the coast of Brazil. Genes for mitochondrial ATPase 8 and 6, a histone intron and sperm bindin were sequenced. Regarding the mitochondrial genes, statistically significant differences are seen between only one color morph and the other four. Essentially no differences were seen in the histone gene among the morphs. However, in bindin all color morphs were differentiated from each other with nine positions under positive selection (Calderon et al. 2010). Whether this represents the first stages of reproductive isolation differentiation in incipient speciation, or merely accelerated evolution of a reproductive protein, remains conjectural at this time.

SELECTION IN THE MAINTENANCE OF RARE SU-BINDIN ALLELES

Fifteen su-bindin alleles are known in *Strongylocentrotus franciscanus*: one represents 58% of alleles, one 15%, one 14%, and the remaining 13% comprise 12 rare alleles (Levitan and Ferrell 2006). When sperm densities were high, embryos that survived were found to have been fertilized with sperm carrying rare su-bindin alleles. Conversely, when sperm densities were low, the common su-bindin alleles were found in the surviving embryos. These results can be explained by sexual conflict, the balance between sperm competition and egg avoidance of polyspermy (Rice and Holland 1997; Frank 2000; Haygood 2004). Theoretically, at high sperm densities, eggs with altered EBR1s survive to become embryos because rare su-bindin alleles fit altered EBR1s poorly, the result being the slowing of sperm fusion and monospermy.

Thus, one explanation for these data is that sexual conflict in polyspermy avoidance maintains the rare alleles in this species and hence explains the extensive su-bindin polymorphism (Levitan and Ferrell 2006). The positive selection seen in su-bindin evolution could result from sperm competition for the "best fit" for binding to an altered egg EBR1 receptor (Kamei and Glabe 2003; Palumbi 2009).

It is difficult to distinguish the above explanation from sexual selection (cryptic female choice) in that eggs that are favored to become embryos are those that have "selected" to fuse with sperm carrying rare su-bindin alleles (Palumbi 1999). Mutants in the tandem repeats of EBR1 could have a better fit with mutant su-bindins and be passed on in the population. Because of its tandem construction, a mutant EBR1 repeat could be spread through the repeat array by concerted evolution, the process mediated by unequal crossing over and gene conversion that homogenizes the sequences of tandem repeats (Elder and Turner 1995). However, the high amount of sequence polymorphism in hyalin repeats of EBR1 shows that concerted evolution has not influenced evolution of this protein (Palumbi 2009).

Thus at low sperm densities, common bindin alleles are used more frequently to produce embryos to avoid egg loss (Levitan and Ferrell 2006). In addition, sea urchin species that are more resistant to polyspermy have fewer bindin alleles than species that are more susceptible to polyspermy (Levitan et al. 2007). However, other work on sexual conflict in polyspermy avoidance shows that even at low sperm densities, a measurable fraction of eggs become polyspermic (Franke et al. 2002). In summary, sexual conflict, resulting from the balance between the egg's interest in monospermy and the sperm's interest to compete with other sperm to be the first sperm to fuse with the egg, is the likely selective force for the rapid evolution, maintenance of polymorphic alleles, and positive selection of su-bindin in *S. franciscanus*. Although sexual conflict in polyspermy prevention is an explanation for maintenance of multiple su-bindin alleles, the possibilities that reinforcement to prevent cross-species fertilizations, and sexual selection, could also be selective forces acting on su-bindin's evolutionary trajectory.

However, the strengths of the blocks against polyspermy, adult sea urchin density, egg reception of bindin, and bindin genotype all influence fertilization success. This is a complex situation that is difficult to quantitatively parse into the components of the selective forces acting on each element. In the sea urchin *Strongylocentrotus purpuratus*, spawned in the sea, both negative and positive selection impinge on the evolution of sperm bindin. This species is sympatric with *S. franciscanus* mentioned above. *S. purpuratus* occurs in higher population densities than *S. franciscanus* and also requires higher sperm densities to achieve fertilization and its eggs are also more resistant to polyspermy. Such complex species differences have been discussed from field experiment fertilization data (Levitan and Stapper 2010) and from the results of theoretical modeling (Tomaiuolo and Levitan 2010).

SU-BINDIN EVOLUTION IN ALLOPATRIC SEA URCHIN SPECIES: SLOW EVOLUTIONARY RATES WITHOUT POSITIVE SELECTION

Four species of sea urchins, genus *Arbacia*, inhabit different coasts, each being separated from another *Arbacia* species by thousands of kilometers. Mitochondrial DNA and su-bindin sequences show they are distinct species with divergence times of ~4–9 million years. Gametes of the Pacific species, *Arbacia incisa*, show no barrier to fertilization when crossed with gametes from the Atlantic species, *Arbacia punctulata* (Metz et al. 1998a). Few amino acid substitutions occur in su-bindins of these four species and only three substitutions out of 235 residues are nonconservative. No evidence of positive selection in su-bindin evolution was found among these four species. Su-bindin evolution has been slow and purifying in the absence of sympatric, congeneric species.

A similar study was performed on su-bindin from three singly occurring (allopatric) species of the genus *Tripnuestes*. Two of these species, that diverged ~3 million years ago, because of the rise of the Isthmus of Panama, differ by only four amino acids and one insertion/deletion. Among these three su-bindins, there is no sign of rapid evolution or positive selection (Zigler and Lessios 2003b). A third study of eight allopatric species of the genus *Lytechinus*, also showed little variation in su-bindin sequences and no signal of rapid evolution by positive selection (Zigler and Lessios 2004).

In all three genera of allopatric sea urchin species mentioned above, su-bindin sequence evolution has been slow and not influenced by positive selection, which is the converse to that of the sympatric species. These findings are indirect evidence supporting the theory that reinforcement to prevent hybrid fertilization among sympatric species might explain the rapid evolution of bindin in sympatric species (Palumbi and Lessios 2005; Lessios 2007). Among the allopatric species of these three genera, egg loss because of inter-species hybrid embryo formation is impossible, so there is no selective pressure to reinforce differences in their gamete recognition systems. Another study of *Echinometra oblonga* supports reinforcement in that sympatric populations of this species have more variation in the su-bindin gene than do allopatric populations (Geyer and Palumbi 2003). Other research on two species of the genus *Heliocidaris* supports the reinforcement hypothesis in that su-bindin evolution may have been driven by the selective pressure to prevent fertilization between the direct developing sea urchin *Heliocidaris erythrogramma* and the indirect developing *Heliocidaris tuberculata* (Zigler et al. 2003).

POSITIVE SELECTION IN SU-BINDIN IN ALLOPATRIC SEA URCHINS

Paracentrotus is the other allopatric species of this genus. The species *Paracentrotus lividus*, in the Mediterranean and west coast of Africa is highly variable in su-bindin sequences. In 63 full-length sequences from different locations, 23 su-bindin haplotypes were found (Calderon et al. 2009). Positive selection was detected in 12 sites on either side of the single intron 5′ to the ~60 residue constant domain. Variation was mainly in the number of glycine rich short repeats flanking the constant domain. Because no other congener is present in the Mediterranean, reinforcement cannot account for the positive selection. Sexual conflict in polyspermy avoidance is probably a more likely explanation, in which a variety of su-bindin alleles are adapting to a variety of EBR1 repeat sequences.

Echinometra lucunter is a sea urchin occurring sympatrically with its sister species, *Echinometra viridis*, in the Caribbean, and allopatrically in other parts of the Atlantic. The sympatric and allopatric populations show mtCO1 sequence differences indicative of ~200,000 years of divergence. However, both populations show equal positive selection in su-bindin, indicating again that reinforcement to prevent intra-species hybrids is probably impossible. More likely hypotheses to account for positive selection in su-bindin are the intra-specific processes of sexual selection, sexual conflict, and sperm competition (Geyer and Lessios 2009).

POSITIVE SELECTION IN THE SEA URCHIN SPERM RECEPTOR FOR EGG JELLY PROTEINS (suREJ)

When sea urchin sperm contact egg jelly (EJ), the fucose sulfate polymer of EJ induces the required acrosome reaction (AR), which consists of the polymerization of actin to form the finger-like acrosomal process coated with su-bindin (Moy and Vacquier 1979; Vacquier and Hirohashi 2004; Darszon et al. 2007; Vilela-Silva et al. 2008). The AR is induced by the opening of two distinct calcium channels and the elevation of sperm pH by ~0.2 units (Darszon et al. 2001; Hirohashi and Vacquier 2002). The sperm plasma membrane has at least two EJ receptors, suREJ-1 and -3, that localize to the plasma membrane covering the acrosome vesicle. suREJ-1 has two carbohydrate recognition domains (CRDs) and completes with EJ for binding to sperm (Moy et al. 1996). suREJ-3 has one CRD and one galactose recognizing lectin domain (Mengerink et al. 2002). Sea urchin REJ proteins are homologs of polycystin-1, the protein mutated in autosomal dominant polycystic kidney disease, the most common monogenic disease of humans (Gunaratne et al. 2007).

suREJ-1 and -3 were sequenced from six *Strongylocentrotus* species. A maximum likelihood analysis showed the CRD modules are subjected to positive selection. The positively selected residues map onto the carbohydrate binding region of each CRD, indicating that they might be important in species-specific induction of the AR by EJ (Mah et al. 2005). Thus, in *Strongylocentrotus*, positive selection can act both on su-bindin, a sperm ligand binding an egg receptor (EBR1) and suREJ-1, a sperm receptor for an egg ligand. Thus, selection can act simultaneously on two distinct sequential events in the fertilization cascade, first acrosome reaction induction, and second, sperm binding to the egg. The reason for the selective pressure underlying the positive selection in the CRDs of the sperm suREJ proteins is not readily apparent. Is it reinforcement to prevent zygote loss, or is it sexual conflict in the balance between sperm competition and egg avoidance of polyspermy?

EXTREME POSITIVE SELECTION IN ABALONE SPERM LYSIN AND 18-kDa PROTEIN

Abalone sperm have two major acrosomal proteins, 16-kDa lysin and 18-kDa protein (sp18) that are single copy genes in the red abalone species. Lysin dissolves the elevated vitelline envelope (VE) surrounding the egg by a nonenzymatic mechanism that probably involves competitively displacing hydrogen bonds that hold the fibrous VE glycoproteins together. sp18 coats the acrosomal process membrane and is thought to mediate sperm-egg fusion (Swanson and Vacquier 1995a; Kresge et al. 2001b). Deduced amino acid sequences of lysin were obtained from 20 abalone species. The amino-terminal sequence of lysin is always species-unique. Other regions are hypervariable, but not necessarily species-unique (Lee and Vacquier 1992; Lee et al. 1995; Kresge et al. 2001). Between species, comparisons show that lysin is extremely divergent in certain regions, yet highly conserved in others. Nucleotide substitution analysis showed that lysin's divergent regions have been subjected to strong positive Darwinian selection (protein adaptive evolution). Using mitochondrial (mt) CO1 as a standard for divergence time, it was estimated that the four most closely related eastern Pacific abalone species diverged from each other 1–2 million years ago. In pairwise comparisons of these four species, both acrosomal proteins appeared to be evolving faster than mtDNA, and sp18 evolves 2.5 times faster than lysin (Metz et al. 1998b). In one comparison between the sympatric white and pinto abalone species, mtCO1 was 1% different in nucleotide sequence, whereas lysin amino acid sequences differed by 15% and sp18 sequences by 23% (Metz et al. 1998b). These are some of fastest rates of divergence known for any metazoan protein. Comparison of three species showed that the five exons of both genes evolve much faster than the four introns. Sequencing ~29,000 base pairs of lysin and sp18 genomic DNA from red abalone individuals separated by ~1200 kilometers showed almost no nucleotide polymorphism, whereas the intron of a G-protein coupled receptor, showed ~7% nucleotide variation among individuals. These data suggest that the lysin and 18-kDa sequences in today's red abalone population are the results of many selective sweeps of favorable mutations in both genes (Metz et al. 1998b). Analysis of maximum likelihood models of 25 lysin sequences showed that the divergence rate (and thus selection pressure) varied in different lineages and that positive selection was most intense among closely related, sympatric species (Yang et al. 2000). Abalone sperm lysin remains one of the most robust examples of positive Darwinian selection known to date (Nunney and Schuenzel 2006).

WORK ON TURBAN SNAILS AND MUSSELS

Sperm of other archeogastropod mollusks, such as turban snails (*Tegula*), also use small cationic acrosomal proteins to dissolve a hole in the egg envelope by a nonenzymatic, species-specific mechanism. Interspecies comparisons show that *Tegula* sperm lysin is also subjected to strong positive selection (Hellberg and Vacquier 1999; Hellberg et al. 2000). The M7 acrosomal lysin of marine mussels, that dissolves the egg envelope, is a C-type lectin that also shows rapid evolution and positive selection (Riginos and McDonald 2003; Springer and Crespi 2007). Two species of mussels, *Mytilus edulis* (Me) and *Mytilus trossulus* (Mt), occur separately or together, and where they co-occur, hybrids exist. Cross species fertilizations of Me eggs and Mt sperm show that where the two species occur together there is greater success of Me eggs being fertilized by Mt sperm as compared to Me eggs from females living in the absence of Mt adults. These data are contrary to what would be expected if reinforcement was acting as a selective force to differentiate the gamete recognition system. These results could be explained by ingression of Mt genes for the egg's sperm receptor system into the Me gene pool in regions of sympatry (Slaughter et al. 2008).

CRYSTAL STRUCTURES OF ABALONE SPERM LYSIN

Lysin crystal structures show that the species unique amino terminus and the hypervariable carboxyl terminus lie on the same face of the molecule away from the helical bundle that comprises most of the protein. Regions of high amino acid substitution appear to be on the outer edges of the helical bundle, whereas conserved residues are on the bundle's inner surfaces (Shaw et al. 1993, 1995). Maximum likelihood analysis of lysin shows that 21 positions out of ~134 have been subjected to positive selection, with the amino terminus being the most highly substituted (Yang et al. 2000). Using recombinant DNA methods to switch the two termini of lysins from two species proved that these regions are indeed involved in species recognition needed for VE dissolution (Lyon and Vacquier 1999).

LYSIN AND THE sp18 AROSE BY GENE DUPLICATION

The sp18 sequence is known for five California abalone species. These five sequences are so different from lysin sequences that BLAST searches of GenBank using sp18 as input do not identify the lysin sequences as homologs. However, secondary structure predictions show that both proteins have the same number, length, and position of α-helices (Swanson and Vacquier 1995b). Both genes have four introns, which are in the same positions and interrupt the codons in the same phase (Metz et al. 1998b). Crystal structures of both proteins showed that both have nearly identical α-carbon ribbon structures, which is excellent evidence that they arose by a gene duplication event followed by their progressive differentiation (Fig. 2) (Kresge et al. 2001a). However, charge distribution and hydrophobic regions of the surfaces of both proteins are very different, possibly reflecting extreme specialization of function. In theory, the primordial abalone had one acrosomal protein that performed both functions of dissolving a hole in the egg VE and fusing the gamete plasma membranes. Because the ancestral gene duplication, lysin has specialized to dissolve egg vitelline envelopes (VE), whereas sp18 has specialized to fuse gamete plasma membranes (Swanson and Vacquier 1995a). The rapid evolution and positive selection seen in these two proteins could be driven by: specialization of function following gene duplication and sperm competition, which would be continuously selecting for the fastest

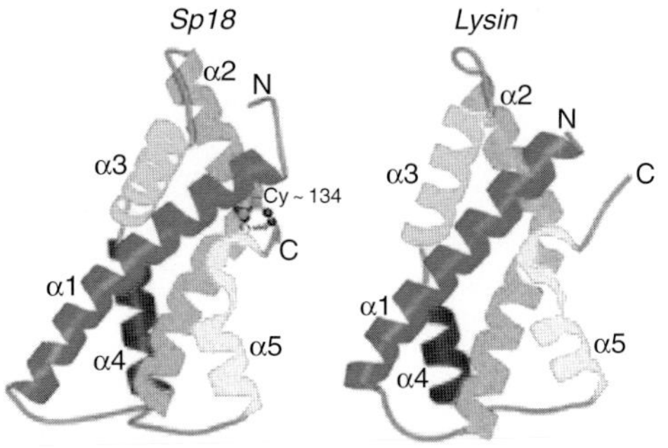

Figure 2. Abalone sperm lysin and sp18 arose from a gene duplication. The α-carbon ribbon diagrams are proof of their common ancestry. Both genes also have the same intron-exon structure. However, the surface properties of the two proteins are very different. (Figure adapted from Kresge et al. [2001a] and reprinted, with permission, from the American Chemical Society © 2001.)

sperm to dissolve a hole in the VE and fuse the gamete plasma membranes.

DUPLICATION AND EVOLUTION OF AN ABALONE LYSIN GENE

One allopatric abalone species, *Haliotis tuberculata coccinea*, lives on the Azores Islands. Another subspecies, *Haliotis tuberculata tuberculata* is allopatric on the Brittany Coast of France. The lysin genes of these subspecies are duplicated and both paralogs are expressed in sperm. The exons of the two paralogs are 83% identical and low polymorphism is found within a small population sampling (Clark et al. 2007). Phylogenetic analysis shows that the gene duplication occurred before the two subspecies arose and that both lysin copies are subjected to positive selection. Because these two subspecies are allopatric, there is no chance for cross-species fertilizations, therefore reinforcement cannot account for the positive selection driving the differentiation of the two paralogous lysins (Clark et al. 2007). Selection pressure on the two lysins might be because they bind different regions of VERL, the egg Vitelline Envelope Receptor for Lysin. Currently, nothing is known about VERL of these subspecies. A second hypothesis is that different alleles of VERL may exist in the population and each lysin paralog is specialized to interact with one VERL allele. Sexual conflict, in the balance between sperm competition and polyspermy avoidance, could explain the positive selection seen in the paralogous lysins as the lysins are continuously adapting to two different alleles of VERL. Such a scenario could result in prezygotic reproductive isolation and speciation (Clark et al. 2007).

VERL, THE ABALONE EGG VITELLINE ENVELOPE RECEPTOR FOR SPERM LYSIN

There are few examples of sperm and egg cognate pairs of interacting fertilization proteins. Abalone sperm lysin and its egg VE receptor, VERL, are the most informative cognate pair. Because sperm lysin dissolves isolated egg VEs, the molecule interacting with lysin was obviously in the VE. Radiolabeled lysin and sucrose density gradients were used to identify VERL, the VE Receptor for Lysin, of mass ~1 million Daltons. Electron microscopy showed VERL to be unbranched rods of undetermined length. VERL bound lysin with high affinity and also species-selectivity (Swanson and Vacquier 1997). Amino acid sequences were obtained and PCR used to amplify the VERL sequence from cDNA and genomic DNA. VERL was composed of tandem repeats of ~153 amino acids. The repeats were very homogeneous within a species, but differed between species. The repeats evolved by "concerted evolution," which homogenizes the sequences so that they are nearly identical (Elder and Turner 1995; Swanson and Vacquier 1998). This finding was contrary to theoretical expectations that, like lysin, VERL should show that positive selection had acted on its divergence (Gavrilets 2000; Gavrilets et al. 2001; Van Doorn et al. 2001; Gavrilets and Waxman 2002; Gavrilets and Hayashi 2005; Hayashi et al. 2007). The most recent work shows that VERL and lysin coevolve at similar rates, that is to say that relatively large changes in lysin sequences are mirrored by similar large changes in VERL sequences suggesting that lysin is "chasing" a rapidly evolving VERL (Clark et al. 2009).

Full-length VERL cDNA (11,166 bp) was cloned from the red abalone (Galindo et al. 2002). About 90% of the amino acid sequence consists of 22 tandem repeats of ~153 residues, which are predicted to be β-sheets (Fig. 3). Following repeat 22 are 353 residues containing a zona pellucida domain (ZP) and a hydrophobic COOH-terminus. The most interesting feature is that repeats 1 and 2 have not been homogenized with repeats 3–22 (Galindo et al. 2002). Sequencing VERL repeats 1–4 from eight species showed that repeats 1 and 2 are subjected to positive selection, and have not been homogenized with repeats 3–22, since the divergence of these eight species (Galindo et al. 2003). The ZP domain was originally found in the three glycoproteins that make up the mammalian egg envelope, the zona pellucida (Wassarman and Litscher 2008). This domain of about 250 amino acids is found in many extracellular proteins and is involved in the polymerization and

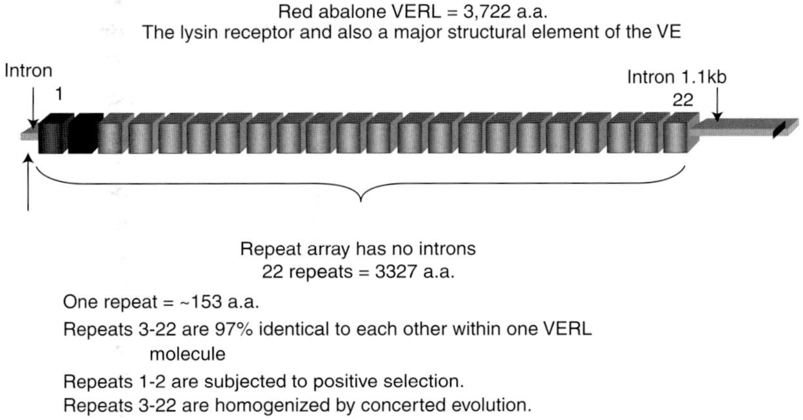

Figure 3. The structure of the red abalone egg vitelline envelope receptor for lysin (VERL). There are 353 residues after the last repeat, which contains one ZP domain. Repeats 1 and 2 evolve by positive selection, whereas repeats 3–22 are subjected to concerted evolution. (Figure adapted from Galindo et al. [2002] and reprinted, with permission, from Elsevier © 2002.)

assembly of structures such as egg envelopes (Jovine et al. 2005; Monné et al. 2008). Further work on the VE has identified at least 30 other ZP domain containing proteins of smaller size than VERL. One of these VEZP proteins, VEZP14, is a paralog of VERL and contains only one VERL repeat that evolves by positive selection twice as rapidly as do repeats 1 and 2 of VERL. VEZP14 also shows preferential binding to the sp18 acrosomal protein and is thus the putative sp18 receptor (Aagaard et al. 2010). Several of these 30 VEZP proteins show positive selection. The selective forces of polyspermy avoidance, specialization following gene duplication and reinforcement of molecular differences between species might account for such strong selection (Aagaard et al. 2006, 2010). What is exciting is that ZP domain proteins are found in both mammalian and abalone egg envelopes. Could this be a case of orthologous evolution from primordial animals, or is this a case of convergent evolution?

COMPARING SEA URCHIN EGG EBR1 AND ABALONE EGG VERL

Sea urchins and abalone diverged at least 500 million years ago. Why should these two distantly related animals have sperm receptors on their egg surfaces that have such similar structures? Sea urchin Sf-EBR1 has 19 tandem CUB/TSP repeats, each dual-domain repeat being ~171 amino acids. In Sf, the last 10 CUB/TSP repeats are species-specific. In Sp, the terminal 11 repeats are hyalin repeats of 81 residues that are not homologous to the CUB/TSP repeats of Sf-EBR1 (Kamei and Glabe 2003). This lack of carboxy-terminal homology might have arisen by exon shuffling that created a new carboxy-terminal EBR1 gene. The hyalin repeats are also found in *Echinometra*, which is far diverged from *Strongylocentrotus* (Palumbi 2009). CUB, TSP, and hyalin domains are all known to be involved in protein–protein binding. The tandem repeats of sea urchin EBR1 are separated by introns, which could promote exon shuffling. The 22 tandem repeats of abalone egg VERL are each ~153 amino acids. VERL repeats 1 and 2 do not homogenize with repeats 3–22, and there are no introns between any of the VERL repeats, which makes exon shuffling improbable (Galindo et al. 2002, 2003). Thus, both these broadcast spawning invertebrates evolved sperm receptors on their eggs that are constructed of repeat arrays of similar sizes and numbers of amino acids. Unequal crossing over was most probably involved in their expansion. The secondary structures of both suggest IgG-like β-sandwich folds.

OYSTER SPERM BINDIN

Oyster (*Crassostrea gigas*) sperm have a ring-shaped acrosomal granule containing a bindin (oy-bindin) that attaches sperm to egg. The rings of insoluble oy-bindin can be isolated (Brandriff et al. 1978). When the acrosomal rings of single males are separated by SDS-PAGE, one or two protein-staining bands appear per individual, of relative molecular masses 35-, 48-, 63-, 75-, and 88-kDa. Translating cDNA sequences of oy-bindins show the protein has a signal sequence of 24 residues, followed by a 97 residue unique amino terminus, which is followed by from one to five tandem fucose-binding lectin repeats (FLR) of 134 residues. Internal FLRs are separated by a spacer sequence of either PGAK or PGAKGK. The differences in SDS-PAGE mobilities of oy-bindin proteins represent different numbers of FLRs, the extremes being the 35-kDa isoform with one FLR and the 88-kDa isoform with five FLRs. The presence of one or two molecular mass classes of bindin per individual represents the two alleles (Moy et al. 2008, Springer et al. 2008).

Southern blots show that oy-bindin is a single copy gene. Genomic DNA contains oy-bindin genes possessing from 1–5 FLRs. Such variation in repeat number is usually created by unequal crossing over. The spermatogenic tissue of individual males having oy-bindin genes with 5 FLRs has oy-bindin mRNAs with 1-5 FLRs; however, only one or two isoforms are translated per individual, indicating that an unexplained mechanism governs translation. The five isoforms of mRNA show that alternative splicing of the primary transcript must occur (Moy et al. 2008, Springer et al. 2008). The amino-terminal 122 residues of the oy-bindin protein is highly conserved in sequence. In 39 full-length sequences from 16 males from five worldwide locations, there were only five substituted amino acids out of 4,315 (Moy et al. 2008). In the exact middle of each FLR is one intron of highly variable size, which is the site of recombination between FLR halves. In 39 full-length oy-bindin cDNA sequences, there were 107 FLRs that contained 41 different FLR sequences, resulting from the recombination of 24 unique amino-terminal FLR halves with 18 unique carboxy-terminal halves. This is the most intra-specific variation that has ever been found in a fertilization protein.

The F-lectin domain is an ancient protein module present in both pro- and eukaryotes. The oy-bindin FLR sequence is about 45% identical to the consensus F-lectin sequence. Maximum likelihood analysis of oy-bindin identifies nine sites subjected to positive section. Threading the oy-bindin F-lectin sequence onto the crystal structure of FLR shows that eight of these sites are clustered in the fucose binding groove (Fig. 4), indicating there is a survival benefit to altering the fucose binding affinity of oy-bindin for its yet unknown egg surface receptor (Moy et al. 2008; Springer et al. 2008; Wu et al. 2010).

As mentioned above, the number of FLRs contained in one oy-bindin gene can vary. One gene with three tandem FLRs had a 3.6 kb retroposon immediately upstream of the first FLR. However, two oy-bindin genes, which

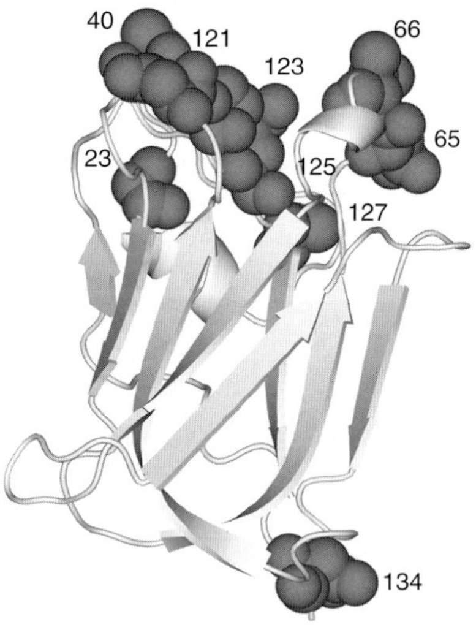

Figure 4. The oyster bindin fucose lectin repeat threads onto the known fucose lectin crystal structure. Dark residues are eight sites under positive selection that surround the fucose-binding groove, indicating the importance of adaptive evolution in binding fucose. Number 138 is the ninth positively selected site, which is not in the fucose binding groove. (Figure adapted from Moy et al. [2008] and reprinted, with permission, from the National Academy of Sciences © 2008.)

both had only one FLR, lacked this retroelement. Thus, insertion of a retroposon might be another way to increase the number of FLRs in an oy-bindin gene. The presence of upstream GA, and complementary downstream CT, microsatellites between FLR repeats also suggests that loopout strand duplex formation, copying and copy integration may be yet another way of altering the number of FLRs in an oy-bindin gene (Moy and Vacquier 2008).

WHY IS OYSTER BINDIN SO VARIABLE?

One male *C. gigas* can synthesize hundreds and maybe thousands of forms of the oy-bindin protein (Moy et al. 2008; Springer et al. 2008). Experimental evidence suggests this species uses every known mechanism to create variation in the FLR sequence. These mechanisms are: alternative splicing to create mRNAs with different numbers and combinations of FLRs, recombination of halves of FLRs to create variation in FLR sequences, retrotransposition, and loopout strand copying to increase the number of FLRs in the tandem repeat array. One big difference between oysters on one hand, and sea urchins and abalone on the other hand, is that oyster eggs do not have an electrical block against polyspermy (Alliegro and Wright 1983; Stephano and Gould 1988; Togo and Morisawa 1999; Gould and Stephano 2003). Because of the lack of an electrical block, polyspermy prevention might be more dependent on the initial events of sperm-egg interaction before the gametes fuse (Moy et al. 2008). We do not know anything about the egg's oy-bindin receptor. We can assume that it is a glycoprotein with exposed fucose residues. Those eggs that survive to pass on their oy-bindin genes might be only those with altered oy-bindin receptors, which slow sperm fusion. However, sperm competition is continuously selecting for the fastest sperm to fuse. To guarantee that fertilization will succeed, males make a wide variety of oy-bindins. These experimental facts support the hypothesis that sexual conflict is the driving force in the great variation seen in oyster sperm bindin. Because the oysters used in these experiments were raised in culture facilities as a single species, reinforcement is not a plausible theory to explain such extreme variation in this fertilization protein. Further knowledge will come from identifying and sequencing the egg receptor for this most interesting protein (Wu et al. 2010).

STRUCTURAL CONSERVATION IN A GAMETE RECOGNITION PROTEIN MODULE IN EUKARYOTES

The mammalian egg envelope, the zona pellucia (ZP), is composed of four glycoproteins (ZP1–4). ZP2 and ZP3 bind sperm and ZP3 can induce the sperm acrosome reaction. ZP proteins are made of two domains, ZP-N and ZP-C. The crystal structure of ZP3's ZP-N domain defines a novel type of immunoglobulin (Ig) superfamily, having: two invariant 1-4 and 2-3 disulfide bonds, a unique E′ strand and a conserved Tyr residue in strand F′. Comparing sequences of ZP-N domains in mammalian ZPs, abalone VERL repeats and yeast mating agglutinin proteins, shows no similarities among these proteins. However, threading the abalone VERL repeat sequences and the amino terminus of abalone VEZP-14 onto the mammalian ZP-N 3-D structure, results in a statistically robust match. The VEZP-14 model was then used to thread the α-agglutinin mating factor of the yeast *Saccharomyces cerevisiae* and the invasion glycoprotein of *Candida albicans*. The yeast IgIII protein identified the ZP-N Ig fold as the top hit. Both yeast proteins show the 1-4, 2-3 disulfide bonding pattern. In conclusion, these sex proteins from mammals, mollusks, and yeast all share the same fold of a novel Ig module. It remains unknown whether this 3-D structural similarity results from an ancestral relationship, or if this is a case of convergent evolution (Swanson et al. 2011). This discovery exemplifies the pitfall of using only primary structure to search for similarities of proteins from diverse species.

CONCLUSION

The quote from Ernst Mayr at the beginning of this review was his guess that selection must be strong on gametes. The rapid rate of sequence divergence, usually driven by positive selection (adaptive protein evolution), witnessed in many fertilization proteins, is testament to his clairvoyance. The factors responsible for such strong selection on fertilization proteins are at present difficult to define with certainty. Sexual conflict, sexual selection, reinforcement, specialization after gene duplication, and yet unknown mechanisms, might work simultaneously in the evolution of gamete recognition systems. Knowledge of sequences of additional cognate pairs of interacting sperm and egg proteins in more animal models should yield the needed information to enable refinement of current hypotheses and the formulation of novel hypotheses to explain this general phenomenon.

ACKNOWLEDGMENTS

This work was funded by NIH Grant HD12986 to V.D.V., NIH Grants HD042563, HD054631, HD057974 and NSF Grants DEB-0743539 and DEB-0918106 to W.J.S.

REFERENCES

Aagaard JE, Yi X, MacCoss MJ, Swanson WJ. 2006. Rapidly evolving zona pellucida domain proteins are a major component of the vitelline envelope of abalone eggs. *Proc Natl Acad Sci* **103:** 17302–17307.

Aagaard JE, Vacquier VD, Maccoss MJ, Swanson WJ. 2010. ZP domain proteins in the abalone egg coat include a paralog of VERL under positive selection that binds lysin and 18-kDa sperm proteins. *Mol Biol Evol* **27:** 193–203.

Alliegro MC, Wright DA. 1983. Polyspermy inhibition in the oyster *Crassostrea virginica*. *J Exp Zool* **227:** 127–137.

Andersson M. 1994. *Sexual selection.* Princeton University Press, Princeton, NJ.

Biermann CH. 1998. The molecular evolution of sperm bindin in sex species of sea urchins (Echinoida: Strongylocentrotidae). *Mol Biol Evol* **15:** 1761–1771.

Biermann CH, Kessing BD, Palumbi SR. 2003. Phylogeny and development of marine model species: Strongylocentrotid sea urchins. *Evol Dev* **5:** 360–371.

Brandriff B, Moy GW, Vacquier VD. 1978. Isolation of sperm bindin from the oyster (*Crassostrea gigas*). *Gamete Res* **1:** 89–99.

Calderón I, Turon X, Lessios HA. 2009. Characterization of the sperm molecule bindin in the sea urchin genus *Paracentrotus*. *J Mol Biol* **68:** 366–376.

Calderón I, Ventura CRR, Turon X, Lessios HA. 2010. Genetic divergence and assortative mating between colour morphs of the sea urchin *Paracentrotus gaimardi*. *Mol Ecol* **19:** 484–493.

Cameron RA, Minor JE, Nishioka D, Britten RJ, Davidson EH. 1990. Locale and level of bindin mRNA in maturing testis of the sea urchin (*Strongylocentrotus purpuratus*). *Dev Biol* **142:** 44–49.

Chuong EB, Tong W, Hoekstra HE. 2010. Maternal-fetal conflict: Rapidly evolving proteins in the rodent placenta. *Mol Biol Evol* **27:** 1221–1225.

Clark NL, Aagaard JE, Swanson WJ. 2006. Evolution of reproductive proteins from animals and plants. *Reproduction* **131:** 11–22.

Clark NL, Findlay GD, Yi X, MacCoss MJ, Swanson WJ. 2007. Duplication and selection on abalone sperm lysin in an allopatric population. *Mol Biol Evol* **24:** 2081–2090.

Clark NL, Gasper J, Sekino M, Springer SA, Aquadro CF, Swanson WJ. 2009. Coevolution of interacting fertilization proteins. *PLoS Genetics* **5:** 1–14.

Coyne JA, Orr HA. 2004. *Speciation.* Sinauer Associates, Sunderland, MA.

Darszon A, Beltrán C, Felix R, Nishigaki T, Treviño CL. 2001. Ion transport in sperm signaling. *Dev Biol* **240:** 1–14.

Darszon A, Treviño CL, Wood C, Galindo B, Rodríguez-Miranda E, Acevedo JJ, Hernandez-González EO, Beltrán C, Martínez-López P, Nishigaki T. 2007. Ion channels in sperm motility and capacitation. *Soc Reprod Fertil Suppl* **65:** 229–244.

Dobzhanski T. 1940. Speciation as a stage in evolutionary divergence. *Am Nat* **74:** 302–321.

Dorus S, Wasbrough ER, Busby J, Wilkin EC, Karr TL. 2010. Sperm proteomics reveals intensified selection on mouse sperm membrane and acrosomal genes. *Mol Biol Evol* **27:** 1235–1246.

Elder JF, Turner BJ. 1995. Concerted evolution of repetitive DNA sequences in eukaryotes. *Q Rev Biol* **70:** 297–320.

Frank SA. 2000. Sperm competition and female avoidance of polyspermy mediated by sperm-egg biochemistry. *Evol Ecol Res* **2:** 613–625.

Franke ES, Babcock RC, Styan CA. 2002. Sexual conflict and polyspermy under sperm-limited conditions: In situ evidence from field simulations with the free spawning marine echinoid (*Evechinus chloroticus*). *Am Nat* **160:** 485–496.

Galindo BE, Moy GW, Swanson WJ, Vacquier VD. 2002. Full-length sequence of VERL, the egg vitelline envelope receptor for abalone sperm lysin. *Gene* **288:** 111–117.

Galindo BE, Vacquier VD, Swanson WJ. 2003. Positive selection in the egg receptor for abalone sperm lysin. *Proc Natl Acad Sci* **100:** 4639–4643.

Gao B, Klein LE, Britten RJ, Davidson EH. 1986. Sequence of mRNA coding for bindin, a species specific sperm protein required for fertilization. *Proc Natl Acad Sci* **83:** 8634–8638.

Gavrilets S. 2000. Rapid evolution of reproductive barriers driven by sexual conflict. *Nature* **403:** 886–889.

Gavrilets S, Hayashi TI. 2005. Speciation and sexual conflict. *Evol Ecol* **19:** 167–198.

Gavrilets S, Waxman D. 2002. Sympatric speciation by sexual conflict. *Proc Natl Acad Sci* **99:** 10533–10538.

Gavrilets S, Arnqvist G, Friberg U. 2001. The evolution of female mate choice by sexual conflict. *Proc Biol Sci* **268:** 531–539.

Geyer LB, Lessios HA. 2009. Lack of character displacement in the male recognition molecule, bindin, in Atlantic sea urchins of the genus *Echinometra*. *Mol Biol Evol* **26:** 2135–2146.

Geyer LB, Palumbi SR. 2003. Reproductive character displacement and the genetics of gamete recognition in tropical sea urchins. *Evolution* **57:** 1049–1060.

Glabe CG, Lennarz WJ. 1979. Species specific sperm adhesion in sea urchins: A quantitative investigation of bindin-mediated egg agglutination. *J Cell Biol* **83:** 595–604.

Glabe CG, Vacquier VD. 1977. Species specific agglutination of eggs by bindin isolated from sea urchin sperm. *Nature* **267:** 836–838.

Gould MC, Stephano JL. 2003. Polyspermy prevention in marine invertebrates. *Microsc Res Technol* **61:** 379–388.

Gunaratne HJ, Moy GW, Kinukawa M, Miyata S, Mah SA, Vacquier VD. 2007. The 10 sea urchin receptor for egg jelly proteins (suREJ) are members of the polycystic kidney disease-1 (PKD1) family. *BMC Genomics* **8:** 235.

Hayashi TI, Vose M, Gavrilets S. 2007. Genetic differentiation by sexual conflict. *Evolution* **61:** 516–529.

Haygood R. 2004. Sexual conflict and protein polymorphism. *Evolution* **58:** 1414–1423.

Hellberg ME, Vacquier VD. 1999. Rapid evolution of fertilization selectivity and lysin cDNA sequences in teguline gastropods. *Mol Biol Evol* **16:** 839–848.

Hellberg ME, Moy GW, Vacquier VD. 2000. Positive selection and propeptide repeats promote rapid interspecies divergence of a gastropod sperm protein. *Mol Biol Evol* **17:** 458–466.

Hirohashi N, Vacquier VD. 2002. High molecular mass egg fucose sulfate polymer is required for opening both calcium channels involved in triggering the sea urchin sperm acrosome reaction. *J Biol Chem* **277:** 1182–1189.

Jovine L, Darie CC, Litscher ES, Wassarman PM. 2005. Zona pellucida domain proteins. *Ann Rev Biochem* **74:** 83–114.

Kamei N, Glabe CG. 2003. The species specific egg receptor for sea urchin sperm adhesion is EBR1, a novel ADAMTS protein. *Genes Dev* **17:** 2502–2507.

Kresge N, Vacquier VD, Stout CD. 2001a. The crystal structure of a fusagenic sperm protein reveals extreme surface properties. *Biochemistry* **40:** 5407–5413.

Kresge N, Vacquier VD, Stout CD. 2001b. Abalone lysin: The dissolving and evolving sperm protein. *BioEssays* **23:** 95–103.

Landry C, Geyer LB, Arakak Y, Uehara T, Palumbi SR. 2003. Recent speciation in the indo-west Pacific: Rapid evolution of gamete recognition and sperm morphology in cryptic species of sea urchin. *Proc R Soc Lond B* **270:** 1839–1847.

Lee Y-H. 2003. Molecular phylogenies and divergence times of sea urchin species of Strongylocentrotidae, Echinoida. *Mol Biol Evol* **20:** 1211–1221.

Lee Y-H, Vacquier VD. 1992. The divergence of species specific abalone sperm lysin is promoted by positive Darwinian selection. *Biol Bull* **182:** 97–104.

Lee Y-H, Ota T, Vacquier VD. 1995. Positive selection is a general phenomenon in the evolution of abalone sperm lysin. *Mol Biol Evol* **12:** 231–238.

Lessios HA. 2007. Reproductive isolation between species of sea urchins. *Bull Mar Sci* **81:** 191–208.

Levitan DR, Ferrell DL. 2006. Selection on gamete recognition proteins depends on sex, density and genotype frequency. *Science* **312:** 267–269.

Levitan DR, Stapper AP. 2010. Simultaneous positive and negative frequency dependent selection on sperm bindin, a gamete recognition protein in the sea urchin *Strongylocentrotus purpuratus*. *Evolution* **64:** 785–797.

Levitan DR, terHorst CP, Fogarty ND. 2007. The risk of polyspermy in three congeneric sea urchins and its implications for gamete incompatibility and reproductive isolation. *Evolution* **61:** 2009–2016.

Lyon JD, Vacquier VD. 1999. Interspecies chimeric sperm lysins identify regions mediating species-specific recognition of the abalone egg vitelline envelope. *Dev Biol* **214:** 151–159.

Mah SA, Swanson WJ, Vacquier VD. 2005. Positive selection in the carbohydrate recognition domains of sea urchin sperm receptor for egg jelly (suREJ) proteins. *Mol Biol Evol* **22:** 533–541.

Manier MK, Palumbi SR. 2008. Intraspecific divergence in sperm morphology of the green sea urchin, *Strongylocentrotus droebachiensis*: Implications for selection in broadcast spawners. *BMC Evol Biol* **8:** 238.

Marks JA, Biermann CH, Eanes WF, Kryvi H. 2008. Sperm polymorphism within the sea urchin *Strongylocentrotus droebachiensis:* Divergence between Pacific and Atlantic Oceans. *Biol Bull* **215:** 115–125.

Mayr E. 1997. Perspective: The objects of selection. *Proc Natl Acad Sci* **94:** 2091–2094.

McCartney MA, Lessios HA. 2002. Quantitative analysis of gamete incompatibility between closely related species of neotropical sea urchins. *Biol Bull* **202:** 166–181.

McCartney MA, Lessios HA. 2004. Adaptive evolution of sperm bindin tracks egg incompatibility in neotropical sea urchins of the genus *Echinometra*. *Mol Biol Evol* **21:** 732–745.

Mengerink KJ, Moy GW, Vacquier VD. 2002. suREJ3, a polycystin-1 protein, is cleaved at the GPS domain and localizes to the acrosomal region of sea urchin sperm. *J Biol Chem* **277:** 943–948.

Metz EC, Palumbi SR. 1996. Positive selection and sequence rearrangements generate extensive polymorphism in the gamete recognition protein bindin. *Mol Biol Evol* **13:** 397–406.

Metz EC, Kane RE, Yanagimachi H, Palumbi SR. 1994. Fertilization between closely related sea urchins is blocked by incompatibilities during sperm-egg attachment and early stages of fusion. *Biol Bull* **187:** 23–34.

Metz EC, Gomez-Gutierrez G, Vacquier VD. 1998a. Mitochondrial DNA and bindin gene sequence evolution among allopatric species of the sea urchin genus *Arbacia*. *Mol Biol Evol* **15:** 185–195.

Metz EC, Robles-Sikisaki R, Vacquier VD. 1998b. Nonsynonymous substitution in abalone sperm fertilization genes exceeds substitution in introns and mitochondrial DNA. *Proc Natl Acad Sci* **95:** 10676–10681.

Minor JE, Fromson DR, Britten RJ, Davidson EH. 1991. Comparison of the bindin proteins of *Strongylocentrotus franciscanus, S. purpuratus* and *Lytechinus variegates*: Sequences involved in the species specificity of fertilization. *Mol Biol Evol* **8:** 781–795.

Monné M, Han L, Schwend T, Burendahl S, Jovine L. 2008. Crystal structure of the ZP-N domain of ZP3 reveals the core fold of animal egg coats. *Nature* **456:** 653–657.

Moy GW, Vacquier VD. 1979. Immunoperoxidase localization of bindin during the adhesion of sperm to sea urchin eggs. *Curr Top Dev Biol* **13:** 31–44.

Moy GW, Vacquier VD. 2008. Bindin genes of the Pacific oyster *Crassostrea gigas. Gene* **423:** 215–220.

Moy GW, Mendoza LM, Swanson WJ, Glabe CG, Vacquier VD. 1996. The sea urchin sperm receptor for egg jelly is a modular protein with extensive homology to the human polycystic kidney disease protein, PKD1. *J Cell Biol* **133:** 809–817.

Moy GW, Springer SA, Adams SL, Swanson WJ, Vacquier VD. 2008. Extraordinary intraspecific diversity in oyster sperm bindin. *Proc Natl Acad Sci* **105:** 1993–1998.

Nunney L, Schuenzel EL. 2006. Detecting natural selection at the molecular level: A reexamination of some classic examples of adaptive evolution. *J Mol Evol* **62:** 176–195.

Palumbi SR. 1999. All males are not created equal: Fertility differences depend on gamete recognition polymorphisms in sea urchins. *Proc Natl Acad Sci* **96:** 12632–12637.

Palumbi SR. 2009. Speciation and the evolution of gamete recognition genes: Pattern and process. *Heredity* **102:** 66–76.

Palumbi SR, Lessios HA. 2005. Evolutionary animation: How do molecular phylogenies compare to Mayr's reconstruction of speciation patterns in the sea? *Proc Natl Acad Sci* **102:** 6566–6572.

Palumbi SR, Metz EC. 1991. Strong reproductive isolation between closely related tropical sea urchins (genus *Echinometra*). *Mol Biol Evol* **8:** 227–239.

Patiño S, Aagaard JE, MacCoss MJ, Swanson WJ, Hart MW. 2009. Bindin from a sea star. *Evol Dev* **11:** 376–381.

Rice WR, Holland B. 1997. The enemies within: Intergenomic conflict. Interlocus contest evolution, ICE and the intraspecific red queen. *Behav Ecol Sociobiol* **41:** 1–10.

Riginos C, McDonald JH. 2003. Positive selection on an acrosomal sperm protein, M17 lysin, in three species of the mussel genus *Mytilus. Mol Biol Evol* **20:** 200–207.

Riginos C, Wang D, Abrams AJ. 2006. Geographic variation and positive selection on M7 lysin, an acrosomal sperm protein in mussels (Mytilus spp). *Mol Biol Evol* **23:** 1952–1965.

Rocha S, Lucio M, Pereira MC, Reis S, Brezesinski G. 2008. The conformation of fusagenic B18 peptide in surfactant solutions. *J Pept Sci* **14:** 436–441.

Shaw A, McRee DE, Vacquier VD, Stout CD. 1993. The crystal structure of lysin, a fertilization protein. *Science* **262:** 1864–1867.

Shaw A, Fortes PAG, Stout CD, Vacquier VD. 1995. Crystal structure and subunit dynamics of the abalone sperm lysin dimer: Egg envelopes dissociate dimers, the monomer is the active species. *J Cell Biol* **130:** 1117–1125.

Slaughter C, McCartney MA, Yund PO. 2008. Comparison of gamete compatibility between two blue mussel species in sympatry and in allopatry. *Biol Bull* **214:** 57–66.

Springer SA, Crespi BJ. 2007. Adaptive gamete recognition divergence in a hybridizing Mytilus population. *Evolution* **61:** 772–783.

Springer SA, Moy GW, Friend DS, Swanson WJ, Vacquier VD. 2008. Oyster sperm bindin is a combinatorial fucose lectin with remarkable intra-species diversity. *Int J Dev Biol* **52:** 759–768.

Stephano JL, Gould M. 1988. Avoiding polyspermy in oyster (Crassostrea gigas). *Aquaculture* **73:** 295–307.

Swanson WJ, Vacquier VD. 1995a. Liposome fusion induced by a M(r) 18,000 protein localized to the acrosomal region of acrosome-reacted abalone spermatozoa. *Biochemistry* **34:** 14202–14208.

Swanson WJ, Vacquier VD. 1995b. Extraordinary divergence and positive Darwinian selection in a fusagenic protein coating the acrosomal process of abalone spermatozoa. *Proc Natl Acad Sci* **92:** 4957–4961.

Swanson WJ, Vacquier VD. 1997. The abalone egg vitelline envelope receptor for sperm lysin is a giant multivalent molecule. *Proc Natl Acad Sci* **94:** 6724–6729.

Swanson WJ, Vacquier VD. 1998. Concerted evolution in an egg receptor for a rapidly evolving abalone sperm protein. *Science* **281:** 710–712.

Swanson WJ, Vacquier VD. 2002. The rapid evolution of reproductive proteins. *Nat Rev Genet* **3:** 137–144.

Swanson WJ, Yang Z, Wolfner MF, Aquadro CF. 2001. Positive Darwinian selection drives the evolution of several female reproductive proteins in mammals. *Proc Natl Acad Sci* **98:** 2509–2514.

Swanson WJ, Aagaard JE, Vacquier VD, Monné M, Sadat Al Hosseini H, Jovine L. 2011. The molecular basis of sex: Linking yeast to humans. *Mol Biol Evol* doi:101093/molbev/msr026.

Togo T, Morisawa M. 1999. Mechanisms for blocking polyspermy in oocytes of the oyster *Crassostrea gigas. J Exp Zool* **283:** 307–314.

Tomaiuolo M, Levitan DR. 2010. Modeling how reproductive ecology can drive protein diversification and result in linkage disequilbrium between sperm and egg proteins. *Am Nat* **176:** 14–25.

Turner LM, Hoekstra HE. 2008. Causes and consequences of evolution of reproductive proteins. *Int J Dev Biol* **52:** 769–780.

Turner LM, Chuong EB, Hoekstra HE. 2008. Comparative analysis of testis protein evolution in rodents. *Genetics* **179:** 2075–2089.

Vacquier VD. 1998. Evolution of gamete recognition proteins. *Science* **281:** 1995–1998.

Vacquier VD, Hirohashi N. 2004. Sea urchin spermatozoa. *Meth Cell Biol* **74:** 524–540.

Vacquier VD, Moy GW. 1977. Isolation of bindin: The protein responsible for adhesion of sperm to sea urchin eggs. *Proc Natl Acad Sci* **74:** 2456–2460.

Vacquier VD, Swanson WJ, Hellberg ME. 1995. What have we learned about sea urchin sperm bindin? *Dev Growth Differ* **37:** 1–10.

Van Doorn GS, Luttikhuizen PC, Weissing FJ. 2001. Sexual selection at the protein level drives the extraordinary divergence of sex related genes during sympatric speciation. *Proc R Soc Lond B* **268:** 2155–2161.

Vilela-Silva A-CES, Hirohashi N, Mourao PAS. 2008. The structure of sulfated polysaccharides ensures a carbohydrate-based mechanism for species recognition during sea urchin fertilization. *Int J Dev Biol* **52:** 551–559.

Wassarman PM, Litscher ES. 2008. Mammalian fertilization: The egg's multifunctional zona pellucida. *Int J Dev Biol* **52:** 665–676.

Wong JL, Wessel GM. 2008. Renovation of the egg extracellular matrix at fertilization. *Int J Dev Biol* **52:** 545–550.

Wu HH, Su B. 2008. Adaptive evolution of SCML1 in primates, a gene involved in male reproduction. *BMC Evol Biol* **8:** 192.

Wu Q, Li L, Zhang G. 2010. *Crassostrea angulata* bindin gene and the divergence of fucose-binding lectin repeats among three species of *Crassostrea. Mar Biotechnol* doi: 10.1007/s10126-010-9304-0.

Xie S, Green J, Bixby JB, Szafranska B, DeMartini JC, Hecht S, Roberts RM. 1997. The diversity and evolutionary relationships of the pregnancy-associated glycoproteins, an aspartic proteinase subfamily consisting of many trophoblast-expressed genes. *Proc Natl Acad Sci* **94:** 12809–12816.

Yang Z, Swanson WJ, Vacquier VD. 2000. Maximum-likelihood analysis of molecular adaptation in abalone sperm lysin reveals variable selective pressures among lineages and sites. *Mol Biol Evol* **17:** 1446–1455.

Zigler KS. 2008. The evolution of sea urchin sperm bindin. *Int J Dev Biol* **52:** 791–796.

Zigler KS, Lessios HA. 2003a. 250 million years of bindin evolution. *Biol Bull* **205:** 8–15.

Zigler KS, Lessios HA. 2003b. Evolution of bindin in the pantropical sea urchin *Tripneustes*: Comparisons to bindin of other genera. *Mol Biol Evol* **20:** 220–231.

Zigler KS, Lessios HA. 2004. Speciation on the coasts of the new world: Phylogeography and the evolution of bindin in the sea urchin genus *Lytechinus. Evolution* **58:** 1225–1241.

Zigler KS, Raff EC, Popodi E, Raff RA, Lessios HA. 2003. Adaptive evolution of bindin in the genus *Heliocidaris* is correlated with the shift to direct development. *Evolution* **57:** 2293–2304.

Zigler KS, McCartney MA, Levitan DR, Lessios HA. 2005. Sea urchin bindin divergence predicts gamete compatibility. *Evolution* **59:** 2399–2404.

Index

K

L

M

N

U

V

X

Y

Z